MATLAB®

A Practical Introduction to Programming and Problem Solving

MATLAB®

A Practical Introduction to Programming and Problem Solving

Fourth Edition

Stormy Attaway

Department of Mechanical Engineering,
Boston University

AMSTERDAM • BOSTON • HEIDELBERG • LONDON
NEW YORK • OXFORD • PARIS • SAN DIEGO
SAN FRANCISCO • SINGAPORE • SYDNEY • TOKYO

Butterworth-Heinemann is an imprint of Elsevier

Butterworth-Heinemann is an imprint of Elsevier
The Boulevard, Langford Lane, Kidlington, Oxford OX5 1GB, UK
50 Hampshire Street, 5th Floor, Cambridge, MA 02139, USA

First published 2009
Second edition 2012
Third edition 2013
Fourth edition 2017

Library of Congress Cataloging-in-Publication Data
A catalog record for this book is available from the Library of Congress

British Library Cataloguing-in-Publication Data
A catalogue record for this book is available from the British Library

ISBN: 978-0-12-804525-1

For information on all Butterworth-Heinemann publications visit our website at https://www.elsevier.com/

Publisher: Todd Green
Acquisition Editor: Stephen Merken
Developmental Editor: Nate McFadden
Production Project Manager: Sujatha Thirugnana Sambandam
Cover Designer: Greg Harris

Typeset by SPi Global, India

Dedication

This book is dedicated to all of my family: my mother Jane Conklin, my sister Catherine Attaway, my brother Banks Attaway, my stepmother Robyn Attaway, and my husband Ted de Winter.

Contents

Preface

MOTIVATION

The purpose of this book is to teach basic programming concepts and skills needed for basic problem solving, all using MATLAB® as the vehicle. MATLAB is a powerful software package that has built-in functions to accomplish a diverse range of tasks, from mathematical operations to three-dimensional imaging. Additionally, MATLAB has a complete set of programming constructs that allow users to customize programs to their own specifications.

There are many books that introduce MATLAB. There are two basic flavors of these books: those that demonstrate the use of the built-in functions in MATLAB, with a chapter or two on some programming concepts, and those that cover only the programming constructs without mentioning many of the built-in functions that make MATLAB efficient to use. Someone who learns just the built-in functions will be well-prepared to use MATLAB, but would not understand basic programming concepts. That person would not be able to then learn a language such as C++ or Java without taking another introductory course, or reading another book, on the programming concepts. Conversely, anyone who learns only programming concepts first (using any language) would tend to write highly inefficient code using control statements to solve problems, not realizing that in many cases these are not necessary in MATLAB.

Instead, this book takes a hybrid approach, introducing both the programming and the efficient uses. The challenge for students is that it is nearly impossible to predict whether they will in fact need to know programming concepts later on or whether a software package such as MATLAB will suffice for their careers. Therefore, the best approach for beginning students is to give them both: the programming concepts, and the efficient built-in functions. Since MATLAB is very easy to use, it is a perfect platform for this approach in teaching programming and problem solving.

As programming concepts are critically important to this book, emphasis is not placed on the time-saving features that evolve with every new MATLAB release.

For example, in most versions of MATLAB, statistics on variables are available readily in the Workspace Window. This is not shown in any detail in the book, as the availability of this feature depends on the version of the software, and because of the desire to explain the concepts in the book.

MODIFICATIONS IN FOURTH EDITION

The changes in the Fourth Edition of this book include the following:

- Use of MATLAB Version R2016a
- A new chapter on Object Oriented Programming, and its use (as of R2014b) in the graphics objects
- The new App Designer, which may eventually replace GUIs and uses object-oriented programming
- The new Live Editor, which creates Live Scripts including text, equations, plots, and code
- Modified and new end-of-chapter exercises
- More conceptual end-of-chapter exercises
- Reorganization of chapters and sections within chapters to make it easier to focus on programming concepts in order to cover more of the book in traditional courses
- More coverage of data structures including categorical arrays and tables
- Increased coverage of built-in functions in MATLAB

KEY FEATURES

Side-by-side Programming Concepts and Built-in Functions

The most important and unique feature of this book is that it teaches programming concepts and the use of the built-in functions in MATLAB, side-by-side. It starts with basic programming concepts such as variables, assignments, input/output, selection, and loop statements. Then, throughout the rest of the book many times a problem will be introduced and then solved using the "programming concept" and also using the "efficient method". This will not be done in every case to the point that it becomes tedious, but just enough to get the ideas across.

Systematic Approach

Another key feature is that the book takes a very systematic, step-by-step approach, building on concepts throughout the book. It is very tempting in a MATLAB text to show built-in functions or features early on with a note that says "we'll do this later". This book does not do that; functions are covered before they are used in examples. Additionally, basic programming concepts will be explained carefully and systematically. Very basic concepts such as looping to calculate a sum, counting in a conditional loop, and error-checking are not found in many texts but are covered here.

File Input/Output

Many applications in engineering and the sciences involve manipulating large data sets that are stored in external files. Most MATLAB texts at least mention the **save** and **load** functions, and in some cases, also some of the lower-level file input/output functions. As file input and output is so fundamental to so many applications, this book will cover several low-level file input/output functions, as well as reading from and writing to spreadsheet files. Later chapters will also deal with audio and image files. These file input/output concepts are introduced gradually: first **load** and **save** in Chapter 3, then lower-level functions in Chapter 9, and finally sound and images in Chapter 13.

User-Defined Functions

User-defined functions are a very important programming concept, and yet many times the nuances and differences between concepts such as types of functions and function calls versus function headers can be very confusing to beginning programmers. Therefore, these concepts are introduced gradually. First, arguably the easiest type of functions to understand, those that calculate and return one single value, are demonstrated in Chapter 3. Later, functions that return no values and functions that return multiple values are introduced in Chapter 6. Finally, advanced function features are shown in Chapter 10.

Advanced Programming Concepts

In addition to the basics, some advanced programming concepts such as string manipulation, data structures (e.g., structures and cell arrays), recursion, anonymous functions, and variable number of arguments to functions are covered. Sorting and indexing are also addressed. All of these are again approached systematically; for example, cell arrays are covered before they are used in file input functions and as labels on pie charts.

Problem Solving Tools

In addition to the programming concepts, some basic mathematics necessary for solving many problems will be introduced. These will include statistical functions, solving sets of linear algebraic equations, and fitting curves to data. The use of complex numbers and some calculus (integration and differentiation) will also be introduced. The built-in functions in MATLAB to perform these tasks will be described.

Plots, Imaging, and Graphical User Interfaces

Simple two-dimensional plots are introduced very early in the book (Chapter 3) so that plot examples can be used throughout. A separate chapter, Chapter 12, shows more plot types, and demonstrates customizing plots and how the graphics properties are handled in MATLAB. This chapter makes use of strings and cell arrays to customize labels. Also, there is an introduction to image processing and the basics necessary to understand programming Graphical User Interfaces (GUIs) in Chapter 13.

Vectorized Code

Efficient uses of the capabilities of the built-in operators and functions in MATLAB are demonstrated throughout the book. In order to emphasize the importance of using MATLAB efficiently, the concepts and built-in functions necessary for writing vectorized code are treated very early in Chapter 2. Techniques such as preallocating vectors and using logical vectors are then covered in Chapter 5 as alternatives to selection statements and looping through vectors and matrices. Methods of determining how efficient the code is, are also covered.

Object-Oriented Programming

Creating objects and classes in MATLAB has been an option for some time, but as of R2014b, all Graphics objects are truly objects. Thus, object-oriented programming (OOP) is now a very important part of MATLAB programming. A new chapter has been introduced on this topic, and applications using App Designer reinforce the concepts.

LAYOUT OF TEXT

This text is divided into two parts: the first part covers programming constructs and demonstrates the programming method versus efficient use of built-in functions to solve problems. The second part covers tools that are used for basic problem solving, including plotting, image processing, and techniques to solve systems of linear algebraic equations, fit curves to data, and perform basic statistical analyses. The first six chapters cover the very basics in MATLAB and in programming, and are all prerequisites for the rest of the book. After that, many chapters in the problem solving section can be introduced when desired, to produce a customized flow of topics in the book. This is true to an extent, although the order of the chapters has been chosen carefully to ensure that the coverage is systematic.

The individual chapters are described here, as well as which topics are required for each chapter.

PART 1: INTRODUCTION TO PROGRAMMING USING MATLAB

Chapter 1: Introduction to MATLAB begins by covering the MATLAB Desktop Environment. Variables, assignment statements, and types are introduced. Mathematical and relational expressions and the operators used in them are covered, as are characters, random numbers, and the use of built-in functions and the Help browser.

Chapter 2: Vectors and Matrices introduces creating and manipulating vectors and matrices. Array operations and matrix operations (such as matrix multiplication) are explained. The use of vectors and matrices as function arguments, and functions that are written specifically for vectors and matrices are covered. Logical vectors and other concepts useful in vectorizing code are emphasized in this chapter.

Chapter 3: Introduction to MATLAB Programming introduces the idea of algorithms and scripts. This includes simple input and output, and commenting. Scripts are then used to create and customize simple plots, and to do file input and output. Finally, the concept of a user-defined function is introduced with only the type of function that calculates and returns a single value.

Chapter 4: Selection Statements introduces the use of logical expressions in **if** statements, with **else** and **elseif** clauses. The **switch** statement is also demonstrated, as is the concept of choosing from a menu. Also, functions that return **logical true** or **false** are covered.

Chapter 5: Loop Statements and Vectorizing Code introduces the concepts of counted (**for**) and conditional (**while**) loops. Many common uses such as summing and counting are covered. Nested loops are also introduced. Some more sophisticated uses of loops such as error-checking and combining loops and selection statements are also covered. Finally, vectorizing code using built-in functions and operators on vectors and matrices instead of looping through them, is demonstrated. Tips for writing efficient code are emphasized, and tools for analyzing code are introduced.

The concepts in the first five chapters are assumed throughout the rest of the book.

Chapter 6: MATLAB Programs covers more on scripts and user-defined functions. User-defined functions that return more than one value and also that do not return anything are introduced. The concept of a program in MATLAB which consists of a script that calls user-defined functions is demonstrated with examples. A longer menu-driven program is shown as a reference, but could be omitted. Subfunctions and scope of variables are also introduced, as are some debugging techniques. The Live Editor is introduced.

The concept of a program is used throughout the rest of the book.

Chapter 7: String Manipulation covers many built-in string manipulation functions as well as converting between string and number types. Several examples include using custom strings in plot labels and input prompts.

Chapter 8: Data Structures: Cell Arrays and Structures introduces two main data structures: cell arrays and structures. Once structures are covered, more complicated data structures such as nested structures and vectors of structures are also introduced. Cell arrays are used in several applications in later chapters, such as file input in Chapter 9, variable number of function arguments in Chapter 10, and plot labels in Chapter 12, and are therefore considered important and are covered first. The section on structures can be omitted, although the use of structure variables to store object properties is shown in Chapter 11. Other data structures such as categorical arrays and tables are also introduced. Methods of sorting are described. Finally, the concept of indexing into a vector is introduced. Sorting a vector of structures and indexing into a vector of structures are described, but these sections can be omitted.

PART II: ADVANCED TOPICS FOR PROBLEM SOLVING WITH MATLAB

Chapter 9: Advanced File Input and Output covers lower-level file input/output statements that require opening and closing the file. Functions that can read the entire file at once as well as those that require reading one line at a time are introduced and examples that demonstrate the differences in their uses are shown. Additionally, techniques for reading from and writing to spreadsheet files and also .mat files that store MATLAB variables are introduced. Cell arrays and string functions are used extensively in this chapter.

Chapter 10: Advanced Functions covers more advanced features of and types of functions, such as anonymous functions, nested functions, and recursive functions. Function handles, and their use both with anonymous functions and function functions are introduced. The concept of having a variable number of input and/or output arguments to a function is introduced; this is implemented using cell arrays. String functions are also used in several examples in this chapter. The section on recursive functions is at the end and may be omitted.

Chapter 11: Introduction to Object-Oriented Programming and Graphics As of version R2014b, all plot objects are actual objects. This chapter introduces Object-Oriented Programming (OOP) concepts and terminology using plot objects, and then expands to how to write your own class definitions and create your own objects.

Chapter 12: Advanced Plotting Techniques continues with more on the plot functions introduced in Chapter 3. Different two-dimensional plot types, such as logarithmic scale plots, pie charts, and histograms are introduced, as is customizing plots using cell arrays and string functions. Three-dimensional plot functions as well as some functions that create the coordinates for specified objects are demonstrated. The notion of Graphics is covered, and some graphics properties such as line width and color are introduced. Core graphics objects and their use by higher-level plotting functions are demonstrated. Applications that involve reading data from files and then plotting, use both cell arrays and string functions.

Chapter 13: Sights and Sounds briefly discusses sound files, and introduces image processing. An introduction to programming Graphical User Interfaces (GUIs) is also given, including the creation of a button group and embedding images in a GUI. Nested functions are used in the GUI examples. The GUI development environment, GUIDE, is introduced. The App Designer is also introduced; it creates OOP code and builds on the concepts from Chapter 11.

Chapter 14: Advanced Mathematics covers six basic topics: it starts with some of the built-in statistical and set operations in MATLAB then curve fitting, complex numbers, solving systems of linear algebraic equations, and integration and differentiation in calculus. Matrix solutions using the Gauss-Jordan and Gauss-Jordan elimination methods are described. Finally, some of the symbolic math toolbox functions are shown, including those that solve equations. This method returns a structure as a result.

PATH THROUGH THE BOOK

It has come to my attention that not all courses that use this text use all sections. In particular, not everyone gets to images and GUIs, which are the cool applications! I have reorganized some of the chapters and sections to make it easier to get to the fun, motivating applications including images and App Designer. What follows is a path through the book to get there, including which sections can be skipped.

Chapter 1: the last two sections, 1.7 and 1.8, can be skipped

Chapter 2: section 2.5 on matrix multiplication can be skipped

Chapters 3-4: are both fundamental

Chapter 5: the last section on Timing can be skipped

Chapter 6: the last two sections can be skipped

Chapter 7: the last section can be skipped

Chapter 8: cell arrays and structures are important, but the last 3 sections can be skipped

Chapter 9: this can be skipped entirely

Chapter 10: Variable number of arguments, nested functions, and anonymous functions are all used in App Designer but the last two sections can be skipped

Chapter 11: the first two sections are fundamental but the last can be skipped

Chapter 12: this can be skipped entirely

Chapter 13: most sections are independent although the concept of callback functions is explained in the GUI section and then used in the GUIDE and App Designer section

Chapter 14: all sections can be skipped

PEDAGOGICAL FEATURES

There are several pedagogical tools that are used throughout this book that are intended to make it easier to learn the material.

First, the book takes a conversational tone with sections called *"Quick Question!"*. These are designed to stimulate thought about the material that has just been covered. The question is posed, and then the answer is given. It will be most beneficial to the reader to try to think about the question before reading the answer! In any case, they should not be skipped over, as the answers often contain very useful information.

"Practice" problems are given throughout the chapters. These are very simple problems that drill the material just covered.

"Explore Other Interesting Features" This book is not intended to be a complete reference book, and cannot possibly cover all of the built-in functions and tools available in MATLAB; however, in every chapter there will be a list of functions and/or commands that are related to the chapter topics, which readers may wish to investigate.

When some problems are introduced, they are solved using both, *"The Programming Concept"* and also *"The Efficient Method"*. This facilitates understanding the built-in functions and operators in MATLAB as well as the underlying programming concepts. *"The Efficient Method"* highlights methods that will save time for the programmer, and in many cases, are also faster to execute in MATLAB.

Additionally, to aid the reader:

- Identifier names are shown in *italic*
- MATLAB function names are shown in **bold**
- Reserved words are shown in **bold and underlined**
- Key important terms are shown in ***bold and italic***

The end of chapter **"Summary"** contains where applicable, several sections:

- **Common Pitfalls**: a list of common mistakes that are made, and how to avoid them
- **Programming Style Guidelines**: In order to encourage "good" programs that others can actually understand, the programming chapters will have guidelines that will make programs easier to read and understand and therefore easier to work with and modify.
- **Key Terms**: a list of the key terms covered in the chapter, in sequence.
- **MATLAB Reserved Words**: a list of the reserved key words in MATLAB. Throughout the text, these are shown in bold, underlined type.
- **MATLAB Functions and Commands**: a list of the MATLAB built-in functions and commands covered in the chapter, in the order covered. Throughout the text, these are shown in bold type.
- **MATLAB Operators**: a list of the MATLAB operators covered in the chapter, in the order covered.
- **Exercises**: a comprehensive set of exercises, ranging from the rote to more engaging applications.

ADDITIONAL BOOK RESOURCES

A companion website with additional teaching resources is available for faculty using this book as a text for their course(s). Please visit www.textbooks.elsevier.com/9780128045251 to register for access to:

- Instructor solutions manual for end of chapter problems
- Instructor solutions manual for "Practice" problems
- Electronic figures from the text for creation of lecture slides
- Downloadable M-files for all examples in the text

Other book-related resources will also be posted there from time to time.

Acknowledgments

I am indebted to many, many family members, colleagues, mentors, and students.

Throughout the last 29 years of coordinating and teaching the basic computation courses for the College of Engineering at Boston University, I have been blessed with many fabulous students as well as graduate teaching fellows and undergraduate teaching assistants (TAs). There have been hundreds of TAs over the years, too many to name individually, but I thank them all for their support. In particular, the following TAs were very helpful in reviewing drafts of the original manuscript and subsequent editions and suggesting examples: Edy Tan, Megan Smith, Brandon Phillips, Carly Sherwood, Ashmita Randhawa, Kevin Ryan, Brian Hsu, Paul Vermilion, Ben Duong, Carlton Duffett, and Raaid Arshad. Kevin Ryan wrote the MATLAB scripts that were used to produce the cover illustrations. Vincent Barilla and Isabella Passaro helped proofread and created problems for the new chapter on OOP for this edition. Carlton and Raaid have been particularly helpful over the last couple of years and have developed many companion Cody Coursework problems.

A number of colleagues have been very encouraging throughout the years. In particular, I would like to thank Tom Bifano and Ron Roy for their support and motivation. I am also indebted to my mentors at Boston University, Bill Henneman of the Computer Science Department and Merrill Ebner of the Department of Manufacturing Engineering as well as Bob Cannon from the University of South Carolina.

I would like to thank all of the reviewers of the proposal and drafts of this book. Their comments have been extremely helpful, and I hope I have incorporated their suggestions to their satisfaction. They include: Dr. Montasir Abbas, Assistant Professor, Virginia Tech; Dr Ruijun Bu, Senior Lecturer, University of Liverpool; Dr. Antonio H. Costa, Professor, University of Massachusetts, Dartmouth; Dr. Patrick Flaherty, Assistant Professor, University of Massachusetts, Amherst; Dr. Daniel Fridline. Assistant Professor, SUNY Maritime College; Dr. Matthias Krauledat, Professor, Hochschule Rhein-Waal;

Dr. Roman Kuc, Professor, Yale University; Dr. Matthew A. Lackner, Associate Professor, University of Massachusetts, Amherst; Ryan Patrick, Lecturer, University of Nebraska, Lincoln; Alison Pechenick, Senior Lecturer, The University of Vermont; Dr. Charles Riedesel, Professor, University of Nebraska, Lincoln; Dr. Daniel F. Ryder, Jr., Associate Professor, Tufts University; Cheryl Schlittler, Instructor, University of Colorado, Colorado Springs; Dr. Geoffrey Shiflett, Associate Professor, University of Southern California; Dr. Hossein Tavana, Assistant Professor, University of Akron; Shaikh Md Rubayiat Tousif, Instructor, The University of Texas at Arlington; Phillip Wong, Lab Coordinator, Portland State University; and Dr. Vladislav V. Yakovlev, Professor, Texas A&M University.

Also, I thank those at Elsevier who helped to make this book possible including: Todd Green, Publisher; Stephen Merken, Acquisitions Editor; Nate McFadden, Sr. Developmental Editor; and Sujatha Thirugnana Sambandam, Production Project Manager.

Finally, thanks go to all of my family, especially my parents Roy Attaway and Jane Conklin, both of whom encouraged me at an early age to read and to write. Thanks also to my husband Ted de Winter for his encouragement and good-natured taking care of the weekend chores while I worked on this project!

The photo used in the image processing section was taken by Ron Roy.

Stormy Attaway
Department of Mechanical Engineering
Boston University

PART 1

Introduction to Programming Using MATLAB

CHAPTER 1

Introduction to MATLAB

KEY TERMS

prompt
programs
script files
toolstrip
variable
assignment statement
assignment operator
user
initializing
incrementing
decrementing
identifier names
reserved words
keywords
mnemonic
types
classes
double precision
floating point
unsigned
characters
strings
default
continuation operator
ellipsis
unary
operand
binary
scientific notation
exponential notation
precedence
associativity
nested parentheses
inner parentheses
help topics
call a function
arguments
returning values
tab completion
constants
random numbers
seed
pseudorandom
open interval
global stream
character encoding
character set
relational expression
Boolean expression
logical expression
relational operators
logical operators
scalars
short-circuit operators
truth table
commutative
roundoff errors
range
casting
type casting
saturation arithmetic
locale setting
logarithm
common logarithm
natural logarithm

CONTENTS

MATLAB® is a very powerful software package that has many built-in tools for solving problems and developing graphical illustrations. The simplest method for using the MATLAB product is interactively; an expression is entered by the user and MATLAB responds immediately with a result. It is also possible to write

MATLAB®. http://dx.doi.org/10.1016/B978-0-12-804525-1.00001-5

scripts and programs in MATLAB, which are essentially groups of commands that are executed sequentially.

This chapter will focus on the basics, including many operators and built-in functions that can be used in interactive expressions.

1.1 GETTING INTO MATLAB

MATLAB is a mathematical and graphical software package with numerical, graphical, and programming capabilities. It includes an integrated development environment, as well as both procedural and object-oriented programming constructs. It has built-in functions to perform many operations, and there are Toolboxes that can be added to augment these functions (e.g., for signal processing). There are versions available for different hardware platforms, in both professional and student editions. The MathWorks releases two versions of MATLAB annually, named by the year and 'a' or 'b'. This book covers the releases through Version R2016a. In cases where there have been changes in recent years, these are noted.

When the MATLAB software is started, a window opens in which the main part is the Command Window (see Fig. 1.1). In the Command Window, you should see:

```
>>
```

The >> is called the ***prompt***. In the Student Edition, the prompt instead is:

```
EDU>>
```

In the Command Window, MATLAB can be used interactively. At the prompt, any MATLAB command or expression can be entered, and MATLAB will respond immediately with the result.

It is also possible to write ***programs*** in MATLAB that are contained in ***script files*** or MATLAB code files. Programs will be introduced in Chapter 3.

The following commands can serve as an introduction to MATLAB and allow you to get help:

- **demo** will bring up MATLAB Examples in the Help Browser, which has examples of some of the features of MATLAB
- **help** will explain any function; **help help** will explain how help works
- **lookfor** searches through the help for a specific word or phrase (Note: this can take a long time)
- **doc** will bring up a documentation page in the Help Browser

To exit from MATLAB, either type **quit** or **exit** at the prompt, or click on MATLAB, then Quit MATLAB from the menu.

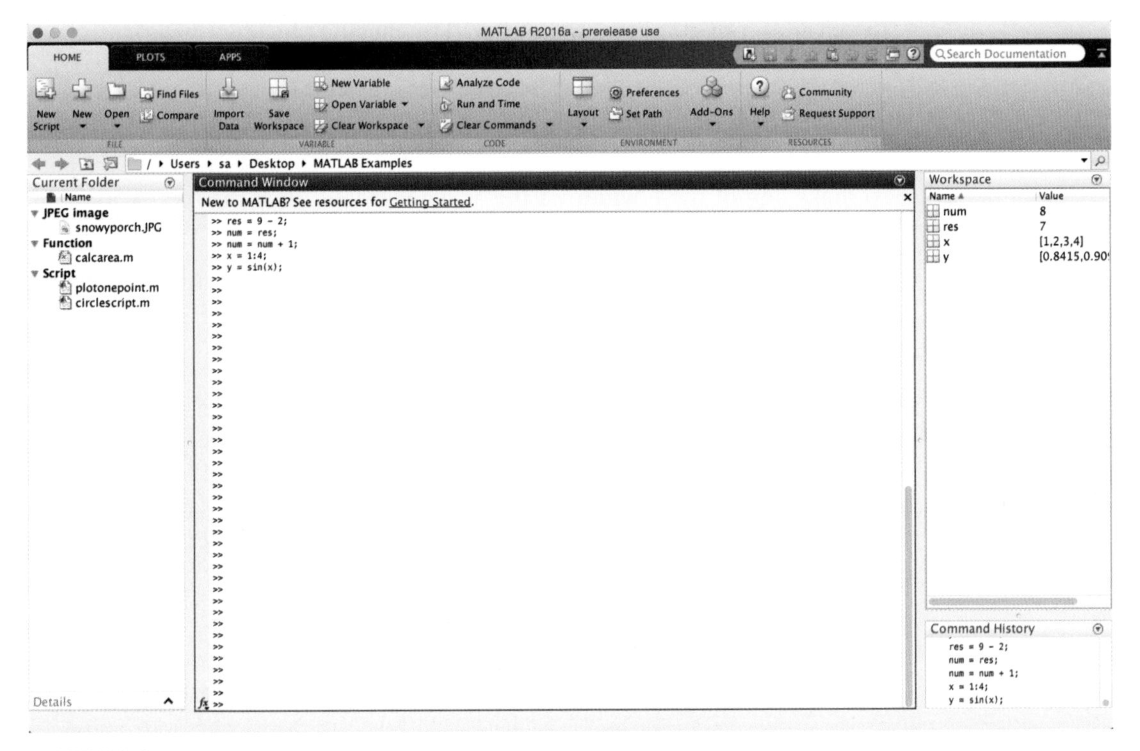

FIGURE 1.1
MATLAB Command Window.

1.2 THE MATLAB DESKTOP ENVIRONMENT

In addition to the Command Window, there are several other windows that can be opened and may be opened by default. What is described here is the default layout for these windows in Version R2016a, although there are other possible configurations. Different versions of MATLAB may show other configurations by default, and the layout can always be customized. Therefore, the main features will be described briefly here.

To the left of the Command Window is the Current Folder Window. The folder that is set as the Current Folder is where files will be saved. This window shows the files that are stored in the Current Folder. These can be grouped in many ways, for example by type, and sorted, for example by name. If a file is selected, information about that file is shown on the bottom where it says "Details."

To the right of the Command Window are the Workspace Window on top and the Command History Window on the bottom. The Command History Window shows commands that have been entered, not just in the current session (in the current Command Window), but previously as well. The Workspace Window will be described in the next section.

This default configuration can be altered by clicking the down arrow at the top right corner of each window. This will show a menu of options (different for each window), including, for example, closing that particular window and undocking that window. Once undocked, bringing up the menu and then clicking on the curled arrow pointing to the lower right will dock the window again. To make any of these windows the active window, click the mouse in it. By default the active window is the Command Window.

Beginning with Version 2012b, the Desktop now has a ***toolstrip***. By default, three tabs are shown ("HOME," "PLOTS," and "APPS"), although another, "SHORTCUTS," can be added.

Under the "HOME" tab there are many useful features, which are divided into functional sections "FILE," "VARIABLE," "CODE," "ENVIRONMENT," and "RESOURCES" (these labels can be seen on the very bottom of the grey toolstrip area). For example, under "ENVIRONMENT," hitting the down arrow under Layout allows for customization of the windows within the Desktop Environment including adding the SHORTCUTS tab. Other toolstrip features will be introduced in later chapters when the relevant material is explained.

1.3 VARIABLES AND ASSIGNMENT STATEMENTS

To store a value in a MATLAB session, or in a program, a ***variable*** is used. The Workspace Window shows variables that have been created and their values. One easy way to create a variable is to use an ***assignment statement***. The format of an assignment statement is

```
variablename = expression
```

The variable is always on the left, followed by the = symbol, which is the ***assignment operator*** (unlike in mathematics, the single equal sign does *not* mean equality), followed by an expression. The expression is evaluated and then that value is stored in the variable. Here is an example of how it would appear in the Command Window:

```
>> mynum = 6
mynum =
     6
>>
```

Here, the ***user*** (the person working in MATLAB) typed "mynum = 6" at the prompt, and MATLAB stored the integer 6 in the variable called *mynum*, and then displayed the result followed by the prompt again. As the equal sign is the assignment operator, and does not mean equality, the statement should be read as "mynum gets the value of 6" (*not* "mynum equals 6").

Note that the variable name must always be on the left, and the expression on the right. An error will occur if these are reversed.

```
>> 6 = mynum
   6 = mynum
     |
 Error: The expression to the left of the equals sign is not a valid target
 for an assignment.
 >>
```

Putting a semicolon at the end of a statement suppresses the output. For example,

```
>> res = 9 – 2;
>>
```

This would assign the result of the expression on the right side, the value 7, to the variable *res*; it just does not show that result. Instead, another prompt appears immediately. However, at this point in the Workspace Window both the variables *mynum* and *res* and their values can be seen.

The spaces in a statement or expression do not affect the result, but make it easier to read. The following statement, which has no spaces, would accomplish exactly the same result as the previous statement:

```
>> res = 9 – 2;
```

Note

In the remainder of the text, the prompt that appears after the result will not be shown.

MATLAB uses a default variable named *ans* if an expression is typed at the prompt and it is not assigned to a variable. For example, the result of the expression 6 + 3 is stored in the variable *ans*:

```
>> 6 + 3
ans =
      9
```

This default variable, *ans*, is reused any time when only an expression, not an assignment statement, is typed at the prompt. Note that it is not a good idea to use *ans* as a name yourself or in expressions.

A shortcut for retyping commands is to hit the up arrow ↑, which will go back to the previously typed command(s). For example, if you decide to assign the result of the expression 6 + 3 to a variable named *result* instead of using the default variable *ans*, you could hit the up arrow and then the left arrow to modify the command rather than retyping the entire statement:

```
>> result = 6 + 3
result =
      9
```

This is very useful, especially if a long expression is entered and it contains an error, and it is desired to go back to correct it.

It is also possible to choose command(s) in the Command History window, and rerun them by right-clicking. Consecutive commands can be chosen by clicking on the first or last and then holding down the Shift and up or down arrows (new in 2014a).

To change a variable, another assignment statement can be used, which assigns the value of a different expression to it. Consider, for example, the following sequence of statements:

```
>> mynum = 3
mynum =
          3
>> mynum = 4 + 2
mynum =
          6
>> mynum = mynum + 1
mynum =
          7
```

In the first assignment statement, the value 3 is assigned to the variable *mynum*. In the next assignment statement, *mynum* is changed to have the value of the expression 4+2, or 6. In the third assignment statement, *mynum* is changed again, to the result of the expression *mynum*+*1*. Since at that time *mynum* had the value 6, the value of the expression was 6+1, or 7.

At that point, if the expression *mynum* + *3* is entered, the default variable *ans* is used since the result of this expression is not assigned to a variable. Thus, the value of *ans* becomes 10 but *mynum* is unchanged (it is still 7). Note that just typing the name of a variable will display its value (the value can also be seen in the Workspace Window).

```
>> mynum + 3
ans =
        10

>> mynum
mynum =
         7
```

1.3.1 Initializing, Incrementing, and Decrementing

Frequently, values of variables change, as shown previously. Putting the first or initial value in a variable is called ***initializing*** the variable.

Adding to a variable is called ***incrementing***. For example, the statement

```
mynum = mynum + 1
```

increments the variable *mynum* by 1.

QUICK QUESTION!

How can 1 be subtracted from the value of a variable called *num*?

Answer:

```
num = num - 1;
```

This is called ***decrementing*** the variable.

1.3.2 Variable Names

Variable names are an example of ***identifier names***. We will see other examples of identifier names, such as function names, in future chapters. The rules for identifier names are as follows:

- The name must begin with a letter of the alphabet. After that, the name can contain letters, digits, and the underscore character (e.g., *value_1*), but it cannot have a space.
- There is a limit to the length of the name; the built-in function **namelengthmax** tells what this maximum length is (any extra characters are truncated).
- MATLAB is case-sensitive, which means that there is a difference between uppercase and lowercase letters. So, variables called *mynum*, *MYNUM*, and *Mynum* are all different (although this would be confusing and should not be done).
- Although underscore characters are valid in a name, their use can cause problems with some programs that interact with MATLAB, so some programmers use mixed case instead (e.g., *partWeights* instead of *part_weights*).
- There are certain words called ***reserved words***, or ***keywords***, that cannot be used as variable names.
- Names of built-in functions (described in the next section) can, but should not, be used as variable names.

Additionally, variable names should always be ***mnemonic***, which means that they should make some sense. For example, if the variable is storing the radius of a circle, a name such as *radius* would make sense; *x* probably wouldn't.

The following commands relate to variables:

- **who** shows variables that have been defined in this Command Window (this just shows the names of the variables)
- **whos** shows variables that have been defined in this Command Window (this shows more information on the variables, similar to what is in the Workspace Window)
- **clear** clears out all variables so they no longer exist

- **clear** *variablename* clears out a particular variable
- **clear** *variablename1 variablename2* ... clears out a list of variables (note: separate the names with spaces)

If nothing appears when **who** or **whos** is entered, that means there aren't any variables! For example, in the beginning of a MATLAB session, variables could be created and then selectively cleared (remember that the semicolon suppresses output).

```
>> who
>> mynum = 3;
>> mynum + 5;
>> who
Your variables are:
ans     mynum
>> clear mynum
>> who
Your variables are:
ans
```

These changes can also be seen in the Workspace Window.

1.3.3 Types

Every variable has a ***type*** associated with it. MATLAB supports many types, which are called ***classes***. (Essentially, a class is a combination of a type and the operations that can be performed on values of that type, but, for simplicity, we will use these terms interchangeably for now. More on classes will be covered in Chapter 11.)

For example, there are types to store different kinds of numbers. For float or real numbers, or in other words, numbers with a decimal place (e.g., 5.3), there are two basic types: **single** and **double**. The name of the type **double** is short for ***double precision***; it stores larger numbers than the **single** type. MATLAB uses a ***floating point*** representation for these numbers.

There are many integer types, such as **int8**, **int16**, **int32**, and **int64**. The numbers in the names represent the number of bits used to store values of that type. For example, the type **int8** uses eight bits altogether to store the integer and its sign. As one bit is used for the sign, this means that seven bits are used to store actual numbers (0s or 1s). There are also ***unsigned*** integer types **uint8**, **uint16**, **uint32**, and **uint64**. For these types, the sign is not stored, meaning that the integer can only be positive (or 0).

The larger the number in the type name, the larger the number that can be stored in it. We will for the most part use the type **int32** when an integer type is required.

The type **char** is used to store either single ***characters*** (e.g., 'x') or ***strings***, which are sequences of characters (e.g., 'cat'). Both characters and strings are enclosed in single quotes.

The type **logical** is used to store **true/false** values.

Variables that have been created in the Command Window can be seen in the Workspace Window. In that window, for every variable, the variable name, value, and class (which is essentially its type) can be seen. Other attributes of variables can also be seen in the Workspace Window. Which attributes are visible by default depends on the version of MATLAB. However, when the Workspace Window is chosen, clicking on the down arrow allows the user to choose which attributes will be displayed by modifying Choose Columns.

By default, numbers are stored as the type **double** in MATLAB. The function **class** can be used to see the type of any variable:

```
>> num = 6 + 3;
>> class(num)
ans =
double
```

1.4 NUMERICAL EXPRESSIONS

Expressions can be created using values, variables that have already been created, operators, built-in functions, and parentheses. For numbers, these can include operators such as multiplication and functions such as trigonometric functions. An example of such an expression is:

```
>> 2 * sin(1.4)
ans =
    1.9709
```

1.4.1 The Format Command and Ellipsis

The ***default*** in MATLAB is to display numbers that have decimal points with four decimal places, as shown in the previous example. (The default means if you do not specify otherwise, this is what you get.) The **format** command can be used to specify the output format of expressions.

There are many options, including making the format **short** (the default) or **long**. For example, changing the format to **long** will result in 15 decimal places.

This will remain in effect until the format is changed back to **short**, as demonstrated in the following.

```
>> format long
>>2 * sin(1.4)
ans =
    1.970899459976920

>> format short
>> 2 * sin(1.4)
ans =
    1.9709
```

The **format** command can also be used to control the spacing between the MATLAB command or expression and the result; it can be either **loose** (the default) or **compact**.

```
>> format loose
>> 5*33
ans =

   165

>> format compact
>> 5*33
ans =
   165
>>
```

Especially long expressions can be continued on the next line by typing three (or more) periods, which is the ***continuation operator***, or the ***ellipsis***. To do this, type part of the expression followed by an ellipsis, then hit the Enter key and continue typing the expression on the next line.

```
>> 3 + 55 - 62 + 4 - 5...
+  22 -1

ans =
    16
```

1.4.2 Operators

There are in general two kinds of operators: ***unary*** operators, which operate on a single value, or ***operand***, and ***binary*** operators, which operate on two values or operands. The symbol "−," for example, is both the unary operator for negation and the binary operator for subtraction.

Here are some of the common operators that can be used with numerical expressions:

```
+  addition
-  negation, subtraction
```

```
*  multiplication
/  division (divided by e.g. 10/5 is 2)
\  division (divided into e.g. 5\10 is 2)
^  exponentiation (e.g. 5^2 is 25)
```

In addition to displaying numbers with decimal points, numbers can also be shown using ***scientific or exponential notation***. This uses ***e*** for the exponent of 10 raised to a power. For example, 2 * 10 ^ 4 could be written two ways:

```
>> 2 * 10^4
ans =
        20000
>> 2e4
ans =
        20000
```

1.4.2.1 Operator Precedence Rules

Some operators have ***precedence*** over others. For example, in the expression 4 + 5 * 3, the multiplication takes precedence over the addition, so first 5 is multiplied by 3, then 4 is added to the result. Using parentheses can change the precedence in an expression:

```
>> 4 + 5 * 3
ans =
    19
>> (4 + 5) * 3
ans =
    27
```

Within a given precedence level, the expressions are evaluated from left to right (this is called ***associativity***).

Nested parentheses are parentheses inside of others; the expression in the ***inner parentheses*** is evaluated first. For example, in the expression `5 – (6 * (4 + 2))`, first the addition is performed, then the multiplication, and finally the subtraction, to result in `–31`. Parentheses can also be used simply to make an expression clearer. For example, in the expression `((4 + (3 * 5)) – 1)`, the parentheses are not necessary, but are used to show the order in which the parts of the expression will be evaluated.

For the operators that have been covered thus far, the following is the precedence (from the highest to the lowest):

```
( )        parentheses
^          exponentiation
-          negation
*, /, \    all multiplication and division
+, -       addition and subtraction
```

PRACTICE 1.1

Think about what the results would be for the following expressions, and then type them in to verify your answers:

```
1\2
-5 ^ 2
(-5) ^ 2
10 - 6/2
5 * 4/2 * 3
```

1.4.3 Built-in Functions and Help

There are many built-in functions in MATLAB. The **help** command can be used to identify MATLAB functions, and also how to use them. For example, typing **help** at the prompt in the Command Window will show a list of ***help topics*** that are groups of related functions. This is a very long list; the most elementary help topics appear at the beginning. Also, if you have any Toolboxes installed, these will be listed.

For example, one of the elementary help topics is listed as **matlab\elfun**; it includes the elementary math functions. Another of the first help topics is **matlab\ops**, which shows the operators that can be used in expressions.

To see a list of the functions contained within a particular help topic, type **help** followed by the name of the topic. For example,

```
>> help elfun
```

will show a list of the elementary math functions. It is a very long list, and it is broken into trigonometric (for which the default is radians, but there are equivalent functions that instead use degrees), exponential, complex, and rounding and remainder functions.

To find out what a particular function does and how to call it, type **help** and then the name of the function. For example, the following will give a description of the **sin** function.

```
>> help sin
```

Note that clicking on the *fx* at the left of the prompt in the Command Window also allows one to browse through the functions in the help topics. Choosing the Help button under Resources to bring up the Documentation page for MATLAB is another method for finding functions by category.

To ***call a function***, the name of the function is given followed by the ***argument(s)*** that are passed to the function in parentheses. Most functions then ***return value(s)***. For example, to find the absolute value of −4, the following expression would be entered:

```
>> abs(-4)
```

which is a ***call*** to the function **abs**. The number in the parentheses, the -4, is the ***argument***. The value 4 would then be ***returned*** as a result.

QUICK QUESTION!

What would happen if you use the name of a function, for example, **abs**, as a variable name?

Answer: This is allowed in MATLAB, but then **abs** could not be used as the built-in function until the variable is cleared. For example, examine the following sequence:

```
>> clear
>> abs(-6)
ans =
     6
>> abs = 11
abs =
    11
>> abs(-6)
Subscript indices must either be real positive
integers or logicals.
>> who
Your variables are:
abs   ans
>> clear abs
>> who
Your variables are:
ans
>> abs(-6)
ans =
     6
```

All of the operators have a functional form. For example, 2 + 5 can be written using the **plus** function as follows.

```
>> plus(2,5)
ans =
     7
```

MATLAB has a useful shortcut that is called the ***tab completion*** feature. If you type the beginning characters in the name of a function, and hit the tab key, a list of functions that begin with the typed characters pops up. This feature has improved over the years; beginning in R2015b, capitalization errors are automatically fixed.

Also, if a function name is typed incorrectly, MATLAB will suggest a correct name.

```
>> abso(-4)
Undefined function or variable 'abso'.
Did you mean:
>> abs(-4)
```

1.4.4 Constants

Variables are used to store values that might change, or for which the values are not known ahead of time. Most languages also have the capacity to store ***constants***, which are values that are known ahead of time and cannot possibly change. An example of a constant value would be **pi**, or π, which is

3.14159... In MATLAB, there are functions that return some of these constant values, some of which include:

pi	3.14159....
i	$\sqrt{-1}$
j	$\sqrt{-1}$
inf	infinity ∞
NaN	stands for "not a number," such as the result of 0/0

1.4.5 Random Numbers

When a program is being written to work with data, and the data are not yet available, it is often useful to test the program first by initializing the data variables to ***random numbers***. Random numbers are also useful in simulations. There are several built-in functions in MATLAB that generate random numbers, some of which will be illustrated in this section.

Random number generators or functions are not truly random. Basically, the way it works is that the process starts with one number, which is called the ***seed***. Frequently, the initial seed is either a predetermined value or it is obtained from the built-in clock in the computer. Then, based on this seed, a process determines the next "random number." Using that number as the seed the next time, another random number is generated, and so forth. These are actually called ***pseudorandom***—they are not truly random because there is a process that determines the next value each time.

The function **rand** can be used to generate uniformly distributed random real numbers; calling it generates one random real number in the ***open interval*** (0,1), which means that the endpoints of the range are not included. There are no arguments passed to the **rand** function in its simplest form. Here are two examples of calling the **rand** function:

```
>> rand
ans =
    0.8147
>> rand
ans =
    0.9058
```

The seed for the **rand** function will always be the same each time MATLAB is started, unless the initial seed is changed. The **rng** function sets the initial seed. There are several ways in which it can be called:

```
>> rng('shuffle')
>> rng(intseed)
>> rng('default')
```

With 'shuffle', the **rng** function uses the current date and time that are returned from the built-in **clock** function to set the seed, so the seed will always be different. An integer can also be passed to be the seed. The 'default' option will set the seed to the default value used when MATLAB starts up. The **rng** function can also be called with no arguments, which will return the current state of the random number generator:

```
>> state_rng = rng;    % gets state
>> randone = rand
randone =
    0.1270
>> rng(state_rng);
>> randtwo = rand    %
randtwo =
    0.1270
```

Note
The words after the % are comments and are ignored by MATLAB.

The random number g [illegible] en-
erates what is called th [illegible] om
functions get their val [illegible]

As **rand** returns a real [illegible] esult
by an integer *N* woul [illegible] , *N*).
For example, multipl [illegible] o the
expression

```
rand*10
```

would return a resu [illegible]

To generate a rando [illegible] ate the
variables *low* and *h* [illegible] ow. For
example, the seque [illegible]

```
>> low = 3;
>> high = 5;
>> rand * (high - lo
```

would generate a random real number in the open inte [illegible]

The function **randn** is used to generate normally distributed random real numbers.

1.4.5.1 Generating Random Integers

As the **rand** function returns a real number, this can be rounded to produce a random integer. For example,

```
>> round(rand*10)
```

would generate one random integer in the range from 0 to 10 inclusive (`rand*10` would generate a random real in the open interval (0, 10); rounding that will return an integer). However, these integers would not be evenly distributed in the range. A better method is to use the function **randi**, which in its simplest form **randi(imax)** returns a random integer in the range from 1 to imax, inclusive. For example, **randi(4)** returns a random integer in the range from 1 to 4. A range can also be passed, for example, **randi ([imin, imax])** returns a random integer in the inclusive range from imin to imax:

```
>> randi([3, 6])
ans =
     4
```

PRACTICE 1.2

Generate a random

- real number in the range (0,1)
- real number in the range (0, 100)
- real number in the range (20, 35)
- integer in the inclusive range from 1 to 100
- integer in the inclusive range from 20 to 35

1.5 CHARACTERS AND STRINGS

A character in MATLAB is represented using single quotes (e.g., 'a' or 'x'). The quotes are necessary to denote a character; without them, a letter would be interpreted as a variable name. Characters are put in an order using what is called a ***character encoding***. In the character encoding, all characters in the computer's ***character set*** are placed in a sequence and given equivalent integer values. The character set includes all letters of the alphabet, digits, and punctuation marks; basically, all of the keys on a keyboard are characters. Special characters, such as the Enter key, are also included. So, 'x', '!', and '3' are all characters. With quotes, '3' is a character, not a number.

Notice the difference in the formatting (the indentation) when a number is displayed versus a character:

```
>> var = 3
var =
     3
>> var = '3'
var =
3
```

MATLAB also handles strings, which are sequences of characters in single quotes.

```
>> myword = 'hello'
myword =
hello
```

The most common character encoding is the American Standard Code for Information Interchange, or ASCII. Standard ASCII has 128 characters, which have equivalent integer values from 0 to 127. The first 32 (integer values 0 through 31) are nonprinting characters. The letters of the alphabet are in order, which means 'a' comes before 'b', then 'c', and so forth. MATLAB actually can use a much larger encoding sequence, which has the same first 128 characters as ASCII. More on the character encoding and converting characters to their numerical values will be covered in Section 1.7.

1.6 RELATIONAL EXPRESSIONS

Expressions that are conceptually either true or false are called ***relational expressions***; they are also sometimes called ***Boolean expressions*** or ***logical expressions***. These expressions can use both ***relational operators***, which relate two expressions of compatible types, and ***logical operators***, which operate on **logical** operands.

The relational operators in MATLAB are:

Operator	Meaning
>	greater than
<	less than
>=	greater than or equals
<=	less than or equals
==	equality
~=	inequality

All these concepts should be familiar, although the actual operators used may be different from those used in other programming languages, or in mathematics classes. In particular, it is important to note that the operator for equality is two consecutive equal signs, not a single equal sign (as the single equal sign is already used as the assignment operator).

For numerical operands, the use of these operators is straightforward. For example, `3 < 5` means "3 less than 5," which is, conceptually, a true expression. In MATLAB, as in many programming languages, "true" is represented by the **logical** value 1, and "false" is represented by the **logical** value 0. So, the expression `3 < 5` actually displays in the Command Window the value 1 (**logical**) in MATLAB. Displaying the result of expressions like this in the Command Window demonstrates the values of the expressions.

```
>> 3 < 5
ans =
      1

>> 2 > 9
ans =
      0
>> class(ans)
ans =
logical
```

The type of the result is **logical**, not **double**. MATLAB also has built-in **true** and **false**.

```
>> true
ans =
      1
```

In other words, **true** is equivalent to **logical(1)** and **false** is equivalent to **logical (0)**. (In some versions of MATLAB, the value shown for the result of these expressions is **true** or **false** in the Workspace Window.) Although these are **logical** values, mathematical operations could be performed on the resulting 1 or 0.

```
>> logresult = 5 < 7
logresult =
    1
>> logresult + 3
ans =
    4
```

Comparing characters (e.g., 'a' < 'c') is also possible. Characters are compared using their ASCII equivalent values in the character encoding. So, 'a' < 'c' is a **true** expression because the character 'a' comes before the character 'c'.

```
>> 'a' < 'c'
ans =
    1
```

The logical operators are:

Operator	Meaning
||	or
&&	and
~	not

All logical operators operate on **logical** or Boolean operands. The **not** operator is a unary operator; the others are binary. The **not** operator will take a **logical** expression, which is **true** or **false**, and give the opposite value. For example, `~(3 <5)` is **false** as `(3 <5)` is **true**. The **or** operator has two **logical** expressions as operands. The result is **true** if either or both of the operands are **true**, and **false** only if both operands are **false**. The **and** operator also operates on two **logical** operands. The result of an **and** expression is **true** only if both operands are **true**; it is **false** if either or both are **false**. The or/and operators shown here are used for *scalars*, or single values. Other or/and operators will be explained in Chapter 2.

The || and && operators in MATLAB are examples of operators that are known as ***short-circuit*** operators. What this means is that if the result of the expression can be determined based on the first part, then the second part will not even be evaluated. For example, in the expression:

```
2 < 4 || 'a' == 'c'
```

the first part, 2 < 4, is **true** so the entire expression is **true**; the second part 'a' == 'c' would not be evaluated.

In addition to these logical operators, MATLAB also has a function **xor**, which is the exclusive or function. It returns **logical true** if one (and only one) of the arguments is **true**. For example, in the following, only the first argument is **true**, so the result is **true**:

```
>> xor(3 < 5, 'a' > 'c')
ans =
     1
```

In this example, both arguments are **true** so the result is **false**:

```
>> xor(3 < 5, 'a' < 'c')
ans =
     0
```

Given the **logical** values of **true** and **false** in variables *x* and *y*, the ***truth table*** (see Table 1.1) shows how the logical operators work for all combinations. Note that the logical operators are ***commutative*** (e.g., *x* || *y* is the same as *y* || *x*).

As with the numerical operators, it is important to know the operator precedence rules. Table 1.2 shows the rules for the operators that have been covered thus far in the order of precedence.

QUICK QUESTION!

Assume that there is a variable *x* that has been initialized. What would be the value of the expression

```
3 < x < 5
```

if the value of *x* is 4? What if the value of *x* is 7?

Answer: The value of this expression will always be **logical true**, or 1, regardless of the value of the variable *x*. Expressions are evaluated from left to right. So, first the expression 3 < *x* will be evaluated. There are only two possibilities: either this will be **true** or **false**, which means that either the expression will have the **logical** value 1 or 0. Then, the rest of the expression will be evaluated, which will be either 1 < 5 or 0 < 5. Both of these expressions are **true**. So, the value of *x* does not matter: the expression 3 < *x* < 5 would be **true** regardless of the value of the variable *x*. This is a logical error; it would not enforce the desired range. If we wanted an expression that was **logical true** only if *x* was in the range from 3 to 5, we could write `3 < x && x < 5` (note that parentheses are not necessary).

Table 1.1 Truth Table for Logical Operators

x	y	~x	x || y	x && y	xor(x,y)
true	true	false	true	true	false
true	false	false	true	false	true
false	false	true	false	false	false

Table 1.2 Operator Precedence Rules

Operators	Precedence
Parentheses: ( )	Highest
Power ^	
Unary: Negation (-), not (~)	
Multiplication, division *, /, \	
Addition, subtraction +, -	
Relational <, <=, >, >=, ==, ~=	
And &&	
Or ||	
Assignment =	Lowest

PRACTICE 1.3

Think about what would be produced by the following expressions, and then type them in to verify your answers.

```
3 == 5 + 2

'b' < 'a' + 1

10 > 5 + 2

(10 > 5) + 2

'c' == 'd' - 1 && 2 < 4

'c' == 'd' - 1  || 2 > 4

xor('c' == 'd' - 1, 2 > 4)

xor('c' == 'd' - 1, 2 < 4)

10 > 5 > 2
```

Note: Be careful about using the equality and inequality operators with numbers. Occasionally, ***roundoff errors*** appear, which means that numbers are close to their correct value but not exact. For example, cos(pi/2) should be 0.

However, because of a roundoff error, it is a very small number but not exactly 0.

```
>> cos(pi/2)
ans =
   6.1232e-17
>> cos(pi/2) == 0
ans =
      0
```

1.7 TYPE RANGES AND TYPE CASTING

The ***range*** of a type, which indicates the smallest and largest numbers that can be stored in the type, can be calculated. For example, the type **uint8** stores 2 ^ 8 or 256 integers, ranging from 0 to 255. The range of values that can be stored in **int8**, however, is from −128 to +127. The range can be found for any type by passing the name of the type as a string (which means in single quotes) to the functions **intmin** and **intmax**. For example,

```
>> intmin('int8')
ans =
  -128
>> intmax('int8')
ans =
  127
```

There are many functions that convert values from one type to another. The names of these functions are the same as the names of the types. These names can be used as functions to convert a value to that type. This is called ***casting*** the value to a different type, or ***type casting***. For example, to convert a value from the type **double**, which is the default, to the type **int32**, the function **int32** would be used. Entering the assignment statement

```
>> val = 6 + 3;
```

would result in the number 9 being stored in the variable *val*, with the default type of **double**, which can be seen in the Workspace Window. Subsequently, the assignment statement

```
>> val = int32(val);
```

would change the type of the variable to **int32**, but would not change its value. Here is another example using two different variables.

```
>> num = 6 + 3;
>> numi = int32(num);
>> whos
  Name     Size     Bytes    Class    Attributes

  num      1x1          8    double
  numi     1x1          4    int32
```

Note that **whos** shows the type (class) of the variables as well as the number of bytes used to store the value of a variable. One byte is equivalent to eight bits, so the type **int32** uses 4 bytes.

One reason for using an integer type for a variable is to save space in memory.

QUICK QUESTION!

What would happen if you go beyond the range for a particular type? For example, the largest integer that can be stored in **int8** is 127, so what would happen if we type cast a larger integer to the type **int8**?

```
>> int8(200)
```

Answer: The value would be the largest in the range, in this case 127. If, instead, we use a negative number that is smaller than the lowest value in the range, its value would be −128. This is an example of what is called ***saturation arithmetic***.

```
>> int8(200)
ans =
  127
>> int8(-130)
ans =
 -128
```

PRACTICE 1.4

- Calculate the range of integers that can be stored in the types **int16** and **uint16**. Use **intmin** and **intmax** to verify your results.
- Enter an assignment statement and view the type of the variable in the Workspace Window. Then, change its type and view it again. View it also using **whos**.

There is also a function **cast** that can cast a variable to a particular type. This has an option to cast a variable to the same type as another, using 'like'.

```
>> a = uint16(43);
>> b = 11;
>> whos
  Name      Size      Bytes     Class     Attributes

  a         1x1           2     uint16
  b         1x1           8     double
>> b = cast(b,'like',a);
>> whos
  Name      Size      Bytes     Class     Attributes

  a         1x1           2     uint16
  b         1x1           2     uint16
```

The numeric functions can also be used to convert a character to its equivalent numerical value (e.g., **double** will convert to a **double** value, and **int32** will convert to an integer value using 32 bits). For example, to convert the character 'a' to its numerical equivalent, the following statement could be used:

```
>> numequiv = double('a')
numequiv =
    97
```

This stores the **double** value 97 in the variable *numequiv*, which shows that the character 'a' is the 98th character in the character encoding (as the equivalent numbers begin at 0). It doesn't matter which number type is used to convert 'a'; for example,

```
>> numequiv = int32('a')
```

would also store the integer value 97 in the variable *numequiv*. The only difference between these will be the type of the resulting variable (**double** in the first case, **int32** in the second).

The function **char** does the reverse; it converts from any number to the equivalent character:

```
>> char(97)
ans =
a
```

Note

Quotes are not shown when the character is displayed.

As the letters of the alphabet are in order, the character 'b' has the equivalent value of 98, 'c' is 99, and so on. Math can be done on characters. For example, to get the next character in the character encoding, 1 can be added either to the integer or the character:

```
>> numequiv = double('a');
>> char(numequiv+1)
ans =
b

>> 'a'+2
ans =
    99
```

The first 128 characters are equivalent to the 128 characters in standard ASCII. MATLAB uses an encoding, however, that has 65,535 characters. The characters from 128 to 65,535 depend on your computer's ***locale setting***, which sets the language for your interface; for example, 'en_US' is the locale for English in the United States.

Converting characters to their equivalent numerical values also works with strings. For example, using the **double** function on a string will show the equivalent numerical value of all characters in the string:

```
>> double('abcd')
ans =
    97   98   99   100
```

To shift the characters of a string "up" in the character encoding, an integer value can be added to a string. For example, the following expression will shift by one:

```
>> char('abcd'+ 1)
ans =
bcde
```

PRACTICE 1.5

- Find the numerical equivalent of the character 'x'.
- Find the character equivalent of 107.

1.8 BUILT-IN NUMERICAL FUNCTIONS

We have seen in Section 1.4.3 that the **help** command can be used to see help topics such as **elfun**, as well as the functions stored in each help topic.

MATLAB has many built-in trigonometric functions for sine, cosine, tangent, and so forth. For example, **sin** is the sine function in radians. The inverse, or arcsine function in radians is **asin**, the hyperbolic sine function in radians is **sinh**, and the inverse hyperbolic sine function is **asinh**. There are also functions that use degrees rather than radians: **sind** and **asind**. Similar variations exist for the other trigonometric functions.

In addition to the trigonometric functions, the **elfun** help topic also has some rounding and remainder functions that are very useful. Some of these include **fix**, **floor**, **ceil**, **round**, **mod**, **rem**, and **sign**.

Both the **rem** and **mod** functions return the remainder from a division; for example, 5 goes into 13 twice with a remainder of 3, so the result of this expression is 3:

```
>> rem(13,5)
ans =
     3
```

QUICK QUESTION!

What would happen if you reversed the order of the arguments by mistake, and typed the following:

```
rem(5,13)
```

Answer: The **rem** function is an example of a function that has two arguments passed to it. In some cases, the order in which the arguments are passed does not matter, but for the **rem** function the order does matter. The **rem** function divides the second argument into the first. In this case, the second argument, 13, goes into 5 zero times with a remainder of 5, so 5 would be returned as a result.

Another function in the **elfun** help topic is the **sign** function, which returns 1 if the argument is positive, 0 if it is 0, and −1 if it is negative. For example,

```
>> sign(-5)
ans =
    -1
>> sign(3)
ans =
     1
```

PRACTICE 1.6

Use the **help** function to find out what the rounding functions **fix**, **floor**, **ceil**, and **round** do. Experiment with them by passing different values to the functions, including some negative, some positive, and some with fractions less than 0.5 and some greater. *It is very important when testing functions that you test thoroughly by trying different kinds of arguments!*

As of R2014b, the **round** function has an option to round to a specified number of digits.

```
>> round(pi,3)
ans =
    3.1420
```

MATLAB has the exponentiation operator ^, and also the function **sqrt** to compute square roots and **nthroot** to find the nth root of a number. For example, the following expression finds the third root of 64.

```
>> nthroot(64,3)
ans =
     4
```

For the case in which $x = b^y$, y is the ***logarithm*** of x to base b, or in other words, $y = \log_b(x)$. Frequently-used bases include b = 10 (called the ***common logarithm***), b = 2 (used in many computing applications), and b = e (the constant e, which equals 2.7183); this is called the ***natural logarithm***. For example,

$100 = 10^2$ so $2 = \log_{10}(100)$
$32 = 2^5$ so $5 = \log_2(32)$

MATLAB has built-in functions to return logarithms:

log(x) returns the natural logarithm
log2(x) returns the base 2 logarithm
log10(x) returns the base 10 logarithm

MATLAB also has a built-in function **exp(n)**, which returns the constant e^n.

QUICK QUESTION!

There is no built-in constant for e (2.718), so how can that value be obtained in MATLAB?

Answer: Use the exponential function **exp**; e or e^1 is equivalent to **exp(1)**.

```
>> exp(1)
ans =
    2.7183
```

Note: Don't confuse the value e with the e used in MATLAB to specify an exponent for scientific notation.

In R2015b, the functions **deg2rad** and **rad2deg** were introduced to convert between degrees and radians, for example,

```
>> deg2rad(180)
ans =
    3.1416
```

■ Explore Other Interesting Features

This section lists some features and functions in MATLAB, related to those explained in this chapter, that you wish to explore on your own.

- Workspace Window: There are many other aspects of the Workspace window to explore. To try this, create some variables. Make the Workspace window the active window by clicking the mouse in it. From there, you can choose which attributes of variables to make visible by choosing Choose Columns from the menu. Also, if you double click on a variable in the Workspace window, this brings up a Variable Editor window that allows you to modify the variable.
- Click on the *fx* next to the prompt in the Command Window, and under MATLAB choose Mathematics, then Elementary Math, then Exponents and Logarithms to see more functions in this category.
- Use **help** to learn about the **path** function and related directory functions such as **addpath** and **which**.
- The **pow2** function.
- Functions related to type casting including **typecast.**
- Find the accuracy of the floating point representation for single and double precision using the **eps** function. ■

SUMMARY

COMMON PITFALLS

It is common when learning to program, to make simple spelling mistakes and to confuse the necessary punctuation. Examples are given here of some very common errors:

- Putting a space in a variable name
- Confusing the format of an assignment statement as

  ```
  expression = variablename
  ```

  rather than

  ```
  variablename = expression
  ```

  The variable name must always be on the left.
- Using a built-in function name as a variable name, and then trying to use the function
- Confusing the two division operators / and \
- Forgetting the operator precedence rules
- Confusing the order of arguments passed to functions—for example, to find the remainder of dividing 3 into 10 using **rem(3,10)** instead of **rem(10,3)**
- Not using different types of arguments when testing functions
- Forgetting to use parentheses to pass an argument to a function (e.g., "fix 2.3" instead of "**fix(2.3)**"). MATLAB returns the ASCII equivalent for each character when this mistake is made (what happens is that is that it is interpreted as the function of a string, "**fix('2.3')**").
- Confusing && and ||
- Confusing || and **xor**
- Putting a space in 2-character operators (e.g., typing "< =" instead of "<=")
- Using = instead of == for equality

PROGRAMMING STYLE GUIDELINES

Following these guidelines will make your code much easier to read and understand, and therefore easier to work with and modify.

- Use mnemonic variable names (names that make sense; for example, *radius* instead of *xyz*)
- Although variables named *result* and *RESULT* are different, avoid this as it would be confusing
- Do not use names of built-in functions as variable names
- Store results in named variables (rather than using *ans*) if they are to be used later
- Do not use *ans* in expressions
- Make sure variable names have fewer characters than **namelengthmax**
- If different sets of random numbers are desired, set the seed for the random functions using **rng**

MATLAB Functions and Commands			
demo	uint16	rand	round
help	uint32	rng	mod
lookfor	uint64	clock	rem
doc	char	randn	sign
quit	logical	randi	sqrt
exit	true	xor	nthroot
namelengthmax	false	intmin	log
who	class	intmax	log2
whos	format	cast	log10
clear	sin	asin	exp
single	abs	sinh	deg2rad
double	plus	asinh	rad2deg
int8	pi	sind	
int16	i	asind	
int32	j	fix	
int64	inf	floor	
uint8	NaN	ceil	

<table>
<tr><th colspan="4">MATLAB Operators</th></tr>
<tr><td>assignment =
ellipsis, or
continuation ...
addition +
negation -
subtraction –</td><td>multiplication *
divided by/
divided into \
exponentiation ^
parentheses ()
greater than ></td><td>less than <
greater than or
equals >=
less than or equals
<=
equality ==</td><td>inequality ~=
or for scalars ||
and for scalars &&
not ~</td></tr>
</table>

Exercises

1. Create a variable *myage* and store your age in it. Subtract 2 from the value of the variable. Add 1 to the value of the variable. Observe the Workspace Window and Command History Window as you do this.
2. Explain the difference between these two statements:

```
result = 9*2
result = 9*2;
```

3. Use the built-in function **namelengthmax** to find out the maximum number of characters that you can have in an identifier name under your version of MATLAB.
4. Create two variables to store a weight in pounds and ounces. Use **who** and **whos** to see the variables. Use **class** to see the types of the variables. Clear one of them and then use **who** and **whos** again.
5. Explore the **format** command in more detail. Use **help format** to find options. Experiment with **format bank** to display dollar values.

6. Find a **format** option that would result in the following output format:

```
>> 5/16 + 2/7
ans =
        67/112
```

7. Think about what the results would be for the following expressions, and then type them in to verify your answers.

```
25 / 5 * 5
4 + 3 ^ 2
(4 + 3) ^ 2
3 \ 12 + 5
4 - 2 * 3
```

As the world becomes more "flat," it is increasingly important for engineers and scientists to be able to work with colleagues in other parts of the world. Correct conversion of data from one system of units to another (e.g., from the metric system to the American system or vice versa) is critically important.

8. Create a variable *pounds* to store a weight in pounds. Convert this to kilograms and assign the result to a variable *kilos*. The conversion factor is 1 kg = 2.2 lb.
9. Create a variable *ftemp* to store a temperature in degrees Fahrenheit (F). Convert this to degrees Celsius (C) and store the result in a variable *ctemp*. The conversion factor is $C = (F - 32) \times 5/9$.
10. The following assignment statements either contain at least one error, or could be improved in some way. Assume that *radius* is a variable that has been initialized. First, identify the problem, and then fix and/or improve them:

```
33 = number
my variable = 11.11;
area = 3.14 * radius ^2;
x = 2 * 3.14 * radius;
```

11. Experiment with the functional form of some operators such as **plus**, **minus**, and **times**.
12. Generate a random
    - real number in the range (0, 20)
    - real number in the range (20, 50)
    - integer in the inclusive range from 1 to 10
    - integer in the inclusive range from 0 to 10
    - integer in the inclusive range from 50 to 100
13. Get into a new Command Window, and type **rand** to get a random real number. Make a note of the number. Then exit MATLAB and repeat this, again making a note of the random number; it should be the same as before. Finally, exit MATLAB and again get into a new Command Window. This time, change the seed before generating a random number; it should be different.
14. What is the difference between x and 'x'?

15. What is the difference between 5 and '5'?
16. The combined resistance R_T of three resistors R_1, R_2, and R_3 in parallel is given by

$$R_T = \frac{1}{\frac{1}{R_1} + \frac{1}{R_2} + \frac{1}{R_3}}$$

    Create variables for the three resistors and store values in each, and then calculate the combined resistance.
17. Explain the difference between constants and variables.
18. What would be the result of the following expressions?

```
'b' >= 'c' - 1
3 == 2 + 1
(3 == 2) + 1
xor(5 < 6, 8 > 4)
10 > 5 > 2
result = 3 ^2 - 20;
0 <= result <= 10
```

19. Create two variables *x* and *y* and store numbers in them. Write an expression that would be **true** if the value of *x* is greater than 5 or if the value of *y* is less than 10, but not if both of those are **true**.
20. Use the equality operator to verify that 3*10 ^5 is equal to 3e5.
21. In the ASCII character encoding, the letters of the alphabet are in order: 'a' comes before 'b' and also 'A' comes before 'B'. However, which comes first—lower or uppercase letters?
22. Are there equivalents to **intmin** and **intmax** for real number types? Use **help** to find out.
23. Use **intmin** and **intmax** to determine the range of values that can be stored in the types **uint32** and **uint64**.
24. Use the **cast** function to cast a variable to be the same type as another variable.
25. Use **help elfun** or experiment to answer the following questions:
    - Is **fix(3.5)** the same as **floor(3.5)**?
    - Is **fix(3.4)** the same as **fix(-3.4)**?
    - Is **fix(3.2)** the same as **floor(3.2)**?
    - Is **fix(-3.2)** the same as **floor(-3.2)**?
    - Is **fix(-3.2)** the same as **ceil(-3.2)**?
26. For what range of values is the function **round** equivalent to the function **floor**? For what range of values is the function **round** equivalent to the function **ceil**?
27. Use **help** to determine the difference between the **rem** and **mod** functions.
28. Find MATLAB expressions for the following

$\sqrt{19}$

$3^{1.2}$

$\tan(\pi)$

29. Using only the integers 2 and 3, write as many expressions as you can that result in 9. Try to come up with at least 10 different expressions (Note: don't just change the order). Be creative! Make sure that you write them as MATLAB expressions. Use operators and/or built-in functions.
30. A vector can be represented by its rectangular coordinates *x* and *y* or by its polar coordinates *r* and θ. Theta is measured in radians. The relationship between them is given by the equations:

    ```
    x = r * cos(θ)
    y = r * sin(θ)
    ```

    Assign values for the polar coordinates to variables *r* and *theta*. Then, using these values, assign the corresponding rectangular coordinates to variables *x* and *y*.
31. In special relativity, the Lorentz factor is a number that describes the effect of speed on various physical properties when the speed is significantly relative to the speed of light. Mathematically, the Lorentz factor is given as:

    $$\gamma = \frac{1}{\sqrt{1 - \frac{v^2}{c^2}}}$$

    Use 3×10^8 m/s for the speed of light, *c*. Create variables for *c* and the speed *v* and from them a variable *lorentz* for the Lorentz factor.
32. A company manufactures a part for which there is a desired weight. There is a tolerance of *N*%, meaning that the range between minus and plus *N*% of the desired weight is acceptable. Create a variable that stores a weight and another variable for *N* (e.g., set it to 2). Create variables that store the minimum and maximum values in the acceptable range of weights for this part.
33. An environmental engineer has determined that the cost *C* of a containment tank will be based on the radius *r* of the tank:

    $$C = \frac{32,430}{r} + 428\pi r$$

    Create a variable for the radius, and then for the cost.
34. A chemical plant releases an amount *A* of pollutant into a stream. The maximum concentration *C* of the pollutant at a point which is at a distance *x* from the plant is:

    $$C = \frac{A}{x}\sqrt{\frac{2}{\Pi e}}$$

    Create variables for the values of *A* and *x*, and then for *C*. Assume that the distance *x* is in meters. Experiment with different values for *x*.

35. The geometric mean g of n numbers x_i is defined as the nth root of the product of x_i:

    $$g = \sqrt[n]{x_1 x_2 x_3 \ldots x_n}$$

    (This is useful, for example, in finding the average rate of return for an investment which is something you'd do in engineering economics.) If an investment returns 15% the first year, 50% the second, and 30% the third year, the average rate of return would be `(1.15*1.50*1.30)`$^{1/3}$. Compute this.

36. Use the **deg2rad** function to convert 180 degrees to radians.

CHAPTER 2

Vectors and Matrices

KEY TERMS

vectors
matrices
row vector
column vector
scalar
elements
array
array operations
colon operator
iterate
step value
concatenating
index
subscript
index vector
transpose
square matrix
subscripted indexing
unwinding a matrix
linear indexing
column major order
columnwise
dimensions
vector of variables
empty vector
deleting elements
three-dimensional matrices
cumulative sum
cumulative product
running sum
nesting calls
scalar multiplication
array operations
array multiplication
array division
logical vector
logical indexing
zero crossings
matrix multiplication
inner dimensions
outer dimensions
dot product or inner product
cross product or outer product

CONTENTS

MATLAB® is short for Matrix Laboratory. Everything in MATLAB is written to work with vectors and matrices. This chapter will introduce vectors and matrices. Operations on vectors and matrices and built-in functions that can be used to simplify code will also be explained. The matrix operations and functions described in this chapter will form the basis for vectorized coding, which will be explained in Chapter 5.

MATLAB®. http://dx.doi.org/10.1016/B978-0-12-804525-1.00002-7

2.1 VECTORS AND MATRICES

Vectors and ***matrices*** are used to store sets of values, all of which are of the same type. A matrix can be visualized as a table of values. The dimensions of a matrix are $r \times c$, where r is the number of rows and c is the number of columns. This is pronounced "r by c." A vector can be either a ***row vector*** or a ***column vector***. If a vector has n elements, a row vector would have the dimensions $1 \times n$, and a column vector would have the dimensions $n \times 1$. A ***scalar*** (one value) has the dimensions 1×1. Therefore, vectors and scalars are actually just special cases of matrices.

Here are some diagrams showing, from left to right, a scalar, a column vector, a row vector, and a matrix:

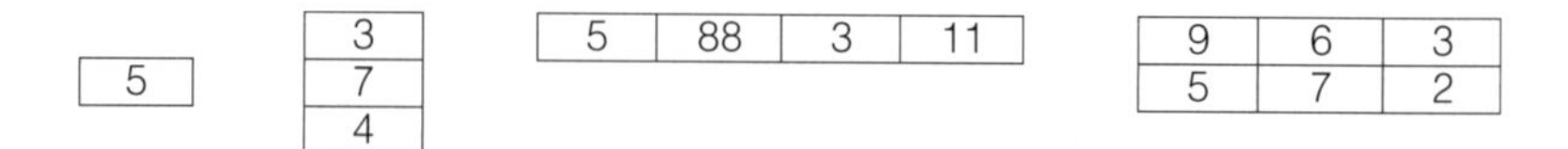

The scalar is 1×1, the column vector is 3×1 (three rows by one column), the row vector is 1×4 (one row by four columns), and the matrix is 2×3 (two rows by three columns). All of the values stored in these matrices are stored in what are called ***elements***.

MATLAB is written to work with matrices and so it is very easy to create vector and matrix variables, and there are many operations and functions that can be used on vectors and matrices.

A vector in MATLAB is equivalent to what is called a one-dimensional ***array*** in other languages. A matrix is equivalent to a two-dimensional array. Usually, even in MATLAB, some operations that can be performed on either vectors or matrices are referred to as ***array operations***. The term array is also frequently used to mean generically either a vector or a matrix.

2.1.1 Creating Row Vectors

There are several ways to create row vector variables. The most direct way is to put the values that you want in the vector in square brackets, separated by either spaces or commas. For example, both of these assignment statements create the same vector v:

```
>> v = [1  2  3  4]
   v =
     1  2  3  4

>> v = [1,2,3,4]
   v =
     1  2  3  4
```

Both of these create a row vector variable that has four elements; each value is stored in a separate element in the vector. The vector is 1×4.

2.1.1.1 The Colon Operator and linspace Function

If, as in the preceding examples, the values in the vector are regularly spaced, the ***colon operator*** can be used to ***iterate*** through these values. For example, 2:6 results in all of the integers from 2 to 6 inclusive:

```
>> vec = 2:6
vec =
        2     3     4     5     6
```

In this vector, there are five elements; the vector is a 1×5 row vector. Note that in this case, the brackets [ ] are not necessary to define the vector.

With the colon operator, a ***step value*** can also be specified by using another colon, in the form (first:step:last). For example, to create a vector with all integers from 1 to 9 in steps of 2:

```
>> nv = 1:2:9
nv =
        1   3   5   7   9
```

QUICK QUESTION!

What happens if adding the step value would go beyond the range specified by the last, for example,

```
1:2:6
```

Answer: This would create a vector containing 1, 3, and 5. Adding 2 to the 5 would go beyond 6, so the vector stops at 5; the result would be

```
1     3     5
```

QUICK QUESTION!

How can you use the colon operator to generate the vector shown below?

```
9   7   5   3   1
```

Answer: `9:-2:1`
The step value can be a negative number, so the resulting sequence is in descending order (from highest to lowest).

The **linspace** function creates a linearly spaced vector; **linspace(x,y,n)** creates a vector with *n* values in the inclusive range of *x* to *y*. If *n* is omitted, the default is 100 points. For example, the following creates a vector with five values linearly spaced between 3 and 15, including the 3 and 15:

```
>> ls = linspace(3,15,5)
ls =
     3     6     9    12    15
```

Similarly, the **logspace** function creates a logarithmically spaced vector; **logspace(x,y,n)** creates a vector with *n* values in the inclusive range from 10^x to 10^y. If *n* is omitted, the default is 50 points. For example, logspace (1,4,4) creates a vector with four elements, logarithmically spaced between 10^1 and 10^4, or in other words 10^1, 10^2, 10^3, and 10^4.

```
>> logspace(1,4,4)
ans =
        10        100        1000        10000
```

Vector variables can also be created using existing variables. For example, a new vector is created here consisting first of all of the values from *nv* followed by all values from *ls*:

```
>> newvec = [nv ls]
newvec =
   1  3  5  7  9  3  6  9  12  15
```

Putting two vectors together like this to create a new one is called ***concatenating*** the vectors.

2.1.1.2 Referring to and Modifying Elements

The elements in a vector are numbered sequentially; each element number is called the ***index***, or ***subscript***. In MATLAB, the indices start at 1. Normally, diagrams of vectors and matrices show the indices. For example, for the variable *newvec* created earlier, the indices 1–10 of the elements are shown above the vector:

newvec

1	2	3	4	5	6	7	8	9	10
1	3	5	7	9	3	6	9	12	15

A particular element in a vector is accessed using the name of the vector variable and the index or subscript in parentheses. For example, the fifth element in the vector *newvec* is a 9.

```
>> newvec(5)
ans =
     9
```

The expression *newvec(5)* would be pronounced "newvec sub 5," where sub is short for the word subscript. A subset of a vector, which would be a vector itself, can also be obtained using the colon operator. For example, the following statement would get the fourth through sixth elements of the vector *newvec*, and store the result in a vector variable *b*:

```
>> b = newvec(4:6)
b =
     7     9     3
```

Any vector can be used for the indices into another vector, not just one created using the colon operator. The indices do not need to be sequential. For example, the following would get the first, tenth, and fifth elements of the vector *newvec*:

```
>> newvec([1 10 5])
ans =
     1    15     9
```

The vector [1 10 5] is called an ***index vector***; it specifies the indices in the original vector that are being referenced.

The value stored in a vector element can be changed by specifying the index or subscript. For example, to change the second element from the preceding vector *b* to now store the value 11 instead of 9:

```
>> b(2) = 11
b =
     7    11     3
```

By referring to an index that does not yet exist, a vector can also be extended. For example, the following creates a vector that has three elements. By then assigning a value to the fourth element, the vector is extended to have four elements.

```
>> rv = [3 55 11]
rv =
    3    55    11
>> rv(4) = 2
rv =
    3    55    11    2
```

If there is a gap between the end of the vector and the specified element, 0s are filled in. For example, the following extends the variable *rv* again:

```
>> rv(6) = 13
rv =
     3    55    11     2     0    13
```

As we will see later, this is actually not a good idea. It is not very efficient because it can take extra time.

PRACTICE 2.1

Think about what would be produced by the following sequence of statements and expressions, and then type them in to verify your answers:

```
pvec = 3:2:10
pvec(2) = 15
pvec(7) = 33
pvec([2:4 7])
linspace(5,11,3)
logspace(2,4,3)
```

2.1.2 Creating Column vectors

One way to create a column vector is to explicitly put the values in square brackets, separated by semicolons (rather than commas or spaces):

```
>> c = [1; 2; 3; 4]
c =
     1
     2
     3
     4
```

There is no direct way to use the colon operator to get a column vector. However, any row vector created using any method can be ***transposed*** to result in a column vector. In general, the transpose of a matrix is a new matrix in which the rows and columns are interchanged. For vectors, transposing a row vector results in a column vector, and transposing a column vector results in a row vector. In MATLAB, the apostrophe is built-in as the transpose operator.

```
>> r = 1:3;
>> c = r'
c =
     1
     2
     3
```

2.1.3 Creating Matrix Variables

Creating a matrix variable is simply a generalization of creating row and column vector variables. That is, the values within a row are separated by either spaces or commas, and the different rows are separated by semicolons. For example, the matrix variable *mat* is created by explicitly entering values:

```
>> mat = [4 3 1; 2 5 6]
mat =
    4  3  1
    2  5  6
```

There must always be the same number of values in each row and each column of a matrix. For example, if you attempt to create a matrix in which there are different numbers of values in the rows, the result will be an error message, such as in the following:

```
>> mat = [3 5 7; 1 2]
Error using vertcat
Dimensions of matrices being concatenated are not consistent.
```

Iterators can be used for the values in the rows using the colon operator. For example:

```
>> mat = [2:4; 3:5]
mat =
     2    3    4
     3    4    5
```

The separate rows in a matrix can also be specified by hitting the Enter key after each row instead of typing a semicolon when entering the matrix values, as in:

```
>> newmat = [2 6 88
33 5 2]
 newmat =
      2    6   88
     33    5    2
```

Matrices of random numbers can be created using the **rand** function. If a single value n is passed to **rand**, an $n \times n$ matrix will be created; this is called a ***square matrix*** (same number of rows and columns).

```
>> rand(2)
ans =
     0.2311    0.4860
     0.6068    0.8913
```

If instead two arguments are passed, they specify the number of rows and columns in that order.

```
>> rand(1,3)
ans =
     0.7621   0.4565   0.0185
```

Matrices of random integers can be generated using **randi**; after the range is passed, the dimensions of the matrix are passed (again, using one value n for an $n \times n$ matrix, or two values for the dimensions):

```
>> randi([5, 10], 2)
ans =
       8    10
       9     5

>> randi([10, 30], 2, 3)
ans =
      21      10      13
      19      17      26
```

Note that the range can be specified for **randi**, but not for **rand**. The format for calling these functions is different. There are a number of ways in which **randi** can be called; use **help** to see them.

MATLAB also has several functions that create special matrices. For example, the **zeros** function creates a matrix of all zeros and the **ones** function creates a matrix of all ones. Like **rand**, either one argument can be passed (which will be both the number of rows and columns), or two arguments (first the number of rows and then the number of columns).

```
>> zeros(3)
ans =
     0     0     0
     0     0     0
     0     0     0

>> ones(2,4)
ans =
     1     1     1     1
     1     1     1     1
```

Note that there is no twos function, or tens, or fifty-threes—just **zeros** and **ones**!

2.1.3.1 Referring to and Modifying Matrix Elements

To refer to matrix elements, the row and then the column subscripts are given in parentheses (always the row first and then the column). For example, this creates a matrix variable *mat* and then refers to the value in the second row, third column of *mat*:

```
>> mat = [2:4; 3:5]
mat =
   2   3   4
   3   4   5

>> mat(2,3)
ans =
   5
```

This is called ***subscripted indexing***; it uses the row and column subscripts. It is also possible to refer to a subset of a matrix. For example, this refers to the first and second rows, second and third columns:

```
>> mat(1:2,2:3)
ans =
   3   4
   4   5
```

Using just one colon by itself for the row subscript means all rows, regardless of how many, and using a colon for the column subscript means all columns. For example, this refers to all columns within the first row or, in other words, the entire first row:

```
>> mat(1,:)
ans =
   2   3   4
```

This refers to the entire second column:

```
>> mat(:, 2)
ans =
     3
     4
```

If a single index is used with a matrix, MATLAB ***unwinds*** the matrix column by column. For example, for the matrix *intmat* created here, the first two elements are from the first column, and the last two are from the second column:

```
>> intmat = [100 77; 28 14]
intmat =
    100    77
     28    14
>> intmat(1)
an =
    100
>> intmat(2)
ans =
     28
>> intmat(3)
ans =
     77
>> intmat(4)
ans =
     14
```

This is called ***linear indexing***. Note that it is usually much better style when working with matrices to use subscripted indexing.

MATLAB stores matrices in memory in ***column major order***, or ***columnwise***, which is why linear indexing refers to the elements in order by columns.

An individual element in a matrix can be modified by assigning a new value to it.

```
>> mat = [2:4; 3:5];
>> mat(1,2) = 11
mat =
       2     11      4
       3      4      5
```

An entire row or column could also be changed. For example, the following replaces the entire second row with values from a vector obtained using the colon operator.

```
>> mat(2,:) = 5:7
mat =
     2    11     4
     5     6     7
```

Notice that as the entire row is being modified, a row vector with the correct length must be assigned.

Any subset of a matrix can be modified, as long as what is being assigned has the same number of rows and columns as the subset being modified.

The exception to this rule is that a scalar can be assigned to any size subset of a vector or matrix; what happens is that the same scalar is assigned to every element referenced. For example,

```
>> m = randi([10 50], 3,5)
m =
    38    11    38    11    41
    11    13    23    27    42
    21    43    48    25    17
>> m(2:3,3:5) = 3
m =
    38    11    38    11    41
    11    13     3     3     3
    21    43     3     3     3
```

To extend a matrix an individual element could not be added as that would mean there would no longer be the same number of values in every row. However, an entire row or column could be added. For example, the following would add a fourth column to the matrix *mat* created previously.

```
>> mat(:,4) = [9 2]'
mat =
     2    11     4     9
     5     6     7     2
```

Just as we saw with vectors, if there is a gap between the current matrix and the row or column being added, MATLAB will fill in with zeros.

```
>> mat(4,:) = 2:2:8
mat =
     2    11     4     9
     5     6     7     2
     0     0     0     0
     2     4     6     8
```

2.1.4 Dimensions

The **length** and **size** functions in MATLAB are used to find ***dimensions*** of vectors and matrices. The **length** function returns the number of elements in a vector. The **size** function returns the number of rows and columns in a vector or matrix. For example, the following vector *vec* has four elements, so its length is 4. It is a row vector, so the size is 1×4.

```
>> vec = -2:1
vec =
   -2   -1   0   1
>> length(vec)
ans =
   4
>> size(vec)
ans =
   1   4
```

To create the following matrix variable *mat*, iterators are used on the two rows and then the matrix is transposed so that it has three rows and two columns or, in other words, the size is 3×2.

```
>> mat = [1:3; 5:7]'
mat =
    1   5
    2   6
    3   7
```

The **size** function returns the number of rows and then the number of columns, so to capture these values in separate variables we put a ***vector of variables*** (two) on the left of the assignment. The variable *r* stores the first value returned, which is the number of rows, and *c* stores the number of columns.

```
>> [r, c] = size(mat)
r =
   3
c =
   2
```

Note that this example demonstrates very important and unique concepts in MATLAB: the ability to have a function return multiple values and the ability to have a vector of variables on the left side of an assignment in which to store the values.

If called as just an expression, the **size** function will return both values in a vector:

```
>> size(mat)
ans =
    3   2
```

For a matrix, the **length** function will return either the number of rows or the number of columns, whichever is largest (in this case the number of rows, 3).

```
>> length(mat)
ans =
      3
```

QUICK QUESTION!

How could you create a matrix of zeros with the same size as another matrix?

Answer: For a matrix variable *mat*, the following expression would accomplish this:

```
zeros(size(mat))
```

The **size** function returns the size of the matrix, which is then passed to the **zeros** function, which then returns a matrix of zeros with the same size as *mat*. It is not necessary in this case to store the values returned from the **size** function in variables.

MATLAB also has a function **numel** which returns the total number of elements in any array (vector or matrix):

```
>> vec = 9:-2:1
vec =
      9     7     5     3     1
>> numel(vec)
ans =
      5

>> mat = [3:2:7; 9 33 11]
mat =
      3     5     7
      9    33    11
>> numel(mat)
ans =
      6
```

For vectors, **numel** is equivalent to the **length** of the vector. For matrices, it is the product of the number of rows and columns.

It is important to note that in programming applications, it is better to not assume that the dimensions of a vector or matrix are known. Instead, to be general, use either the **length** or **numel** function to determine the number of elements in a vector, and use **size** (and store the result in two variables) for a matrix.

MATLAB also has a built-in expression **end** that can be used to refer to the last element in a vector; for example, *v(end)* is equivalent to *v(length(v))*. For matrices, it can refer to the last row or column. So, for example, using **end** for the row index would refer to the last row.

In this case, the element referred to is in the first column of the last row:

```
>> mat = [1:3; 4:6]'
mat =
      1     4
      2     5
      3     6
>> mat(end,1)
ans =
      3
```

Using **end** for the column index would refer to a value in the last column (e.g., the last column of the second row):

```
>> mat(2,end)
ans =
      5
```

The expression **end** can only be used as an index.

2.1.4.1 Changing Dimensions

In addition to the transpose operator, MATLAB has several built-in functions that change the dimensions or configuration of matrices (or in many cases vectors), including **reshape**, **fliplr**, **flipud**, **flip**, and **rot90**.

The **reshape** function changes the dimensions of a matrix. The following matrix variable *mat* is 3×4 or, in other words, it has 12 elements (each in the range from 1 to 100).

```
>> mat = randi(100, 3, 4)
    14    61     2    94
    21    28    75    47
    20    20    45    42
```

These 12 values could instead be arranged as a 2×6 matrix, 6×2, 4×3, 1×12, or 12×1. The **reshape** function iterates through the matrix columnwise. For example, when reshaping *mat* into a 2×6 matrix, the values from the first column in the original matrix (14, 21, and 20) are used first, then the values from the second column (61, 28, and 20), and so forth.

```
>> reshape(mat,2,6)
ans =
    14    20    28     2    45    47
    21    61    20    75    94    42
```

Note that in these examples *mat* is unchanged; instead, the results are stored in the default variable *ans* each time.

There are several functions that flip arrays. The **fliplr** function "flips" the matrix from left to right (in other words the left-most column, the first column, becomes the last column and so forth), and the **flipud** function flips up to down.

```
>> mat
mat =
    14    61    2    94
    21    28   75    47
    20    20   45    42
>> fliplr(mat)
ans =
    94     2   61    14
    47    75   28    21
    42    45   20    20

>> mat
mat =
    14    61    2    94
    21    28   75    47
    20    20   45    42
>> flipud(mat)
ans =
    20    20    45    42
    21    28    75    47
    14    61     2    94
```

The **flip** function, introduced in R2013b, flips any array; it flips a vector (left to right if it is a row vector or up to down if it is a column vector) or a matrix (up to down by default).

The **rot90** function rotates the matrix counterclockwise 90 degrees, so for example, the value in the top right corner becomes instead the top left corner and the last column becomes the first row.

```
>> mat
mat =
    14    61     2    94
    21    28    75    47
    20    20    45    42
>> rot90(mat)
ans =
    94    47    42
     2    75    45
    61    28    20
    14    21    20
```

QUICK QUESTION!

Is there a ***rot180*** function? Is there a ***rot-90*** function (to rotate clockwise)?

Answer: Not exactly, but a second argument can be passed to the **rot90** function which is an integer *n*; the function will rotate 90**n* degrees. The integer can be positive or negative. For example, if 2 is passed, the function will rotate the matrix 180 degrees (so, it would be the same as rotating the result of **rot90** another 90 degrees).

```
>> mat
mat =
    14    61     2    94
    21    28    75    47
    20    20    45    42
>> rot90(mat,2)
ans =
    42    45    20    20
    47    75    28    21
    94     2    61    14
```

If a negative number is passed for *n*, the rotation would be in the opposite direction, that is, clockwise.

```
>> mat
mat =
    14    61     2    94
    21    28    75    47
    20    20    45    42
>> rot90(mat,-1)
ans =
    20    21    14
    20    28    61
    45    75     2
    42    47    94
```

The function **repmat** can be used to create a matrix; **repmat(mat,m,n)** creates a larger matrix that consists of an $m \times n$ matrix of copies of *mat*. For example, here is a 2×2 random matrix:

```
>> intmat = randi(100,2)
intmat =
    50    34
    96    59
```

Replicating this matrix six times as a 3×2 matrix would produce copies of *intmat* in this form:

intmat	intmat
intmat	intmat
intmat	intmat

```
>> repmat(intmat,3,2)
ans =
    50  34  50  34
    96  59  96  59
    50  34  50  34
    96  59  96  59
    50  34  50  34
    96  59  96  59
```

The function **repelem**, on the other hand, replicates each element from a matrix in the dimensions specified; this function was introduced in R2015a.

```
>> repelem(intmat,3,2)
ans =
    50  50  34  34
    50  50  34  34
    50  50  34  34
    96  96  59  59
    96  96  59  59
    96  96  59  59
```

2.1.5 Empty Vectors

An ***empty vector*** (a vector that stores no values) can be created using empty square brackets:

```
>> evec = []
evec =
      []
>> length(evec)
ans =
    0
```

Values can then be added to an empty vector by concatenating, or adding, values to the existing vector. The following statement takes what is currently in *evec*, which is nothing, and adds a 4 to it.

Note

There is a difference between having an empty vector variable and not having the variable at all.

```
>> evec = [evec 4]
evec =
      4
```

The following statement takes what is currently in *evec*, which is 4, and adds an 11 to it.

```
>> evec = [evec 11]
evec =
      4  11
```

This can be continued as many times as desired, to build a vector up from nothing. Sometimes this is necessary, although generally it is not a good idea if it can be avoided because it can be quite time consuming.

Empty vectors can also be used to ***delete elements*** from vectors. For example, to remove the third element from a vector, the empty vector is assigned to it:

```
>> vec = 4:8
vec =
      4        5        6        7        8
>> vec(3) = []
vec =
      4        5        7        8
```

The elements in this vector are now numbered 1 through 4. Note that the variable *vec* has actually changed.

Subsets of a vector could also be removed. For example:

```
>> vec = 3:10
vec =
      3       4       5       6       7       8       9       10
>> vec(2:4) = []
vec =
      3       7       8       9      10
```

Individual elements cannot be removed from matrices, as matrices always need to have the same number of elements in every row.

```
>> mat = [7 9 8; 4 6 5]
mat =
      7       9       8
      4       6       5
>> mat(1,2) = [];
Subscripted assignment dimension mismatch.
```

However, entire rows or columns could be removed from a matrix. For example, to remove the second column:

```
>> mat(:,2) = []
mat =
      7       8
      4       5
```

Also, if linear indexing is used with a matrix to delete an element, the matrix will be reshaped into a row vector.

```
>> mat = [7 9 8; 4 6 5]
mat =
       7     9     8
       4     6     5
>> mat(3) = []
mat =
       7     4     6     8     5
```

(Again, using linear indexing is not a good idea.)

PRACTICE 2.2

Think about what would be produced by the following sequence of statements and expressions, and then type them in to verify your answers.

```
mat = [1:3; 44 9  2; 5:-1:3]
mat(3,2)
mat(2,:)
size(mat)
mat(:,4) = [8;11;33]
numel(mat)
v = mat(3,:)
v(v(2))
v(1) = []
reshape(mat,2,6)
```

2.1.6 Three-Dimensional Matrices

The matrices that have been shown so far have been two-dimensional; these matrices have rows and columns. Matrices in MATLAB are not limited to two dimensions, however. In fact, in Chapter 13 we will see image applications in which ***three-dimensional matrices*** are used. For a three-dimensional matrix, imagine a two-dimensional matrix as being flat on a page, and then the third dimension consists of more pages on top of that one (so, they are stacked on top of each other).

Here is an example of creating a three-dimensional matrix. First, two two-dimensional matrices *layerone* and *layertwo* are created; it is important that they have the same dimensions (in this case, 3×5). Then, these are made into "layers" in a three-dimensional matrix *mat*. Note that we end up with a matrix that has two layers, each of which is 3×5. The resulting three-dimensional matrix has dimensions $3 \times 5 \times 2$.

```
>> layerone = reshape(1:15,3,5)
layerone =
       1     4     7    10    13
       2     5     8    11    14
       3     6     9    12    15
```

```
>> layertwo = fliplr(flipud(layerone))
layertwo =
    15    12     9     6     3
    14    11     8     5     2
    13    10     7     4     1

>> mat(:,:,1) = layerone
mat =
     1     4     7    10    13
     2     5     8    11    14
     3     6     9    12    15
>> mat(:,:,2) = layertwo
mat(:,:,1) =
     1     4     7    10    13
     2     5     8    11    14
     3     6     9    12    15
mat(:,:,2) =
    15    12     9     6     3
    14    11     8     5     2
    13    10     7     4     1
>> size(mat)
ans =
     3     5     2
```

Three-dimensional matrices can also be created using the **zeros**, **ones**, and **rand** functions by specifying three dimensions to begin with. For example, **zeros(2,4,3)** will create a $2 \times 4 \times 3$ matrix of all 0s.

Unless specified otherwise, in the remainder of this book "matrices" will be assumed to be two-dimensional.

2.2 VECTORS AND MATRICES AS FUNCTION ARGUMENTS

In MATLAB an entire vector or matrix can be passed as an argument to a function; the function will be evaluated on every element. This means that the result will be of the same size as the input argument.

For example, let us find the absolute value of every element of a vector *vec*. The **abs** function will automatically return the absolute value of each individual element and the result will be a vector with the same length as the input vector.

```
>> vec = -2:1
vec =
    -2    -1     0     1
>> absvec = abs(vec)
absvec =
     2     1     0     1
```

For a matrix, the resulting matrix will have the same size as the input argument matrix. For example, the **sign** function will find the sign of each element in a matrix:

```
>> mat = [0 4 -3; -1 0 2]
mat =
     0     4    -3
    -1     0     2
>> sign(mat)
ans =
     0     1    -1
    -1     0     1
```

Functions such as **abs** and **sign** can have either scalars or arrays (vectors or matrices) passed to them. There are a number of functions that are written specifically to operate on vectors or on columns of matrices; these include the functions **min**, **max**, **sum**, and **prod**. These functions will be demonstrated first with vectors, and then with matrices.

For example, assume that we have the following vector variables:

```
>> vec1 = 1:5;
>> vec2 = [3 5 8 2];
```

The function **min** will return the minimum value from a vector, and the function **max** will return the maximum value.

```
>> min(vec1)
ans =
     1
>> max(vec2)
ans =
     8
```

The function **sum** will sum all of the elements in a vector. For example, for *vec1* it will return 1+2+3+4+5 or 15:

```
>> sum(vec1)
ans =
    15
```

The function **prod** will return the product of all of the elements in a vector; for example, for *vec2* it will return 3*5*8*2 or 240:

```
>> prod(vec2)
ans =
   240
```

There are also functions that return cumulative results; the functions **cumsum** and **cumprod** return the ***cumulative sum*** or ***cumulative product***, respectively. A cumulative or ***running sum*** stores the sum so far at each step as it adds the elements from the vector. For example, for *vec1*, it would store the first element,

1, then 3 (1 + 2), then 6 (1 + 2 + 3), then 10 (1 + 2 + 3 + 4), then, finally, 15 (1 + 2 + 3 + 4 + 5). The result is a vector that has as many elements as the input argument vector that is passed to it:

```
>> cumsum(vec1)
ans =
     1     3     6     10     15
>> cumsum(vec2)
ans =
     3     8    16     18
```

The **cumprod** function stores the cumulative products as it multiplies the elements in the vector together; again, the resulting vector will have the same length as the input vector:

```
>> cumprod(vec1)
ans =
     1     2     6     24     120
```

Similarly, there are **cummin** and **cummax** functions, which were introduced in R2014b. Also, in R2014b, a 'reverse' option was introduced for all of the cumulative functions. For example,

```
>> cumsum(vec1,'reverse')
ans =
    15    14    12     9     5
```

For matrices, all of these functions operate on every individual column. If a matrix has dimensions $r \times c$, the result for the **min**, **max**, **sum**, and **prod** functions will be a $1 \times c$ row vector, as they return the minimum, maximum, sum, or product respectively for every column. For example, assume the following matrix:

```
>> mat = randi([1 20], 3, 5)
mat =
     3    16     1    14     8
     9    20    17    16    14
    19    14    19    15     4
```

The following are the results for the **max** and **sum** functions:

```
>> max(mat)
ans =
    19    20    19    16    14
>> sum(mat)
ans =
    31    50    37    45    26
```

To find a function for every row, instead of every column, one method would be to transpose the matrix.

```
>> max(mat')
ans =
     16     20     19
>> sum(mat')
ans =
     42     76     71
```

QUICK QUESTION!

As these functions operate columnwise, how can we get an overall result for the matrix? For example, how would we determine the overall maximum in the matrix?

Answer: We would have to get the maximum from the row vector of column maxima, in other words, ***nest the calls*** to the **max** function:

```
>> max(max(mat))
ans =
    20
```

For the **cumsum** and **cumprod** functions, again they return the cumulative sum or product of every column. The resulting matrix will have the same dimensions as the input matrix:

```
>> mat
mat =
       3    16     1    14     8
       9    20    17    16    14
      19    14    19    15     4
>> cumsum(mat)
ans =
       3    16     1    14     8
      12    36    18    30    22
      31    50    37    45    26
```

Note that the first row in the resulting matrix is the same as the first row in the input matrix. After that, the values in the rows accumulate. Similarly, the **cummin** and **cummax** functions find the cumulative minima and maxima for every column.

```
>> cummin(mat)
ans =
     3  16   1  14   8
     3  16   1  14   8
     3  14   1  14   4
>> cummax(mat,'reverse')
 ans =
     19  20  19  16  14
     19  20  19  16  14
     19  14  19  15   4
```

Another useful function that can be used with vectors and matrices is **diff**. The function **diff** returns the differences between consecutive elements in a vector. For example,

```
>> diff([4 7 15 32])
ans =
     3     8    17

>> diff([4 7 2 32])
ans =
     3    -5    30
```

For a vector *v* with a length *n*, the length of **diff(v)** will be *n*-1. For a matrix, the **diff** function will operate on each column.

```
>> mat = randi(20, 2,3)
mat =
    17     3    13
    19    19     2
>> diff(mat)
ans =
     2    16   -11
```

2.3 SCALAR AND ARRAY OPERATIONS ON VECTORS AND MATRICES

Numerical operations can be done on entire vectors or matrices. For example, let's say that we want to multiply every element of a vector *v* by 3.

In MATLAB, we can simply multiply *v* by 3 and store the result back in *v* in an assignment statement:

```
>> v= [3 7 2 1];
>> v = v*3
v =
      9  21   6   3
```

As another example, we can divide every element by 2:

```
>> v= [3 7 2 1];
>> v/2
ans =
    1.5000 3.5000 1.0000 0.5000
```

To multiply every element in a matrix by 2:

```
>> mat = [4:6; 3:-1:1]
mat =
    4   5   6
    3   2   1
```

```
>> mat * 2
ans =
     8   10   12
     6    4    2
```

This operation is referred to as ***scalar multiplication***. We are multiplying every element in a vector or matrix by a scalar (or, for scalar division, dividing every element in a vector or a matrix by a scalar).

QUICK QUESTION!

There is no tens function to create a matrix of all tens, so how could we accomplish that?

Answer: We can either use the **ones** function and multiply by ten, or the **zeros** function and add ten:

```
>> ones(1,5) * 10
ans =
    10   10   10   10   10
>> zeros(2) + 10
ans =
    10   10
    10   10
```

Array operations are operations that are performed on vectors or matrices term by term, or element by element. This means that the two arrays (vectors or matrices) must be of the same size to begin with. The following examples demonstrate the array addition and subtraction operators.

```
>> v1 = 2:5
v1 =
     2     3     4     5
>> v2 = [33 11 5 1]
v2 =
    33    11     5     1
>> v1 + v2
ans =
    35    14     9     6

>> mata = [5:8; 9:-2:3]
mata =
     5     6     7     8
     9     7     5     3

>> matb = reshape(1:8,2,4)
matb =
     1     3     5     7
     2     4     6     8

>> mata - matb
ans =
     4     3     2     1
     7     3    -1    -5
```

However, for any operation that is based on multiplication (which means multiplication, division, and exponentiation), a dot must be placed in front of the operator for array operations. For example, for the exponentiation operator `.^` must be used when working with vectors and matrices, rather than just the `^` operator. Squaring a vector, for example, means multiplying each element by itself so the `.^` operator must be used.

```
>> v= [3 7 2 1];
>> v ^ 2
Error using ^
Inputs must be a scalar and a square matrix.
To compute elementwise POWER, use POWER (.^) instead.

>> v .^ 2
ans =
      9    49    4    1
```

Similarly, the operator .* must be used for ***array multiplication*** and ./ or .\ for ***array division***. The following examples demonstrate array multiplication and array division.

```
>> v1 = 2:5
v1 =
      2     3     4     5
>> v2 = [33 11 5 1]
v2 =
     33    11     5     1

>> v1 .* v2
ans =
     66    33    20     5

>> mata = [5:8; 9:-2:3]
mata =
      5     6     7     8
      9     7     5     3
>> matb = reshape(1:8, 2,4)
matb =
      1     3     5     7
      2     4     6     8

>> mata ./ matb
ans =
    5.0000    2.0000    1.4000    1.1429
    4.5000    1.7500    0.8333    0.3750
```

The operators `.^`, `.*`, `./`, and `.\` are called array operators and are used when multiplying or dividing vectors or matrices of the same size term by term. Note that matrix multiplication is a very different operation, and will be covered in Section 2.5.

PRACTICE 2.3

Create a vector variable and subtract 3 from every element in it.
Create a matrix variable and divide every element by 3.
Create a matrix variable and square every element.

2.4 LOGICAL VECTORS

Logical vectors use relational expressions that result in **true/false** values.

2.4.1 Relational Expressions with Vectors and Matrices

Relational operators can be used with vectors and matrices. For example, let's say that there is a vector *vec*, and we want to compare every element in the vector to 5 to determine whether it is greater than 5 or not. The result would be a vector (with the same length as the original) with **logical true** or **false** values.

```
>> vec = [5 9 3 4 6 11];
>> isg = vec > 5
isg =
     0   1   0   0   1   1
```

Note that this creates a vector consisting of all **logical true** or **false** values. Although the result is a vector of ones and zeros, and numerical operations can be done on the vector *isg*, its type is **logical** rather than **double**.

```
>> doubres = isg + 5
doubres =
     5     6     5     5     6     6

>> whos
  Name        Size      Bytes  Class

  doubres     1x6          48  double array
  isg         1x6           6  logical array
  vec         1x6          48  double array
```

To determine how many of the elements in the vector *vec* were greater than 5, the **sum** function could be used on the resulting vector *isg*:

```
>> sum(isg)
ans =
     3
```

What we have done is to create a ***logical vector*** *isg*. This logical vector can be used to index into the original vector. For example, if only the elements from the vector that are greater than 5 are desired:

```
>> vec(isg)
ans =
      9     6     11
```

This is called ***logical indexing***. Only the elements from *vec* for which the corresponding element in the logical vector *isg* is **logical true** are returned.

QUICK QUESTION!

Why doesn't the following work?

```
>> vec = [5 9 3 4 6 11];
>> v = [0 1 0 0 1 1];
>> vec(v)
Subscript indices must either be real positive
integers or logicals.
```

Answer: The difference between the vector in this example and *isg* is that *isg* is a vector of logicals (**logical** 1s and 0s), whereas [0 1 0 0 1 1] by default is a vector of **double** values. ***Only logical 1s and 0s can be used to index into a vector.*** So, type casting the variable *v* would work:

```
>> v = logical(v);
>> vec(v)
ans =
      9     6     11
```

To create a vector or matrix of all **logical** 1s or 0s, the functions **true** and **false** can be used.

```
>> false(2)
ans =
      0   0
      0   0
>> true(1,5)
ans =
      1   1   1   1   1
```

Beginning with R2016a, the **ones** and **zeros** functions can also create **logical** arrays directly.

```
>> logzer = ones(1,5, 'logical')
logzer =
      1     1     1     1     1
>> class(logzer)
ans =
logical
```

2.4.2 Logical Built-In Functions

There are built-in functions in MATLAB which are useful in conjunction with **logical** vectors or matrices; two of these are the functions **any** and **all**. The function **any** returns **logical true** if any element in a vector represents **true**, and **false** if not. The function **all** returns **logical true** only if all elements represent **true**. Here are some examples.

```
>> any(isg)
ans =
     1
>> all(true(1,3))
ans =
     1
```

For the following variable *vec2*, some, but not all, elements are **true**; consequently, **any** returns **true** but **all** returns **false**.

```
>> vec2 = logical([1 1 0 1])
vec2 =
     1    1    0    1
>> any(vec2)
ans =
     1
>> all(vec2)
ans =
     0
```

The function **find** returns the indices of a vector that meet the given criteria. For example, to find all of the elements in a vector that are greater than 5:

```
>> vec = [5 3 6 7 2]
vec =
     5    3    6    7    2
>> find(vec > 5)
ans =
     3    4
```

For matrices, the **find** function will use linear indexing when returning the indices that meet the specified criteria. For example:

```
>> mata = randi(10,2,4)
mata =
     5   6   7   8
     9   7   5   3
>> find(mata == 5)
ans =
     1
     6
```

For both vectors and matrices, an empty vector will be returned if no elements match the criterion. For example,

```
>> find(mata == 11)
ans =
   Empty matrix: 0-by-1
```

The function **isequal** is useful in comparing arrays. In MATLAB, using the equality operator with arrays will return 1 or 0 for each element; the **all** function could then be used on the resulting array to determine whether all elements were equal or not. The built-in function **isequal** also accomplishes this:

```
>> vec1 = [1 3 -4 2 99];
>> vec2 = [1 2 -4 3 99];
>> vec1 == vec2
ans =
      1   0   1   0   1
>> all(vec1 == vec2)
ans =
      0
>> isequal(vec1,vec2)
ans =
      0
```

This works with strings, also.

```
>> str1 = 'hello';
>> str2 = 'howdy';
>> str1 == str2
ans =
      1   0   0   0   0
>> isequal(str1, str2)
ans =
      0
>> isequal(str1, 'hello')
ans =
      1
```

However, one difference is that if the two arrays are not of the same dimensions, the **isequal** function will return **logical** 0, whereas using the equality operator will result in an error message.

QUICK QUESTION!

If we have a vector *vec* that erroneously stores negative values, how can we eliminate those negative values?

Answer: One method is to determine where they are and delete these elements:

```
>> vec = [11 -5 33 2 8 -4 25];
>> neg = find(vec < 0)
neg =
     2   6
>> vec(neg) = []
vec =
     11  33   2   8  25
```

Alternatively, we can just use a logical vector rather than **find**:

```
>> vec = [11 -5 33 2 8 -4 25];
>> vec(vec < 0) = []
vec =
     11  33   2   8  25
```

PRACTICE 2.4

Modify the result seen in the previous Quick Question!. Instead of deleting the negative elements, retain only the positive ones. (Hint: Do it two ways, using **find** and using a logical vector with the expression *vec>=0.*)

The following is an example of an application of several of the functions mentioned here. A vector that stores a signal can contain both positive and negative values. (For simplicity, we will assume no zeros, however.) For many applications it is useful to find the ***zero crossings***, or where the signal goes from being positive to negative or vice versa. This can be accomplished using the functions **sign**, **diff**, and **find**.

```
>> vec = [0.2 -0.1 -0.2 -0.1 0.1 0.3 -0.2];
>> sv = sign(vec)
sv =
     1  -1  -1  -1   1   1  -1

>> dsv = diff(sv)
dsv =
    -2   0   0   2   0  -2

>> find(dsv ~= 0)
ans =
     1   4   6
```

This shows that the signal crossings are between elements 1 and 2, 4 and 5, and 6 and 7.

MATLAB also has or and and operators that work elementwise for arrays:

Operator	Meaning
|	elementwise or for arrays
&	elementwise and for arrays

These operators will compare any two vectors or matrices, as long as they are of the same size, element by element and return a vector or matrix of the same size of **logical** 1s and 0s. The operators || and && are only used with scalars, not arrays. For example:

```
>> v1 = logical([1 0 1 1]);
>> v2 = logical([0 0 1 0]);
>> v1 & v2
ans =
     0   0   1   0
>> v1 | v2
ans =
     1   0   1   1
>> v1 && v2
Operands to the || and && operators must be convertible to logical scalar
values.
```

As with the numerical operators, it is important to know the operator precedence rules. Table 2.1 shows the rules for the operators that have been covered so far, in the order of precedence.

Table 2.1 Operator Precedence Rules

Operators	Precedence
Parentheses: ( )	Highest
Transpose and power: ', ^, .^	
Unary: negation (−), not (~)	
Multiplication, division *,/,\,.*,./,.\	
Addition, subtraction +, −	
Relational <, <=, >, >=, ==, ~=	
Element-wise and &	
Element-wise or		
And && (scalars)	
Or || (scalars)	
Assignment =	Lowest

2.5 MATRIX MULTIPLICATION

Matrix multiplication does *not* mean multiplying term by term; it is not an array operation. Matrix multiplication has a very specific meaning. First of all, to multiply a matrix A by a matrix B to result in a matrix C, the number of columns of A must be the same as the number of rows of B. If the matrix A has dimensions $m \times n$, that means that matrix B must have dimensions $n \times$ *something*; we'll call it p.

We say that the ***inner dimensions*** (the ns) must be the same. The resulting matrix C has the same number of rows as A and the same number of columns as B (i.e., the ***outer dimensions*** $m \times p$). In mathematical notation,

$$[A]_{m \times n}[B]_{n \times p} = [C]_{m \times p}$$

This only defines the size of C, not how to find the elements of C.

The elements of the matrix C are defined as the sum of products of corresponding elements in the rows of A and columns of B, or in other words,

$$c_{ij} = \sum_{k=1}^{n} a_{ik} b_{kj}$$

In the following example, A is 2×3 and B is 3×4; the inner dimensions are both 3, so performing the matrix multiplication A*B is possible (note that B*A would not be possible). C will have as its size the outer dimensions 2×4. The elements in C are obtained using the summation just described. The first row of C is obtained using the first row of A and in succession the columns of B. For example, C(1,1) is 3*1+8*4+0*0 or 35. C(1,2) is 3*2+8*5+0*2 or 46.

$$\underset{A}{\begin{bmatrix} 3 & 8 & 0 \\ 1 & 2 & 5 \end{bmatrix}} * \underset{B}{\begin{bmatrix} 1 & 2 & 3 & 1 \\ 4 & 5 & 1 & 2 \\ 0 & 2 & 3 & 0 \end{bmatrix}} = \underset{C}{\begin{bmatrix} 35 & 46 & 17 & 19 \\ 9 & 22 & 20 & 5 \end{bmatrix}}$$

In MATLAB, the * operator will perform this matrix multiplication:

```
>> A = [3 8 0; 1 2 5];
>> B = [1 2 3 1; 4 5 1 2; 0 2 3 0];
>> C = A*B
C =
     35    46    17    19
      9    22    20     5
```

PRACTICE 2.5

When two matrices have the same dimensions and are square, both array and matrix multiplication can be performed on them. For the following two matrices, perform A.*B, A*B, and B*A by hand and then verify the results in MATLAB.

$$\underset{A}{\begin{bmatrix} 1 & 4 \\ 3 & 3 \end{bmatrix}} \quad \underset{B}{\begin{bmatrix} 1 & 2 \\ -1 & 0 \end{bmatrix}}$$

2.5.1 Matrix Multiplication for Vectors

As vectors are just special cases of matrices, the matrix operations described previously (addition, subtraction, scalar multiplication, multiplication, and transpose) also work on vectors, as long as the dimensions are correct.

For vectors, we have already seen that the transpose of a row vector is a column vector, and the transpose of a column vector is a row vector.

To multiply vectors, they must have the same number of elements, but one must be a row vector and the other a column vector. For example, for a column vector c and row vector r:

$$c = \begin{bmatrix} 5 \\ 3 \\ 7 \\ 1 \end{bmatrix} \quad r = \begin{bmatrix} 6 & 2 & 3 & 4 \end{bmatrix}$$

Note that r is 1×4, and c is 4×1, so

$$[r]_{1\times4}[c]_{4\times1} = [s]_{1\times1}$$

or, in other words, a scalar:

$$[6\ \ 2\ \ 3\ \ 4]\begin{bmatrix}5\\3\\7\\1\end{bmatrix} = 6*5+2*3+3*7+4*1 = 61$$

whereas $[c]_{4\times1}[r]_{1\times4} = [\mathrm{M}]_{4\times4}$, or in other words a 4×4 matrix:

$$\begin{bmatrix}5\\3\\7\\1\end{bmatrix}[6\ \ 2\ \ 3\ \ 4] = \begin{bmatrix}30 & 10 & 15 & 20\\18 & 6 & 9 & 12\\42 & 14 & 21 & 28\\6 & 2 & 3 & 4\end{bmatrix}$$

In MATLAB, these operations are accomplished using the * operator, which is the matrix multiplication operator. First, the column vector c and row vector r are created.

```
>> c = [5 3 7 1]';
>> r = [6 2 3 4];
>> r*c
ans =
     61
>> c*r
ans =
     30  10  15  20
     18   6   9  12
     42  14  21  28
      6   2   3   4
```

There are also operations specific to vectors: the ***dot product*** and ***cross product***. The ***dot* product** or ***inner product*** of two vectors a and b is written as $a \bullet b$ and is defined as

$$a_1b_1 + a_2b_2 + a_3b_3 + \ldots + a_nb_n = \sum_{i=1}^{n} a_ib_i$$

where both a and b have n elements and a_i and b_i represent elements in the vectors. In other words, this is like matrix multiplication when multiplying a row vector a by a column vector b; the result is a scalar. This can be accomplished using the * operator and transposing the second vector, or by using the **dot** function in MATLAB:

```
>> vec1 = [4 2 5 1];
>> vec2 = [3 6 1 2];
>> vec1*vec2'
ans =
     31
>> dot(vec1,vec2)
ans =
     31
```

The ***cross product*** or ***outer product*** $a \times b$ of two vectors a and b is defined only when both a and b have three elements. It can be defined as a matrix multiplication of a matrix composed of the elements from a in a particular manner shown here and the column vector b.

$$a \times b = \begin{bmatrix} 0 & -a_3 & a_2 \\ a_3 & 0 & -a_1 \\ -a_2 & a_1 & 0 \end{bmatrix} \begin{bmatrix} b_1 \\ b_2 \\ b_3 \end{bmatrix} = [a_2b_3 - a_3b_2, \;\; a_3b_1 - a_1b_3, \;\; a_1b_2 - a_2b_1]$$

MATLAB has a built-in function **cross** to accomplish this.

```
>> vec1 = [4 2 5];
>> vec2 = [3 6 1];

>> cross(vec1,vec2)
ans =
   -28  11  18
```

Explore Other Interesting Features

- There are many functions that create special matrices, for example, **hilb** for a Hilbert matrix, **magic**, and **pascal**.
- The **gallery** function, which can return many different types of test matrices for problems.
- The **ndims** function to find the number of dimensions of an argument.
- The **shiftdim** function.
- The **circshift** function. How can you get it to shift a row vector, resulting in another row vector?
- How to reshape a three-dimensional matrix.
- The **range** function
- Passing 3D matrices to functions. For example, if you pass a $3 \times 5 \times 2$ matrix to the **sum** function, what would be the size of the result?
- The **meshgrid** function can specify the x and y coordinates of points in images, or can be used to calculate functions on two variables x and y. It receives as input arguments two vectors, and returns as output arguments two matrices that specify separately x and y values. ■

SUMMARY

COMMON PITFALLS

- Attempting to create a matrix that does not have the same number of values in each row
- Confusing matrix multiplication and array multiplication. Array operations, including multiplication, division, and exponentiation, are performed term by term (so the arrays must have the same size); the

operators are .*, ./, .\, and .^. For matrix multiplication to be possible, the inner dimensions must agree and the operator is *.

- Attempting to use an array of **double** 1s and 0s to index into an array (must be **logical**, instead)
- Forgetting that for array operations based on multiplication, the dot must be used in the operator. In other words, for multiplying, dividing by, dividing into, or raising to an exponent term by term, the operators are .*, ./, .\, and .^.
- Attempting to use || or && with arrays. Always use | and & when working with arrays; || and && are only used with scalars.

PROGRAMMING STYLE GUIDELINES

- If possible, try not to extend vectors or matrices, as it is not very efficient.
- Do not use just a single index when referring to elements in a matrix; instead, use both the row and column subscripts (use subscripted indexing rather than linear indexing)
- To be general, never assume that the dimensions of any array (vector or matrix) are known. Instead, use the function **length** or **numel** to determine the number of elements in a vector, and the function **size** for a matrix:

```
len = length(vec);
[r, c] = size(mat);
```

- Use **true** instead of **logical(1)** and **false** instead of **logical(0)**, especially when creating vectors or matrices.

MATLAB Functions and Commands			
linspace	reshape	max	any
logspace	fliplr	sum	all
zeros	flipud	prod	find
ones	flip	cumsum	isequal
length	rot90	cumprod	dot
size	repmat	cummin	cross
numel	repelem	cummax	
end	min	diff	

MATLAB Operators	
colon	:
transpose	'
array operators	.^, .*, ./, .\
elementwise or for matrices	|
elementwise and for matrices	&
matrix multiplication	*

Exercises

1. If a variable has the dimensions 3 × 4, could it be considered to be (check all that apply):
   a matrix
   a row vector
   a column vector
   a scalar
2. If a variable has the dimensions 1 × 5, could it be considered to be (check all that apply):
   a matrix
   a row vector
   a column vector
   a scalar
3. If a variable has the dimensions 5 × 1, could it be considered to be (check all that apply):
   a matrix
   a row vector
   a column vector
   a scalar
4. If a variable has the dimensions 1 × 1, could it be considered to be (check all that apply):
   a matrix
   a row vector
   a column vector
   a scalar
5. Using the colon operator, create the following row vectors

```
2      3      4      5      6      7
1.1000     1.3000     1.5000     1.7000
8      6      4      2
```

6. Using a built-in function, create a vector *vec* which consists of 20 equally spaced points in the range of – pi to +pi.
7. Write an expression using linspace that will result in the same as 2: 0.2: 3
8. Using the colon operator and also the **linspace** function, create the following row vectors:

```
-5   -4   -3   -2   -1
 5    7    9
 8    6    4
```

9. How many elements would be in the vectors created by the following expressions?

```
linspace(3,2000)
logspace(3,2000)
```

10. Create a variable *myend* which stores a random integer in the inclusive range of 5 to 9. Using the colon operator, create a vector that iterates from 1 to *myend* in steps of 3.
11. Using the colon operator and the transpose operator, create a column vector *myvec* that has the values −1 to 1 in steps of 0.5.
12. Write an expression that refers to only the elements that have odd-numbered subscripts in a vector, regardless of the length of the vector. Test your expression on vectors that have both an odd and even number of elements.
13. Generate a 2 × 4 matrix variable *mat*. Replace the first row with 1:4. Replace the third column (you decide with which values).
14. Generate a 2 × 4 matrix variable *mat*. Verify that the number of elements is the product of the number of rows and columns.
15. Which would you normally use for a matrix: **length** or **size**? Why?
16. When would you use **length** vs. **size** for a vector?
17. Generate a 2 × 3 matrix of random
    - real numbers, each in the range (0, 1)
    - real numbers, each in the range (0, 10)
    - integers, each in the inclusive range from 5 to 20
18. Create a variable *rows* that is a random integer in the inclusive range from 1 to 5. Create a variable *cols* that is a random integer in the inclusive range from 1 to 5. Create a matrix of all zeros with the dimensions given by the values of *rows* and *cols*.
19. Create a matrix variable *mat*. Find as many expressions as you can that would refer to the last element in the matrix, without assuming that you know how many elements or rows or columns it has (i.e., make your expressions general).
20. Create a vector variable *vec*. Find as many expressions as you can that would refer to the last element in the vector, without assuming that you know how many elements it has (i.e., make your expressions general).
21. Create a 2 × 3 matrix variable *mat*. Pass this matrix variable to each of the following functions and make sure you understand the result: **flip**, **fliplr**, **flipud**, and **rot90**. In how many different ways can you **reshape** it?
22. What is the difference between `fliplr(mat)` and `mat = fliplr(mat)`?
23. Use **reshape** to reshape the row vector 1:4 into a 2x2 matrix; store this in a variable named mat. Next, make 2 × 3 copies of mat using both **repelem** and **repmat**.
24. Create a 3 × 5 matrix of random real numbers. Delete the third row.
25. Given the matrix:

```
>> mat = randi([1 20], 3,5)
mat =
     6  17   7  13  17
    17   5   4  10  12
     6  19   6   8  11
```

Why wouldn't this work:

```
mat(2:3, 1:3) = ones(2)
```

26. Create a three-dimensional matrix with dimensions $2 \times 4 \times 3$ in which the first "layer" is all 0 s, the second is all 1 s and the third is all 5 s. Use **size** to verify the dimensions.
27. Create a vector x which consists of 20 equally spaced points in the range from $-\pi$ to $+\pi$. Create a y vector which is **sin(x)**.
28. Create a 3×5 matrix of random integers, each in the inclusive range from -5 to 5. Get the **sign** of every element.
29. Find the sum $3+5+7+9+11$.
30. Find the sum of the first *n* terms of the harmonic series where *n* is an integer variable greater than one.

$$1+\frac{1}{2}+\frac{1}{3}+\frac{1}{4}+\frac{1}{5}+\cdots$$

31. Find the following sum by first creating vectors for the numerators and denominators:

$$\frac{3}{1}+\frac{5}{2}+\frac{7}{3}+\frac{9}{4}$$

32. Create a matrix and find the product of each row and column using **prod**.
33. Create a 1×6 vector of random integers, each in the inclusive range from 1 to 20. Use built-in functions to find the minimum and maximum values in the vector. Also create a vector of cumulative sums using **cumsum**.
34. Write a relational expression for a vector variable that will verify whether the last value in a vector created by **cumsum** is the same as the result returned by **sum**.
35. Create a vector of five random integers, each in the inclusive range from -10 to 10. Perform each of the following:
    - subtract 3 from each element
    - count how many are positive
    - get the cumulative minimum
36. Create a 3×5 matrix. Perform each of the following:
    - Find the maximum value in each column.
    - Find the maximum value in each row.
    - Find the maximum value in the entire matrix.
    - Find the cumulative maxima.
37. Find two ways to create a 3×5 matrix of all 100s (Hint: use **ones** and **zeros**).
38. Given the two matrices:

$$A = \begin{bmatrix} 1 & 2 & 3 \\ 4 & -1 & 6 \end{bmatrix} \quad B = \begin{bmatrix} 2 & 4 & 1 \\ 1 & 3 & 0 \end{bmatrix}$$

Perform the following operations:

```
A + B
A - B
A * B
```

39. The built-in function **clock** returns a vector that contains 6 elements: the first three are the current date (year, month, and day) and the last three represent the current time in hours, minutes, and seconds. The seconds is a real number, but all others are integers. Store the result from clock in a variable called *myc*. Then, store the first three elements from this variable in a variable *today* and the last three elements in a variable *now*. Use the fix function on the vector variable *now* to get just the integer part of the current time.
40. A vector *v* stores for several employees of the Green Fuel Cells Corporation, their hours worked one week followed for each by the hourly pay rate. For example, if the variable stores

```
>> v
v =
33.000010.500040.000018.000020.00007.5000
```

that means the first employee worked 33 hours at $10.50 per hour, the second worked 40 hours at $18 an hour, and so on. Write code that will separate this into two vectors, one that stores the hours worked and another that stores the hourly rates. Then, use the array multiplication operator to create a vector, storing in the new vector, the total pay for every employee.
41. A company is calibrating some measuring instrumentation and has measured the radius and height of one cylinder 8 separate times; they are in vector variables *r* and *h*. Find the volume from each trial, which is given by $\pi r^2 h$. Also use logical indexing first to make sure that all measurements were valid (>0).

```
>> r = [5.499 5.498 5.5 5.5 5.52 5.51 5.5 5.48];
>> h = [11.1 11.12 11.09 11.11 11.11 11.1 11.08 11.11];
```

42. For the following matrices A, B, and C:

$$A=\begin{bmatrix}1 & 4\\ 3 & 2\end{bmatrix} \quad B=\begin{bmatrix}2 & 1 & 3\\ 1 & 5 & 6\\ 3 & 6 & 0\end{bmatrix} \quad C=\begin{bmatrix}3 & 2 & 5\\ 4 & 1 & 2\end{bmatrix}$$

- Give the result of 3*A.
- Give the result of A*C.
- Are there any other matrix multiplications that can be performed? If so, list them.

43. For the following vectors and matrices A, B, and C:

$$A=\begin{bmatrix}4 & 1 & -1\\ 2 & 3 & 0\end{bmatrix} \quad B=\begin{bmatrix}1 & 4\end{bmatrix} \quad C=\begin{bmatrix}2\\ 3\end{bmatrix}$$

Perform the following operations, if possible. If not, just say it can't be done!

```
A * B
B * C
C * B
```

44. The matrix variable *rainmat* stores the total rainfall in inches for some districts for the years 2010–13. Each row has the rainfall amounts for a given district. For example, if *rainmat* has the value:

```
>> rainmat
ans =
25 33 29 42
53 44 40 56
etc.
```

district 1 had 25 in. in 2010, 33 in 2011, etc. Write expression(s) that will find the number of the districts that had the highest total rainfall for the entire four year period.

45. Generate a vector of 20 random integers, each in the range from 50 to 100. Create a variable *evens* that stores all of the even numbers from the vector, and a variable *odds* that stores the odd numbers.
46. Assume that the function **diff** does not exist. Write your own expression(s) to accomplish the same thing for a vector.
47. Create a vector variable *vec*; it can have any length. Then, write assignment statements that would store the first half of the vector in one variable and the second half in another. Make sure that your assignment statements are general, and work whether *vec* has an even or odd number of elements (Hint: use a rounding function such as **fix**).

CHAPTER 3

Introduction to MATLAB Programming

KEY TERMS

computer program
scripts
live script
algorithm
modular program
top-down design
external file
default input device
prompting
default output device
execute/run
high-level languages
machine language
executable
compiler
source code
object code
interpreter
documentation
comments
block comment
comment blocks
input/output (I/O)
user
empty string
error message
formatting
format string
place holder
conversion characters
newline character
field width
leading blanks
trailing zeros
plot symbols
markers
line types
toggle
modes
writing to a file
appending to a file
reading from a file
user-defined functions
function call
argument
control
return value
function header
output arguments
input arguments
function body
function definition
local variables
scope of variables
base workspace

CONTENTS

We have now used the MATLAB® product interactively in the Command Window. That is sufficient when all one needs is a simple calculation. However, in many cases, quite a few steps are required before the final result can be obtained. In these cases, it is more convenient to group statements together in what is called a ***computer program***.

In this chapter, we will introduce the simplest MATLAB programs, which are called ***scripts***. Examples of scripts that customize simple plots will illustrate

MATLAB®. http://dx.doi.org/10.1016/B978-0-12-804525-1.00003-9

the concept. Input will be introduced, both from files and from the user. Output to files and to the screen will also be introduced. Finally, user-defined functions that calculate and return a single value will be described. These topics serve as an introduction to programming, which will be expanded on in Chapter 6.

As of Version R2016a, there are two types of scripts. A new, richer script type called a ***live script*** has been created in MATLAB. Live scripts will be introduced in Chapter 6, in which programming concepts will be covered in more depth. In this chapter, we will create simple scripts stored in MATLAB code files, which have an extension of .m.

3.1 ALGORITHMS

Before writing any computer program, it is useful to first outline the steps that will be necessary. An ***algorithm*** is the sequence of steps needed to solve a problem. In a ***modular*** approach to programming, the problem solution is broken down into separate steps, and then each step is further refined until the resulting steps are small enough to be manageable tasks. This is called the ***top-down design*** approach.

As a simple example, consider the problem of calculating the area of a circle. First, it is necessary to determine what information is needed to solve the problem, which in this case is the radius of the circle. Next, given the radius of the circle, the area of the circle would be calculated. Finally, once the area has been calculated, it has to be displayed in some way. The basic algorithm then is three steps:

- Get the input: the radius
- Calculate the result: the area
- Display the output

Even with an algorithm this simple, it is possible to further refine each of the steps. When a program is written to implement this algorithm, the steps would be as follows:

- Where does the input come from? Two possible choices would be from an ***external file*** or from the user (the person who is running the program) who enters the number by typing it using the keyboard. For every system, one of these will be the ***default input device*** (which means, if not specified otherwise, this is where the input comes from!). If the user is supposed to enter the radius, the user has to be told to type in the radius (and, in what units). Telling the user what to enter is called ***prompting***. So, the input step actually includes two steps: prompt the user to enter a radius and then read it into the program.

- To calculate the area, the formula is needed. In this case, the area of the circle is π multiplied by the square of the radius. So, that means the value of the constant for π is needed in the program.
- Where does the output go? Two possibilities are: (1) to an external file, or (2) to the screen. Depending on the system, one of these will be the ***default output device***. When displaying the output from the program, it should always be as informative as possible. In other words, instead of just printing the area (just the number), it should be printed in a nice sentence format. Also, to make the output even more clear, the input should be printed. For example, the output might be the sentence: "For a circle with a radius of 1 inch, the area is 3.1416 inches squared."

For most programs, the basic algorithm consists of three steps that have been outlined below:

1. Get the input(s)
2. Calculate the result(s)
3. Display the result(s)

As can be seen here, even the simplest solution to a problem can then be refined further. This is top-down design.

3.2 MATLAB SCRIPTS

Once a problem has been analyzed, and the algorithm for its solution written and refined, the solution to the problem is then written in a particular programming language. A computer program is a sequence of instructions, in a given language, that accomplishes a task. To ***execute*** or ***run*** a program, is to have the computer actually follow these instructions sequentially.

High-level languages have English-like commands and functions, such as "print this" or "if $x<5$ do something." The computer, however, can only interpret commands written in its ***machine language***. Programs that are written in high-level languages must therefore be translated into machine language before the computer can actually execute the sequence of instructions in the program. A program that does this translation from a high-level language to an ***executable*** file is called a ***compiler***. The original program is called the ***source code***, and the resulting executable program is called the ***object code***. Compilers translate from the source code to object code; this is then executed as a separate step.

By contrast, an ***interpreter*** goes through the code line-by-line, translating and executing each command as it goes. MATLAB uses what are called either script files or MATLAB code files, which have an extension .m on the file name. These script files are interpreted, rather than compiled. Therefore, the correct terminology is that these are scripts, and not programs. However, the terms are used somewhat loosely by many people, and documentation in MATLAB itself refers to scripts as programs. In this book, we will reserve the use of the word "program" to mean a set of scripts and functions, as described briefly in Section 3.7 and then in more detail in Chapter 6.

A script is a sequence of MATLAB instructions that is stored in a file with an extension of .m and saved. The contents of a script can be displayed in the Command Window using the **type** command. The script can be executed, or run, by simply entering the name of the file (without the .m extension).

Before creating a script, make sure the Current Folder is set to the folder in which you want to save your files.

The steps involved in creating a script depend on the version of MATLAB. The easiest method is to click on "New Script" under the HOME tab. Alternatively, one can click on the down arrow under "New" and then choose Script (see Fig. 3.1)

A new window will appear called the Editor (which can be docked). In the latest versions of MATLAB, this window has three tabs: "EDITOR," "PUBLISH," and "VIEW." Next, simply type the sequence of statements (note that line numbers will appear on the left).

When finished, save the file by choosing the Save down arrow under the EDITOR tab. Make sure that the extension of .m is on the file name (this should be the default). The rules for file names are the same as for variables (they must start with a letter; after that there can be letters, digits, or the underscore.)

If you have entered commands in the Command Window, and decide that you would like to put them into a script, an alternate method for creating a script is

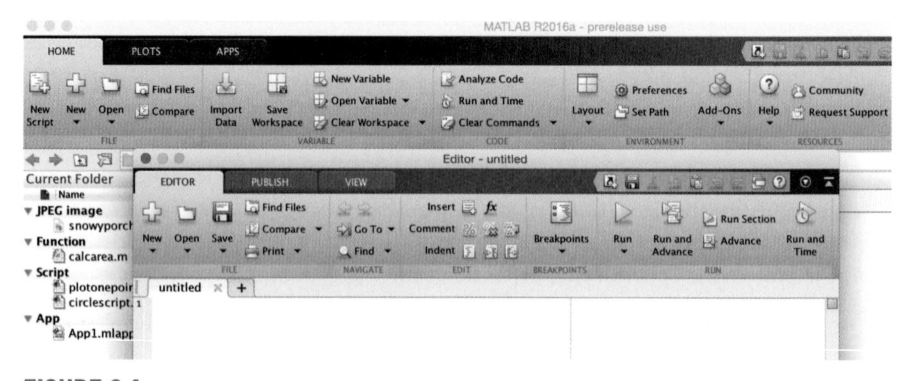

FIGURE 3.1
Toolstrip and Editor.

to select the commands in the Command History window, and then right click. This will give options for creating a script or live script, and will then prepopulate the editor with those commands.

In our first example, we will now create a script called *script1.m* that calculates the area of a circle. It assigns a value for the radius, and then calculates the area based on that radius.

In this text, scripts will be displayed in a box with the name of the file on top.

script1.m

```
radius = 5
area = pi * (radius ^2)
```

There are two ways to view a script once it has been written: either open the Editor Window to view it or use the **type** command, as shown here, to display it in the Command Window. The **type** command shows the contents of the file named *script1.m*; notice that the .m is not included:

```
>> type script1
radius = 5
area = pi * (radius ^2)
```

To actually run or execute the script from the Command Window, the name of the file is entered at the prompt (again, without the .m). When executed, the results of the two assignment statements are displayed, as the output was not suppressed for either statement.

```
>> script1
radius =
     5
area =
     78.5398
```

Once the script has been executed, you may find that you want to make changes to it (especially if there are errors!). To edit an existing file, there are several methods to open it. The easiest are:

- Within the Current Folder Window, double-click on the name of the file in the list of files.
- Choosing the Open down arrow will show a list of Recent Files

3.2.1 Documentation

It is very important that all scripts be ***documented*** well, so that people can understand what the script does and how it accomplishes its task. One way of documenting a script is to put ***comments*** in it. In MATLAB, a comment is anything from a % to the end of that particular line. Comments are completely ignored when the script is executed. To put in a comment, simply type the % symbol at the beginning of a line, or select the comment lines and then click on the Edit down arrow and click on the % symbol, and the Editor will put in the % symbols at the beginning of those lines for the comments.

For example, the previous script to calculate the area of a circle could be modified to have comments:

circlescript.m

```
% This script calculates the area of a circle

% First the radius is assigned
radius = 5
% The area is calculated based on the radius
area = pi * (radius ^2)
```

The first comment at the beginning of the script describes what the script does; this is sometimes called a ***block comment***. Then, throughout the script, comments describe different parts of the script (not usually a comment for every line, however!). Comments don't affect what a script does, so the output from this script would be the same as for the previous version.

The **help** command in MATLAB works with scripts as well as with built-in functions. The first block of comments (defined as contiguous lines at the beginning) will be displayed. For example, for *circlescript*:

```
>> help circlescript
  This script calculates the area of a circle
```

The reason that a blank line was inserted in the script between the first two comments is that otherwise both would have been interpreted as one contiguous comment, and both lines would have been displayed with **help**. The very first comment line is called the "H1 line"; it is what the function **lookfor** searches through.

PRACTICE 3.1

Write a script to calculate the circumference of a circle ($C=2\pi r$). Comment the script.

Longer comments, called ***comment blocks***, consist of everything in between %{ and %}, which must be alone on separate lines. For example:

```
%{
  this is used for a really
  Really
  REALLY
  long comment
%}
```

3.3 INPUT AND OUTPUT

The previous script would be much more useful if it were more general; for example, if the value of the radius could be read from an external source rather than being assigned in the script. Also, it would be better to have the script print the output in a nice, informative way. Statements that accomplish these tasks are called ***input/output*** statements or ***I/O*** for short. Although, for simplicity, examples of input and output statements will be shown here in the Command Window, these statements will make the most sense in scripts.

3.3.1 Input Function

Input statements read in values from the default or standard input device. In most systems, the default input device is the keyboard, so the input statement reads in values that have been entered by the ***user***, or the person who is running the script. To let the user know what he or she is supposed to enter, the script must first prompt the user for the specified values.

The simplest input function in MATLAB is called **input**. The **input** function is used in an assignment statement. To call it, a string is passed that is the prompt that will appear on the screen, and whatever the user types will be stored in the variable named on the left of the assignment statement. For ease of reading the prompt, it is useful to put a colon and then a space after the prompt. For example,

```
>> rad = input('Enter the radius: ')
Enter the radius: 5
rad =
     5
```

If character or string input is desired, 's' must be added as a second argument to the **input** function:

```
>> letter = input('Enter a char: ','s')
Enter a char: g
letter =
g
```

If the user enters only spaces or tabs before hitting the Enter key, they are ignored and an ***empty string*** is stored in the variable:

```
>> mychar = input('Enter a character: ', 's')
Enter a character:
mychar =
     ''
```

Note

Although normally the quotes are not shown around a character or string, in this case they are shown to demonstrate that there is nothing inside of the string.

However, if blank spaces are entered before other characters, they are included in the string. In the next example, the user hits the space bar four times before entering "go." The **length** function returns the number of characters in the string.

```
>> mystr = input('Enter a string: ', 's')
Enter a string:     go
mystr =
    go
>> length(mystr)
ans =
     6
```

QUICK QUESTION!

What would be the result if the user enters blank spaces after other characters? For example, the user here entered "xyz   " (four blank spaces):

```
>> mychar = input('Enter chars: ', 's')
Enter chars: xyz
mychar =
xyz
```

Answer: The space characters would be stored in the string variable. It is difficult to see earlier, but is clear from the length of the string.

```
>> length(mychar)
ans =
     7
```

The string can actually be seen in the Command Window by using the mouse to highlight the value of the variable; the xyz and four spaces will be highlighted.

It is also possible for the user to type quotation marks around the string rather than including the second argument 's' in the call to the **input** function.

```
>> name = input('Enter your name: ')
Enter your name: 'Stormy'
name =
Stormy
```

However, this assumes that the user would know to do this, so it is better to signify that character input is desired in the **input** function itself. Also, if the 's' is specified and the user enters quotation marks, these would become part of the string.

```
>> name = input('Enter your name: ','s')
Enter your name: 'Stormy'
name =
'Stormy'
>> length(name)
ans =
     8
```

Note what happens if string input has not been specified, but the user enters a letter rather than a number.

```
>> num = input('Enter a number: ')
Enter a number: t
Error using input
Undefined function or variable 't'.

Enter a number: 3
num =
     3
```

MATLAB gave an ***error message*** and repeated the prompt. However, if *t* is the name of a variable, MATLAB will take its value as the input.

```
>> t = 11;
>> num = input('Enter a number: ')
Enter a number: t
num =
    11
```

Separate **input** statements are necessary if more than one input is desired. For example,

```
>> x = input('Enter the x coordinate: ');
>> y = input('Enter the y coordinate: ');
```

Normally in a script the results from **input** statements are suppressed with a semicolon at the end of the assignment statements.

PRACTICE 3.2

Create a script that would prompt the user for a length, and then 'f' for feet or 'm' for meters, and store both inputs in variables. For example, when executed it would look like this (assuming the user enters 12.3 and then m):

```
Enter the length: 12.3
Is that f(eet) or m(eters)?: m
```

It is also possible to enter a vector. The user can enter any valid vector, using any valid syntax such as square brackets, the colon operator, or functions such as **linspace**.

```
>> v = input('Enter a vector: ')
Enter a vector: [3  8  22]
v =
     3     8    22
```

3.3.2 Output Statements: disp and fprintf

Output statements display strings and/or the results of expressions, and can allow for ***formatting*** or customizing how they are displayed. The simplest output function in MATLAB is **disp**, which is used to display the result of an expression or a string without assigning any value to the default variable *ans*. However, **disp** does not allow formatting. For example,

```
>> disp('Hello')
Hello

>> disp(4^3)
    64
```

Formatted output can be printed to the screen using the **fprintf** function. For example,

```
>> fprintf('The value is %d, for sure!\n',4^3)
The value is 64, for sure!
>>
```

To the **fprintf** function, first a string (called the ***format string***) is passed that contains any text to be printed, as well as formatting information for the expressions to be printed. In this case, the %d is an example of format information.

The %d is sometimes called a ***place holder*** because it specifies where the value of the expression that is after the string, is to be printed. The character in the place holder is called the ***conversion character***, and it specifies the type of value that is being printed. There are others, but what follows is a list of the simple place holders:

```
%d    integer (it stands for decimal integer)
%f    float (real number)
%c    character (one character)
%s    string of characters
```

Note
Don't confuse the % in the place holder with the symbol used to designate a comment.

The character '\n' at the end of the string is a special character called the ***newline character***; what happens when it is printed is that the output that follows moves down to the next line.

QUICK QUESTION!

What do you think would happen if the newline character is omitted from the end of an **fprintf** statement?

Answer: Without it, the next prompt would end up on the same line as the output. It is still a prompt, and so an expression can be entered, but it looks messy as shown here.

```
>> fprintf('The value is %d, surely!', 4^3)
The value is 64, surely!>> 5 + 3
ans =
     8
```

Note that with the **disp** function, however, the prompt will always appear on the next line:

```
>> disp('Hi')
Hi
>>
```

Also, note that an ellipsis can be used after a string but not in the middle.

QUICK QUESTION!

How can you get a blank line in the output?

Answer: Have two newline characters in a row.

```
>> fprintf('The value is %d,\n\nOK!\n',4 ^3)
The value is 64,

OK!
```

This also points out that the newline character can be anywhere in the string; when it is printed, the output moves down to the next line.

Note that the newline character can also be used in the prompt in the **input** statement; for example:

```
>> x = input('Enter the \nx coordinate: ');
Enter the
x coordinate: 4
```

However, the newline is the ONLY formatting character allowed in the prompt in **input**.

To print two values, there would be two place holders in the format string, and two expressions after the format string. The expressions fill in for the place holders in sequence.

```
>> fprintf('The int is %d and the char is %c\n', ...
    33 - 2, 'x')
The int is 31 and the char is x
```

A ***field width*** can also be included in the place holder in **fprintf**, which specifies how many characters in total are to be used in printing. For example, %5d would indicate a field width of 5 for printing an integer and %10s would indicate a field width of 10 for a string. For floats, the number of decimal places can also be specified; for example, %6.2f means a field width of 6 (including the decimal point and the two decimal places) with 2 decimal places. For floats, just the number of decimal places can be specified; for example, %.3f indicates 3 decimal places, regardless of the field width.

```
>> fprintf('The int is %3d and the float is %6.2f\n',...
            5,4.9)
The int is  5 and the float is   4.90
```

Note

If the field width is wider than necessary, ***leading blanks*** are printed, and if more decimal places are specified than necessary, ***trailing zeros*** are printed.

There are many other options for the format string. For example, the value being printed can be left-justified within the field width using a minus sign. The following example shows the difference between printing the

QUICK QUESTION!

What do you think would happen if you tried to print 1234.5678 in a field width of 3 with 2 decimal places?

```
>> fprintf('%3.2f\n', 1234.5678)
```

Answer: It would print the entire 1234, but round the decimals to two places, that is,

1234.57

If the field width is not large enough to print the number, the field width will be increased. Basically, to cut the number off would give a misleading result, but rounding the decimal places does not change the number significantly.

QUICK QUESTION!

What would happen if you use the %d conversion character but you're trying to print a real number?

Answer: MATLAB will show the result using exponential notation

```
>> fprintf('%d\n',1234567.89)
1.234568e+006
```

Note that if you want exponential notation, this is not the correct way to get it; instead, there are conversion characters that can be used. Use the **help** browser to see this option, as well as many others!

integer 3 using %5d and using %-5d. The x's below are used to show the spacing.

```
>> fprintf('The integer is xx%5dxx and xx%-5dxx\n',3,3)
The integer is xx    3xx and xx3    xx
```

Also, strings can be truncated by specifying "decimal places":

```
>> fprintf('The string is %s or %.2s\n', 'street', 'street')
The string is street or st
```

There are several special characters that can be printed in the format string in addition to the newline character. To print a slash, two slashes in a row are used, and also to print a single quote two single quotes in a row are used. Additionally, '\t' is the tab character.

```
>> fprintf('Try this out: tab\t quote '' slash \\ \n')
Try this out: tab    quote ' slash \
```

3.3.2.1 Printing Vectors and Matrices

For a vector, if a conversion character and the newline character are in the format string, it will print in a column regardless of whether the vector itself is a row vector or a column vector.

```
>> vec = 2:5;
>> fprintf('%d\n', vec)
2
3
4
5
```

Without the newline character, it would print in a row but the next prompt would appear on the same line:

```
>> fprintf('%d', vec)
2345>>
```

However, in a script, a separate newline character could be printed to avoid this problem. It is also much better to separate the numbers with spaces.

printvec.m

```
% This demonstrates printing a vector

vec = 2:5;
fprintf('%d ',vec)
fprintf('\n')
```

```
>> printvec
2 3 4 5
```

If the number of elements in the vector is known, that many conversion characters can be specified and then the newline:

```
>> fprintf('%d %d %d %d\n', vec)
2 3 4 5
```

This is not very general however, and is therefore not preferable.

For matrices, MATLAB unwinds the matrix column-by-column. For example, consider the following 2×3 matrix:

```
>> mat = [5 9 8; 4 1 10]
mat =
     5     9     8
     4     1    10
```

Specifying one conversion character and then the newline character will print the elements from the matrix in one column. The first values printed are from the first column, then the second column, and so on.

```
>> fprintf('%d\n', mat)
5
4
9
1
8
10
```

If three of the %d conversion characters are specified, the **fprintf** will print three numbers across on each line of output, but again the matrix is unwound column-by-column. It again prints first the two numbers from the first column (across on the first row of output), then the first value from the second column, and so on.

```
>> fprintf('%d %d %d\n', mat)
5   4   9
1   8   10
```

If the transpose of the matrix is printed, however, using the three %d conversion characters, the matrix is printed as it appears when created.

```
>> fprintf('%d %d %d\n', mat')   % Note the transpose
5   9   8
4   1   10
```

For vectors and matrices, even though formatting cannot be specified, the **disp** function may be easier to use in general than **fprintf** because it displays the result in a straight-forward manner. For example,

```
>> mat = [15 11 14; 7 10 13]
mat =
    15    11    14
     7    10    13
>> disp(mat)
    15    11    14
     7    10    13

>> vec = 2:5
vec =
     2     3     4     5
>> disp(vec)
     2     3     4     5
```

Note that when loops are covered in Chapter 5, formatting the output of matrices will be easier. For now, however, **disp** works well.

3.4 SCRIPTS WITH INPUT AND OUTPUT

Putting all of this together now, we can implement the algorithm from the beginning of this chapter. The following script calculates and prints the area of a circle. It first prompts the user for a radius, reads in the radius, and then calculates and prints the area of the circle based on this radius.

circleIO.m

```
% This script calculates the area of a circle
% It prompts the user for the radius

% Prompt the user for the radius and calculate
% the area based on that radius
fprintf('Note: the units will be inches.\n')
radius = input('Please enter the radius: ');
area = pi * (radius ^2);

% Print all variables in a sentence format
fprintf('For a circle with a radius of %.2f inches,\n',...
    radius)
fprintf('the area is %.2f inches squared\n',area)
```

Executing the script produces the following output:

```
>> circleIO
Note: the units will be inches.
Please enter the radius: 3.9
For a circle with a radius of 3.90 inches,
the area is 47.78 inches squared
```

Note that the output from the first two assignment statements (including the **input**) is suppressed by putting semicolons at the end. That is usually done in scripts, so that the exact format of what is displayed by the program is controlled by the **fprintf** functions.

PRACTICE 3.3

Write a script to prompt the user separately for a character and a number, and print the character in a field width of 3 and the number left-justified in a field width of 8 with 3 decimal places. Test this by entering numbers with varying widths.

3.5 SCRIPTS TO PRODUCE AND CUSTOMIZE SIMPLE PLOTS

MATLAB has many graphing capabilities. Customizing plots is often desired and this is easiest to accomplish by creating a script rather than typing one command at a time in the Command Window. For that reason, simple plots and how to customize them will be introduced in this chapter on MATLAB programming.

The help topics that contain graph functions include **graph2d** and **graph3d**. Typing **help graph2d** would display some of the two dimensional graph functions, as well as functions to manipulate the axes and to put labels and titles on the graphs. The Search Documentation under MATLAB Graphics also has a section on two- and three-dimensional plots.

3.5.1 The Plot Function

For now, we'll start with a very simple graph of one point using the **plot** function.

The following script, *plotonepoint*, plots one point. To do this, first values are given for the *x* and *y* coordinates of the point in separate variables. The point is plotted using a red star ('*'). The plot is then customized by specifying the minimum and maximum values on first the *x* and then *y*-axes. Labels are then put on the *x*-axis, the *y*-axis, and the graph itself using the functions **xlabel**, **ylabel**, and **title**. (Note: there are no default labels for the axes.)

All of this can be done from the Command Window, but it is much easier to use a script. The following shows the contents of the script *plotonepoint* that accomplishes this. The *x* coordinate represents the time of day (e.g. 11 am) and the *y* coordinate represents the temperature (e.g. in degrees Fahrenheit) at that time.

plotonepoint.m

```
% This is a really simple plot of just one point!

% Create coordinate variables and plot a red '*'
x = 11;
y = 48;
plot(x,y,'r*')

% Change the axes and label them
axis([9 12 35 55])
xlabel('Time')
ylabel('Temperature')

% Put a title on the plot
title('Time and Temp')
```

In the call to the **axis** function, one vector is passed. The first two values are the minimum and maximum for the *x*-axis, and the last two are the minimum and maximum for the *y*-axis. Executing this script brings up a Figure Window with the plot (see Fig. 3.2).

In general, the script could prompt the user for the time and temperature, rather than just assigning values. Then, the axis function could be used based on whatever the values of *x* and *y* are, as in the following example:

```
axis([x-2  x+2  y-10  y+10])
```

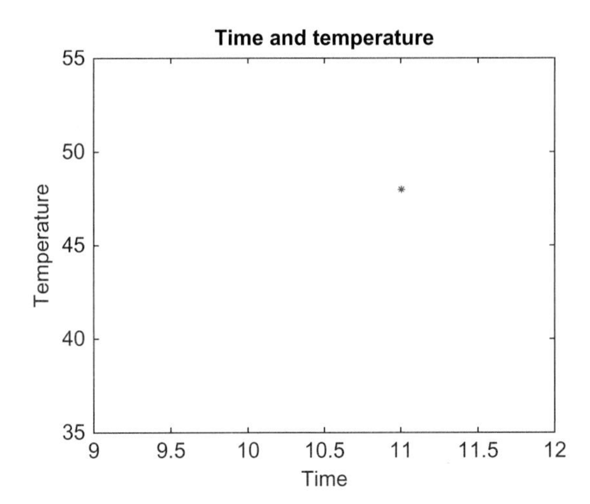

FIGURE 3.2
Plot of one data point.

In addition, although they are the x and y coordinates of a point, variables named *time* and *temp* might be more mnemonic than x and y.

PRACTICE 3.4

Modify the script *plotonepoint* to prompt the user for the time and temperature, and set the axes based on these values.

To plot more than one point, x and y vectors are created to store the values of the (x,y) points. For example, to plot the points

```
(1,1)
(2,5)
(3,3)
(4,9)
(5,11)
(6,8)
```

first an x vector is created that has the x values (as they range from 1 to 6 in steps of 1, the colon operator can be used) and then a y vector is created with the y values. The following will create (in the Command Window) x and y vectors and then plot them (see Fig. 3.3).

```
>> x = 1:6;
>> y = [1 5 3 9 11 8];
>> plot(x,y)
```

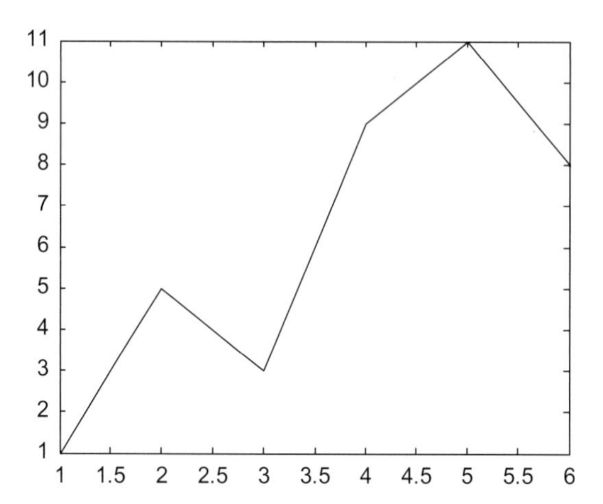

FIGURE 3.3
Plot of data points from vectors.

Note that the points are plotted with straight lines drawn in between. Also, the axes are set up according to the data; for example, the *x* values range from 1 to 6 and the *y* values from 1 to 11, so that is how the axes are set up. There are many options for the **axis** function; for example, just calling it with no arguments returns the values used for the *x* and *y*-axes ranges.

```
>> arang = axis
arang =
     1   6   1  11
```

Axes can also be turned on and off, and they can be made square or equal to each other. A subset of the data can also be shown by limiting the extent of the axes.

Also, note that in this case the *x* values are the indices of the *y* vector (the *y* vector has six values in it, so the indices iterate from 1 to 6). When this is the case, it is not necessary to create the *x* vector. For example,

```
>> plot(y)
```

will plot exactly the same figure without using an *x* vector.

3.5.1.1 Customizing a Plot: Color, Line Types, Marker Types

Plots can be done in the Command Window, as shown here, if they are really simple. However, at many times it is desired to customize the plot with labels, titles, and so on, so it makes more sense to do this in a script. Using the **help** function for **plot** will show the many options such as the line types and colors. In the previous script *plotonepoint*, the string 'r*' specified a red star for the point type. The LineSpec, or line specification, can specify up to three different properties in a string, including the color, line type, and the symbol or marker used for the data points.

The possible colors are:

```
b blue
g green
r red
c cyan
m magenta
y yellow
k black
w white
```

Either the single character listed above or the full name of the color can be used in the string to specify the color. The ***plot symbols***, or ***markers***, that can be used are:

```
.  point
o  circle
x  x-mark
+  plus
*  star
s  square
d  diamond
v  down triangle
^  up triangle
<  left triangle
>  right triangle
p  pentagram
h  hexagram
```

Line types can also be specified by the following:

```
-        solid
:        dotted
-.       dash dot
--       dashed
(none)   no line
```

If no line type is specified and no marker type is specified, a solid line is drawn between the points, as seen in the last example.

3.5.2 Simple Related Plot Functions

Other functions that are useful in customizing plots include **clf**, **figure**, **hold**, **legend**, and **grid**. Brief descriptions of these functions are given here; use **help** to find out more about them:

`clf:` clears the Figure Window by removing everything from it.

`figure:` creates a new, empty Figure Window when called without any arguments. Calling it as **`figure(n)`** where *n* is an integer, is a way of creating and maintaining multiple Figure Windows, and of referring to each individually.

hold: is a toggle that freezes the current graph in the Figure Window, so that new plots will be superimposed on the current one. Just **hold** by itself is a ***toggle***, so calling this function once turns the hold on, and then the next time turns it off. Alternatively, the commands **hold on** and **hold off** can be used.

legend: displays strings passed to it, in a legend box in the Figure Window, in order of the plots in the Figure Window

grid: displays grid lines on a graph. Called by itself, it is a toggle that turns the grid lines on and off. Alternatively, the commands **grid on** and **grid off** can be used.

Also, there are many plot types. We will see more in Chapter 12, but another simple plot type is a **bar** chart.

For example, the following script creates two separate Figure Windows. First, it clears the Figure Window. Then, it creates an *x* vector and two different *y* vectors (*y1* and *y2*). In the first Figure Window, it plots the *y1* values using a bar chart. In the second Figure Window, it plots the *y1* values as black lines, puts **hold on** so that the next graph will be superimposed, and plots the *y2* values as black circles. It also puts a legend on this graph and uses a grid. Labels and titles are omitted in this case as it is generic data.

plot2figs.m

```
% This creates 2 different plots, in 2 different
%  Figure Windows, to demonstrate some plot features

clf
x = 1:5; % Not necessary
y1 = [2 11 6 9 3];
y2 = [4 5 8 6 2];
% Put a bar chart in Figure 1
figure(1)
bar(x,y1)
% Put plots using different y values on one plot
% with a legend
figure(2)
plot(x,y1,'k')
hold on
plot(x,y2,'ko')
grid on
legend('y1','y2')
```

Running this script will produce two separate Figure Windows. If there are no other active Figure Windows, the first, which is the bar chart, will be in the one numbered "Figure 1" in MATLAB. The second will be "Figure 2." See Fig. 3.4 for both plots.

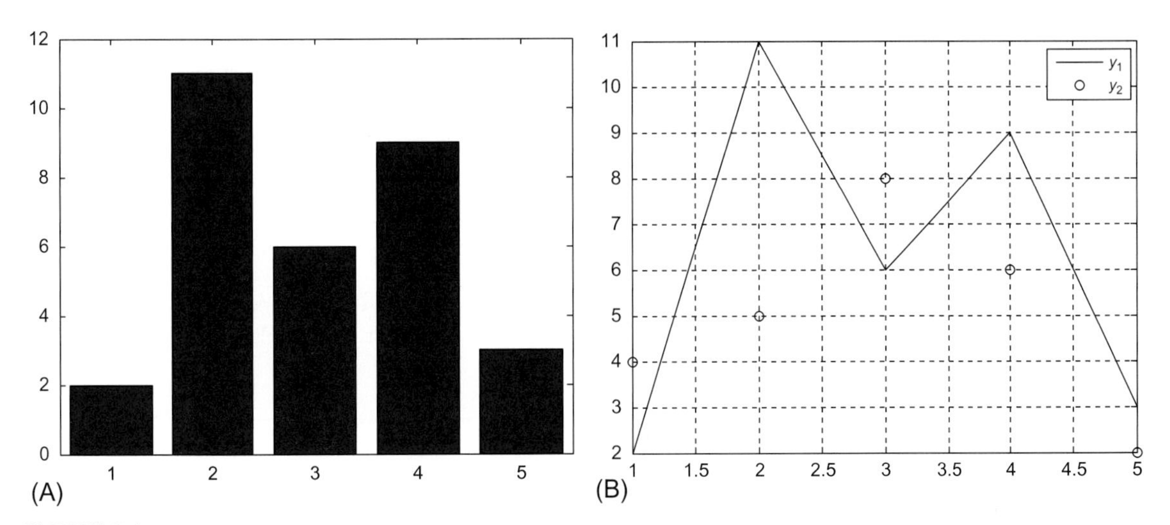

FIGURE 3.4
(A) Bar chart produced by script. (B) Plot produced by script, with a grid and legend.

Note that the first and last points are on the axes, which makes them difficult to be seen. That is why the **axis** function is used frequently, as it creates space around the points so that they are all visible.

PRACTICE 3.5

Modify the *plot2figs* script using the **axis** function so that all points are easily seen.

The ability to pass a vector to a function and have the function evaluate every element of the vector can be very useful in creating plots. For example, the following script graphically displays the difference between the **sin** and **cos** functions:

sinncos.m

```
% This script plots sin(x) and cos(x) in the same Figure Window
%  for values of x ranging from 0 to 2*pi

clf
x = 0: 2*pi/40: 2*pi;
y = sin(x);
plot(x,y,'ro')
hold on
y = cos(x);
plot(x,y,'b+')
legend('sin', 'cos')
xlabel('x')
ylabel('sin(x) or cos(x)')
title('sin and cos on one graph')
```

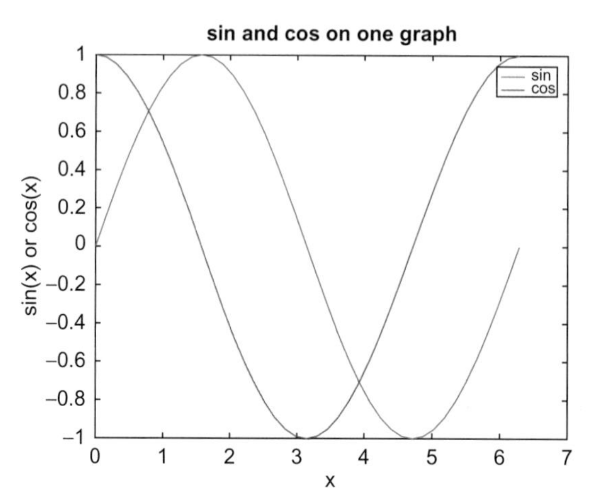

FIGURE 3.5
Plot of **sin** and **cos** in one Figure Window with a legend.

The script creates an *x* vector; iterating through all of the values from 0 to 2*π in steps of 2*π/40 gives enough points to get a good graph. It then finds the sine of each *x* value, and plots these. The command **hold on** freezes this in the Figure Window so the next plot will be superimposed. Next, it finds the cosine of each *x* value and plots these points. The **legend** function creates a legend; the first string is paired with the first plot, and the second string with the second plot. Running this script produces the plot seen in Fig. 3.5.

Beginning with Version R2014b, when **hold on** is used, MATLAB uses a sequence of colors for the plots, rather than using the default color for each. Of course, colors can also be specified as was done in this script.

Note that instead of using **hold on**, both functions could have been plotted using one call to the **plot** function:

```
plot(x,sin(x),x,cos(x))
```

PRACTICE 3.6

Write a script that plots exp(x) and log(x) for values of x ranging from 0 to 3.5.

3.6 INTRODUCTION TO FILE INPUT/OUTPUT (LOAD AND SAVE)

In many cases, input to a script will come from a data file that has been created by another source. Also, it is useful to be able to store output in an external file that can be manipulated and/or printed later. In this section, the simplest methods used to read from an external data file and also to write to an external data file will be demonstrated.

There are basically three different operations, or ***modes*** on files. Files can be:

- read from
- written to
- appended to

Writing to a file means writing to a file from the beginning. ***Appending to a file*** is also writing, but starting at the end of the file rather than the beginning. In other words, appending to a file means adding to what was already there.

There are many different file types, which use different filename extensions. For now, we will keep it simple and just work with .dat or .txt files when working with data or text files. There are several methods for reading from files and writing to files; we will, for now, use the **load** function to read and the **save** function to write to files. More file types and functions for manipulating them will be discussed in Chapter 9.

3.6.1 Writing Data to a File

The **save** command can be used to write data from a matrix to a data file, or to append to a data file. The format is:

```
save filename matrixvariablename -ascii
```

The "-ascii" qualifier is used when creating a text or data file. For example, the following creates a matrix and then saves the values from the matrix variable to a data file called *testfile.dat*:

```
>> mymat = rand(2,3)
mymat =
     0.4565   0.8214   0.6154
     0.0185   0.4447   0.7919

>> save testfile.dat mymat -ascii
```

This creates a file called "testfile.dat" that stores the numbers:

```
0.4565   0.8214   0.6154
0.0185   0.4447   0.7919
```

The **type** command can be used to display the contents of the file; note that scientific notation is used:

```
>> type testfile.dat

  4.5646767e-001   8.2140716e-001   6.1543235e-001
  1.8503643e-002   4.4470336e-001   7.9193704e-001
```

Note that if the file already exists, the **save** command will overwrite the file; **save** always writes from the beginning of a file.

3.6.2 Appending Data to a Data File

Once a text file is created, data can be appended to it. The format is the same as the preceding, with the addition of the qualifier "-append." For example, the following creates a new random matrix and appends it to the file that was just created:

```
>> mat2 = rand(3,3)
mymat =
       0.9218    0.4057    0.4103
       0.7382    0.9355    0.8936
       0.1763    0.9169    0.0579
>> save testfile.dat mat2 -ascii -append
```

This results in the file "testfile.dat" containing the following:

```
     0.4565    0.8214    0.6154
     0.0185    0.4447    0.7919
     0.9218    0.4057    0.4103
     0.7382    0.9355    0.8936
     0.1763    0.9169    0.0579
```

Note

Although technically any size matrix could be appended to this data file, to be able to read it back into a matrix later there would have to be the same number of values on every row (or, in other words, the same number of columns).

PRACTICE 3.7

Prompt the user for the number of rows and columns of a matrix, create a matrix with that many rows and columns of random integers, and write it to a file.

3.6.3 Reading from a File

Reading from a file is accomplished using **load**. Once a file has been created (as in the preceding), it can be read into a matrix variable. If the file is a data file, the **load** command will read from the file "filename.ext" (e.g. the extension might be .dat) and create a matrix with the same name as the file. For example, if the data file "testfile.dat" had been created as shown in the previous section, this would read from it, and store the result in a matrix variable called *testfile*:

```
>> clear
>> load testfile.dat
>> who
Your variables are:
testfile
>> testfile
testfile =
```

```
0.4565    0.8214    0.6154
0.0185    0.4447    0.7919
0.9218    0.4057    0.4103
0.7382    0.9355    0.8936
0.1763    0.9169    0.0579
```

The **load** command works only if there are the same number of values in each line, so that the data can be stored in a matrix, and the **save** command only writes from a matrix to a file. If this is not the case, lower-level file I/O functions must be used; these will be discussed in Chapter 9.

3.6.3.1 Example: Load from a File and Plot the Data

As an example, a file called "timetemp.dat" stores two lines of data. The first line is the times of day, and the second line is the recorded temperature at each of those times. The first value of 0 for the time represents midnight. For example, the contents of the file might be:

```
0      3      6      9      12     15     18     21
55.5   52.4   52.6   55.7   75.6   77.7   70.3   66.6
```

The following script loads the data from the file into a matrix called *timetemp*. It then separates the matrix into vectors for the time and temperature, and then plots the data using black star (*) symbols.

timetempprob.m

```
% This reads time and temperature data for an afternoon
% from a file and plots the data

load timetemp.dat

% The times are in the first row, temps in the second row
time = timetemp(1,:);
temp = timetemp(2,:);

% Plot the data and label the plot
plot(time,temp,'k*')
xlabel('Time')
ylabel('Temperature')
title('Temperatures one afternoon')
```

Running the script produces the plot seen in Fig. 3.6.

Note that it is difficult to see the point at time 0 as it falls on the *y*-axis. The **axis** function could be used to change the axes from the defaults shown here.

To create the data file, the Editor in MATLAB can be used; it is not necessary to create a matrix and **save** it to a file. Instead, just enter the numbers in a new script file, and Save As *timetemp.dat,* making sure that the Current Folder is set.

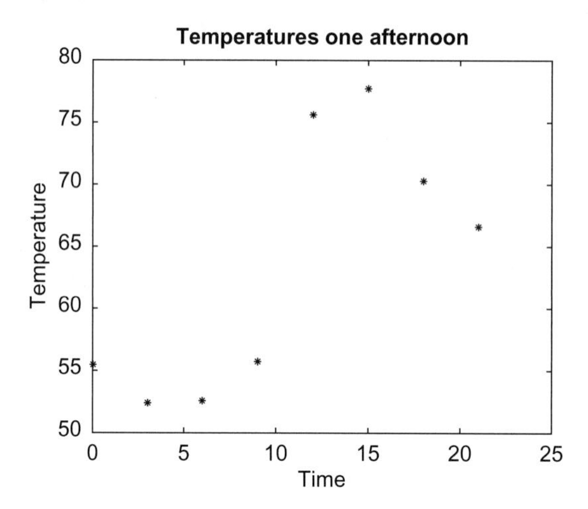

FIGURE 3.6
Plot of temperature data from a file.

PRACTICE 3.8

The sales (in billions) for two separate divisions of the ABC Corporation for each of the four quarters of 2013 are stored in a file called "salesfigs.dat":

```
1.2 1.4 1.8 1.3
2.2 2.5 1.7 2.9
```

- First, create this file (just type the numbers in the Editor, and Save As "salesfigs.dat").
- Then, write a script that will
  - load the data from the file into a matrix
  - separate this matrix into 2 vectors.
  - create the plot seen in Fig. 3.7 (which uses black circles and stars as the plot symbols).

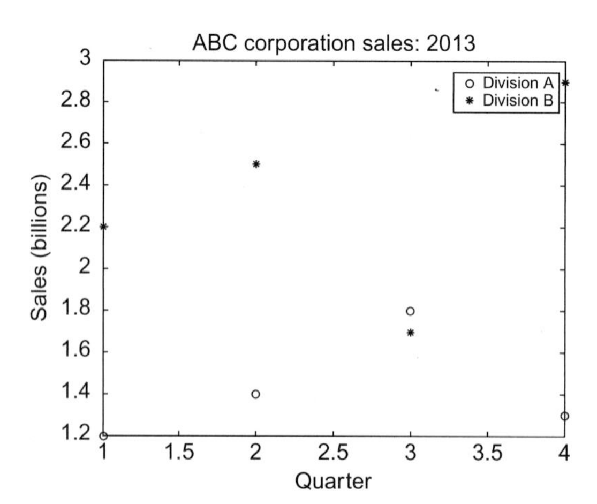

FIGURE 3.7
Plot of sales data from file.

QUICK QUESTION!

Sometimes files are not in the format that is desired. For example, a file "expresults.dat" has been created that has some experimental results, but the order of the values is reversed in the file:

```
4  53.4
3  44.3
2  50.0
1  55.5
```

How could we create a new file that reverses the order?

Answer: We can **load** from this file into a matrix, use the **flipud** function to "flip" the matrix up to down, and then **save** this matrix to a new file:

```
>> load expresults.dat
>> expresults
expresults =
    4.0000   53.4000
    3.0000   44.3000
    2.0000   50.0000
    1.0000   55.5000
>> correctorder = flipud(expresults)
correctorder =
    1.0000   55.5000
    2.0000   50.0000
    3.0000   44.3000
    4.0000   53.4000
>> save neworder.dat correctorder - ascii
```

3.7 USER-DEFINED FUNCTIONS THAT RETURN A SINGLE VALUE

We have already seen the use of many functions in MATLAB. We have used many built-in functions such as **sin**, **fix**, **abs**, and **double**. In this section, ***user-defined functions*** will be introduced. These are functions that the programmer defines, and then uses, in either the Command Window or in a script.

There are several different types of functions. For now, we will concentrate on the kind of function that calculates and returns a single result. Other types of functions will be introduced in Chapter 6.

First, let us review some of what we already know about functions, including the use of built-in functions. Although by now the use of these functions is straightforward, explanations will be given in some detail here in order to compare and contrast the use of user-defined functions.

The **length** function is an example of a built-in function that calculates a single value; it returns the length of a vector. As an example,

```
length(vec)
```

is an expression that represents the number of elements in the vector *vec*. This expression could be used in the Command Window or in a script. Typically, the value returned from this expression might be assigned to a variable:

```
>> vec = 1:3:10;
>> lv = length(vec)
lv =
      4
```

Alternatively, the length of the vector could be printed:

```
>> fprintf('The length of the vector is %d\n', length(vec))
The length of the vector is 4
```

The ***function call*** to the **length** function consists of the name of the function, followed by the ***argument*** in parentheses. The function receives as input the argument, and returns a result. What happens when the call to the function is encountered is that ***control*** is passed to the function itself (in other words, the function begins executing). The argument(s) are also passed to the function.

The function executes its statements and does whatever is necessary (the actual contents of the built-in functions are not generally known or seen by the user) to determine the number of elements in the vector. As the function is calculating a single value, this result is then ***returned*** and it becomes the value of the expression. Control is also passed back to the expression that called it in the first place, which then continues (e.g. in the first example the value would then be assigned to the variable *lv* and in the second example the value was printed).

3.7.1 Function Definitions

There are different ways to organize scripts and functions, but, for now, every function that we write will be stored in a separate file. Like scripts, function files have an extension of .m. Although to enter function definitions in the Editor it is possible to choose the New down arrow and then Function, it will be easier for now to type in the function by choosing New Script (this ignores the defaults that are provided when you choose Function).

A function in MATLAB that returns a single result consists of the following.

- The ***function header*** (the first line), comprised of:
  - the reserved word **function**
  - the name of the ***output argument*** followed by the assignment operator (=), as the function ***returns*** a result
  - the name of the function (*important*—This should be the same as the name of the file in which this function is stored to avoid confusion)
  - the ***input arguments*** in parentheses, which correspond to the arguments that are passed to the function in the function call

- A comment that describes what the function does (this is printed when **help** is used)
- The ***body*** of the function, which includes all statements and eventually must put a value in the output argument
- **end** at the end of the function (note that this is not necessary in many cases in current versions of MATLAB, but it is considered a good style anyway)

The general form of a ***function definition*** for a function that calculates and returns one value looks like this:

functionname.m

```
function outputargument = functionname(input arguments)
% Comment describing the function

Statements here; these must include putting a value in the output
argument

end % of the function
```

For example, the following is a function called *calcarea* that calculates and returns the area of a circle; it is stored in a file called *calcarea.m*.

calcarea.m

```
function area = calcarea(rad)
% calcarea calculates the area of a circle
% Format of call: calcarea(radius)
% Returns the area

area = pi * rad * rad;
end
```

The radius of a circle is passed to the function to the input argument *rad*; the function calculates the area of this circle and stores it in the output argument *area*.

In the function header, we have the reserved word **function**, then the output argument *area* followed by the assignment operator =, then the name of the function (the same as the name of the file), and then the input argument *rad*, which is the radius. As there is an output argument in the function header, somewhere in the body of the function we must put a value in this output argument. This is how a value is returned from the function. In this case, the function is simple and all we have to do is assign to the output argument *area* the value of the built-in constant **pi** multiplied by the square of the input argument *rad*.

The function can be displayed in the Command Window using the **type** command.

```
>> type calcarea

    function area = calcarea(rad)
    % calcarea calculates the area of a circle
    % Format of call: calcarea(radius)
    % Returns the area

    area = pi * rad * rad;
    end
```

Note

Many of the functions in MATLAB are implemented as functions that are stored in files with an extension of .m; these can also be displayed using **type**.

3.7.2 Calling a Function

The following is an example of a call to this function in which the value returned is stored in the default variable *ans*:

```
>> calcarea(4)
ans =
   50.2655
```

Technically, calling the function is done with the name of the file in which the function resides. To avoid confusion, it is easiest to give the function the same name as the file name, so that is how it will be presented in this book. In this example, the function name is *calcarea* and the name of the file is *calcarea.m*. The result returned from this function can also be stored in a variable in an assignment statement; the name could be the same as the name of the output argument in the function itself, but that is not necessary. So, for example, either of these assignments would be fine:

```
>> area = calcarea(5)
area =
   78.5398

>> myarea = calcarea(6)
myarea =
  113.0973
```

The output could also be suppressed when calling the function:

```
>> mya = calcarea(5.2);
```

Note

The printing is not done in the function itself; rather, the function returns the area and then an output statement can print or display it.

The value returned from the *calcarea* function could also be printed using either **disp** or **fprintf**:

```
>> disp(calcarea(4))
   50.2655
>> fprintf('The area is %.1f\n', calcarea(4))
The area is 50.3
```

QUICK QUESTION!

Could we pass a vector of radii to the *calcarea* function?

Answer: This function was written assuming that the argument was a scalar, so calling it with a vector instead would produce an error message:

```
>> calcarea(1:3)
   Error using *
   Inner matrix dimensions must agree.

   Error in calcarea (line 6)
        area = pi * rad * rad;
```

This is because the * was used for multiplication in the function, but .* must be used when multiplying vectors term by term. Changing this in the function would allow either scalars or vectors to be passed to this function:

calcareaii.m

```
function area = calcareaii(rad)
% calcareaii returns the area of a circle
% The input argument can be a vector of radii
% Format: calcareaii(radiiVector)

area = pi * rad .* rad;
end
```

```
>> calcareaii(1:3)
ans =
  3.1416 12.5664 28.2743

>> calcareaii(4)
ans =
  50.2655
```

Note that the .* operator is only necessary when multiplying the radius vector by itself. Multiplying by **pi** is scalar multiplication, so the .* operator is not needed there. We could have also used:

```
area = pi * rad .^ 2;
```

Using **help** with either of these functions displays the contiguous block of comments under the function header (the block comment). It is useful to put the format of the call to the function in this block comment:

```
>> help calcarea
  calcarea calculates the area of a circle
  Format of call: calcarea(radius)
  Returns the area
```

The suggested corrections for invalid filenames in the Command Window works for user-defined files as of Version R2014b.

```
>> clacarea(3)
Undefined function or variable 'clacarea'.
Did you mean:
>> calcarea(3)
```

Many organizations have standards regarding what information should be included in the block comment in a function. These can include:

- Name of the function
- Description of what the function does

- Format of the function call
- Description of input arguments
- Description of output argument
- Description of variables used in function
- Programmer name and date written
- Information on revisions

Although this is an excellent programming style, for the most part of this book these will be omitted simply to save space. Also, documentation in MATLAB suggests that the name of the function should be in all uppercase letters in the beginning of the block comment. However, this can be somewhat misleading in that MATLAB is case-sensitive and typically lowercase letters are used for the actual function name.

3.7.3 Calling a User-Defined Function from a Script

Now, we will modify our script that prompts the user for the radius and calculates the area of a circle to call our function *calcarea* to calculate the area of the circle rather than doing this in the script.

circleCallFn.m

```
% This script calculates the area of a circle
% It prompts the user for the radius
radius = input('Please enter the radius: ');
% It then calls our function to calculate the
%  area and then prints the result
area = calcarea(radius);
fprintf('For a circle with a radius of %.2f,',radius)
fprintf(' the area is %.2f\n',area)
```

Running this will produce the following:

```
>> circleCallFn
Please enter the radius: 5
For a circle with a radius of 5.00, the area is 78.54
```

3.7.3.1 Simple Programs

In this book, a script that calls function(s) is what we will call a MATLAB program. In the previous example, the program consisted of the script *circleCallFn* and the function it calls, *calcarea*. A simple program, consisting of a script that calls a function to calculate and return a value, looks like the format shown in Fig. 3.8.

It is also possible for a function to call another (whether built-in or user-defined).

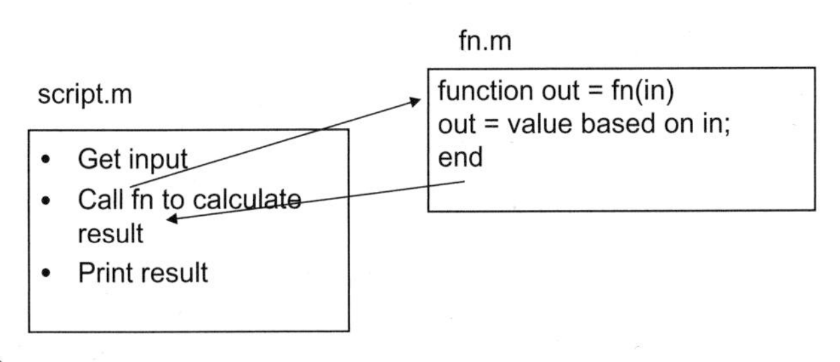

FIGURE 3.8
General form of a simple program.

3.7.4 Passing Multiple Arguments

In many cases it is necessary to pass more than one argument to a function. For example, the volume of a cone is given by

$$V = \frac{1}{3}\pi r^2 h$$

where r is the radius of the circular base and h is the height of the cone. Therefore, a function that calculates the volume of a cone needs both the radius and the height:

conevol.m

```
function outarg = conevol(radius, height)
% conevol calculates the volume of a cone
% Format of call: conevol(radius, height)
% Returns the volume

outarg = (pi/3) * radius .^ 2 .* height;
end
```

As the function has two input arguments in the function header, two values must be passed to the function when it is called. The order makes a difference. The first value that is passed to the function is stored in the first input argument (in this case, *radius*) and the second argument in the function call is passed to the second input argument in the function header.

This is very important: the arguments in the function call must correspond one-to-one with the input arguments in the function header.

Here is an example of calling this function. The result returned from the function is simply stored in the default variable *ans*.

```
>> conevol(4,6.1)
ans =
   102.2065
```

In the next example, the result is instead printed with a format of two decimal places.

```
>> fprintf('The cone volume is %.2f\n',conevol(3, 5.5))
The cone volume is 51.84
```

Note that by using the array exponentiation and multiplication operators, it would be possible to pass arrays for the input arguments, as long as the dimensions are the same.

QUICK QUESTION!

Nothing is technically wrong with the following function, but what about it does not make sense?

fun.m

```
function out = fun(a,b,c)
out = a*b;
end
```

Answer; Why pass the third argument if it is not used?

PRACTICE 3.9

Write a script that will prompt the user for the radius and height, call the function *conevol* to calculate the cone volume, and print the result in a nice sentence format. So, the program will consist of a script and the *conevol* function that it calls.

PRACTICE 3.10

For a project, we need some material to form a rectangle. Write a function *calcrectarea* that will receive the length and width of a rectangle in inches as input arguments, and will return the area of the rectangle. For example, the function could be called as shown, in which the result is stored in a variable and then the amount of material required is printed, rounded up to the nearest square inch.

```
>> ra = calcrectarea(3.1, 4.4)
ra =
   13.6400

>> fprintf('We need %d sq in.\n', ceil(ra))
We need 14 sq in.
```

3.7.5 Functions With Local Variables

The functions discussed thus far have been very simple. However, in many cases the calculations in a function are more complicated, and may require the use of extra variables within the function; these are called ***local variables***.

For example, a closed cylinder is being constructed of a material that costs a certain dollar amount per square foot. We will write a function that will calculate and return the cost of the material, rounded up to the nearest square foot, for a cylinder with a given radius and a given height. The total surface area for the closed cylinder is

$$SA = 2\pi rh + 2\pi r^2$$

For a cylinder with a radius of 32 in., height of 73 in., and cost per square foot of the material of $4.50, the calculation would be given by the following algorithm.

- Calculate the surface area $SA = 2*\pi*32*73 + 2*\pi*32*32\ in^2$.
- Convert the SA from square inches to square feet = SA/144.
- Calculate the total cost = SA in square feet * cost per square foot.

The function includes local variables to store the intermediate results.

cylcost.m

```
function outcost = cylcost(radius, height, cost)
% cylcost calculates the cost of constructing a closed
%    cylinder
% Format of call: cylcost(radius, height, cost)
% Returns the total cost

% The radius and height are in inches
% The cost is per square foot

% Calculate surface area in square inches
surf_area = 2 * pi * radius .* height + 2 * pi * radius . ^ 2;

% Convert surface area in square feet and round up
surf_areasf = ceil(surf_area/144);

% Calculate cost
outcost = surf_areasf .* cost;
end
```

The following shows examples of calling the function:

```
>> cylcost(32,73,4.50)
ans =
   661.5000
```

```
>> fprintf('The cost would be $%.2f\n', cylcost(32,73,4.50))
The cost would be $661.50
```

3.7.6 Introduction to Scope

It is important to understand the ***scope of variables***, which is where they are valid. More will be described in Chapter 6, but, basically, variables used in a script are also known in the Command Window and vice versa. All variables used in a function, however, are local to that function. Both the Command Window and scripts use a common workspace, the ***base workspace***. Functions, however, have their own workspaces. This means that when a script is executed, the variables can subsequently be seen in the Workspace Window and can be used from the Command Window. This is not the case with functions, however.

3.8 COMMANDS AND FUNCTIONS

Some of the commands that we have used (e.g. **format**, **type**, **save**, and **load**) are just shortcuts for function calls. If all of the arguments to be passed to a function are strings, and the function does not return any values, it can be used as a command. For example, the following produce the same results:

```
>> type script1

radius = 5
area = pi * (radius ^ 2)

>> type('script1')

radius = 5
area = pi * (radius ^ 2)
```

Using **load** as a command creates a variable with the same name as the file. If a different variable name is desired, it is easiesr to use the functional form of **load**. For example,

```
>> type pointcoords.dat

3.3    1.2
4      5.3

>> points = load('pointcoords.dat')
points =
    3.3000    1.2000
    4.0000    5.3000
```

This stores the result in a variable *points* rather than *pointcoords*.

■ Explore Other Interesting Features

Note that this chapter serves as an introduction to several topics, most of which will be covered in more detail in future chapters. Before getting to those chapters, the following are some things you may wish to explore.

- The **help** command can be used to see short explanations of built-in functions. At the end of this, a doc page link is also listed. These documentation pages frequently have much more information and useful examples. They can also be reached by typing "doc fnname" where fnname is the name of the function.
- Look at formatSpec on the doc page on the **fprintf** function for more ways in which expressions can be formatted, for example, padding numbers with zeros and printing the sign of a number.
- Use the Search Documentation to find the conversion characters used to print other types, such as unsigned integers and exponential notation. ■

SUMMARY

COMMON PITFALLS

- Spelling a variable name different ways in different places in a script or function.
- Forgetting to add the second 's' argument to the **input** function when character input is desired.
- Not using the correct conversion character when printing.
- Confusing **fprintf** and **disp.** Remember that only **fprintf** can format.

PROGRAMMING STYLE GUIDELINES

- Especially for longer scripts and functions, start by writing an algorithm.
- Use comments to document scripts and functions, as follows:
  - a block of contiguous comments at the top to describe a script
  - a block of contiguous comments under the function header for functions
  - comments throughout any code file (script or function) to describe each section
- Make sure that the "H1" comment line has useful information.
- Use your organization's standard style guidelines for block comments.
- Use mnemonic identifier names (names that make sense, e.g. *radius* instead of *xyz*) for variable names and for file names.
- Make all output easy to read and informative.

- Put a newline character at the end of every string printed by **fprintf** so that the next output or the prompt appears on the line below.
- Put informative labels on the *x*- and *y*-axes, and a title on all plots.
- Keep functions short—typically no longer than one page in length.
- Suppress the output from all assignment statements in functions and scripts.
- Functions that return a value do not normally print the value; it should simply be returned by the function.
- Use the array operators .*, ./, .\, and .^ in functions so that the input arguments can be arrays and not just scalars.

MATLAB Reserved Words	
function	end

MATLAB Functions and Commands			
type	xlabel	figure	load
input	ylabel	hold	save
disp	title	legend	
fprintf	axis	grid	
plot	clf	bar	

MATLAB Operators	
comment %	comment block %{, %}

Exercises

1. Using the top-down design approach, write an algorithm for making a sandwich.
2. Write a simple script that will calculate the volume of a hollow sphere,

$$\frac{4\pi}{3}(r_o^3 - r_i^3)$$

where r_i is the inner radius and r_o is the outer radius. Assign a value to a variable for the inner radius, and also assign a value to another variable for the outer radius. Then, using these variables, assign the volume to a third variable. Include comments in the script. Use **help** to view the comments in your script.

3. Write a statement that prompts the user for his/her favorite number.

4. Write a statement that prompts the user for his/her name.
5. Write an **input** statement that will prompt the user for a real number, and store it in a variable. Then, use the **fprintf** function to print the value of this variable using 2 decimal places.
6. Experiment, in the Command Window, with the **fprintf** function for real numbers. Make a note of what happens for each. Use **fprintf** to print the real number 12345.6789.
   - without specifying any field width
   - in a field width of 10 with 4 decimal places
   - in a field width of 10 with 2 decimal places
   - in a field width of 6 with 4 decimal places
   - in a field width of 2 with 4 decimal places
7. Experiment, in the Command Window, with the **fprintf** function for integers. Make a note of what happens for each. Use **fprintf** to print the integer 12345.
   - without specifying any field width
   - in a field width of 5
   - in a field width of 8
   - in a field width of 3
8. When would you use **disp** instead of **fprintf**? When would you use **fprintf** instead of **disp**?
9. Write a script called *echostring* that will prompt the user for a string, and will echo print the string in quotes:

```
>> echostring
Enter your string: hi there
Your string was: 'hi there'
```

10. If the lengths of two sides of a triangle and the angle between them are known, the length of the third side can be calculated. Given the lengths of two sides (b and c) of a triangle, and the angle between them α in degrees, the third side a is calculated as follows:

$$a^2 = b^2 + c^2 - 2\,b\,c\,\cos(\alpha)$$

Write a script *thirdside* that will prompt the user and read in values for b, c, and α (in degrees), and then calculate and print the value of a, with 3 decimal places. The format of the output from the script should look exactly like this:

```
>> thirdside
Enter the first side: 2.2
Enter the second side: 4.4
Enter the angle between them: 50

The third side is 3.429
```

For more practice, write a function to calculate the third side, so the script will call this function.

11. Write a script that will prompt the user for a character, and will print it twice; once left-justified in a field width of 5, and again right-justified in a field width of 3.
12. Write a script *lumin* that will calculate and print the luminosity *L* of a star in Watts. The luminosity *L* is given by $L=4\pi d^2 b$ where *d* is the distance from the sun in meters and *b* is the brightness in Watts/meters^2. Here is an example of executing the script:

```
>> lumin
This script will calculate the luminosity of a star.
When prompted, enter the star's distance from the sun
  in meters, and its brightness in W/meters squared.

Enter the distance: 1.26e12
Enter the brightness: 2e-17
The luminosity of this star is 399007399.75 watts
```

13. A script "iotrace" has been written. Here's what the desired output looks like:

```
>> iotrace
Please enter a number: 33
Please enter a character: x
Your number is 33.00
Your char is       x!
```

Fix this script so that it works as shown previously:

```
mynum = input('Please enter a number:\n ');
mychar = input('Please enter a character: ');
fprintf('Your number is %.2f, mynum)
fprintf('Your char is %c!\n', mychar)
```

14. Write a script that assigns values for the *x* coordinate and then *y* coordinate of a point, and then plot this using a green+.
15. Plot **sin(x)** for *x* values ranging from 0 to π (in separate Figure Windows):
    - using 10 points in this range
    - using 100 points in this range
16. When would it be important to use **legend** in a plot?
17. Why do we always suppress all assignment statements in scripts?
18. Atmospheric properties such as temperature, air density, and air pressure are important in aviation. Create a file that stores temperatures in degrees Kelvin at various altitudes. The altitudes are in the first column and the temperatures in the second. For example, it may look like this:

```
1000    288
2000    281
3000    269
```

19. Generate a random integer *n*, create a vector of the integers 1 through *n* in steps of 2, square them, and plot the squares.

20. Create a 3×6 matrix of random integers, each in the range of 50–100. Write this to a file called *randfile.dat*. Then, create a new matrix of random integers, but this time make it a 2×6 matrix of random integers, each in the range of 50–100. Append this matrix to the original file. Then, read the file (which will be to a variable called *randfile*) just to make sure that it worked!
21. A particular part is being turned on a lathe. The diameter of the part is supposed to be 20,000 mm. The diameter is measured every 10 min and the results are stored in a file called *partdiam.dat*. Create a data file to simulate this. The file will store the time in minutes and the diameter at each time. Plot the data.
22. Create a file called "testtan.dat" comprised of two lines with three real numbers on each line (some negative, some positive, in the—1 to 3 range). The file can be created from the Editor, or saved from a matrix. Then, **load** the file into a matrix and calculate the tangent of every element in the resulting matrix.
23. Write a function *calcrectarea* that will calculate and return the area of a rectangle. Pass the length and width to the function as input arguments.

Renewable energy sources such as biomass are gaining increasing attention. Biomass energy units include megawatt hours (MWh) and gigajoules (GJ). One MWh is equivalent to 3.6 GJ. For example, one cubic meter of wood chips produces 1 MWh.

24. Write a function *mwh_to_gj* that will convert MWh to GJ.
25. List some differences between a script and a function.
26. In quantum mechanics, the angular wavelength for a wavelength λ is defined as $\lambda/2\pi$. Write a function named *makeitangular* that will receive the wavelength as an input argument, and will return the angular wavelength.
27. Write a *fives* function that will receive two arguments for the number of rows and columns, and will return a matrix with that size of all fives.
28. Write a function *isdivby4* that will receive an integer input argument, and will return **logical** 1 for **true** if the input argument is divisible by 4, or **logical false** if it is not.
29. Write a function *isint* that will receive a number input argument *innum*, and will return 1 for **true** if this number is an integer, or 0 for **false** if not. Use the fact-that *innum* should be equal to **int32(innum)** if it is an integer. Unfortunately, due to round-off errors, it should be noted that it is possible to get **logical** 1 for **true** if the input argument is close to an integer. Therefore the output may not be what you might expect, as shown here.

```
>> isint(4)
ans =
     1
>> isint(4.9999)
ans =
     0
>> isint(4.9999999999999999999999999999)
ans =
     1
```

30. A Pythagorean triple is a set of positive integers (a,b,c) such that $a^2+b^2=c^2$. Write a function *ispythag* that will receive three positive integers (a, b, c in that order) and will return **logical** 1 for **true** if they form a Pythagorean triple, or 0 for **false** if not.
31. A function can return a vector as a result. Write a function *vecout* that will receive one integer argument and will return a vector that increments from the value of the input argument to its value plus 5, using the colon operator. For example,

```
>> vecout(4)
ans =
     4   5   6   7   8   9
```

32. Write a function that is called *pickone*, which will receive one input argument *x*, which is a vector, and will return one random element from the vector. For example,

```
>> pickone(4:7)
ans =
     5
>> disp(pickone(-2:0))
-1
>> help pickone
pickone(x) returns a random element from vector x
```

33. The conversion depends on the temperature and other factors, but an approximation is that 1 in. of rain is equivalent to 6.5 in. of snow. Write a script that prompts the user for the number of inches of rain, calls a function to return the equivalent amount of snow, and prints this result. Write the function, as well!
34. In thermodynamics, the Carnot efficiency is the maximum possible efficiency of a heat engine operating between two reservoirs at different temperatures. The Carnot efficiency is given as

$$\eta = 1 - \frac{T_C}{T_H}$$

where T_C and T_H are the absolute temperatures at the cold and hot reservoirs, respectively. Write a script *carnot* that will prompt the user for the two reservoir temperatures in Kelvin, call a function to calculate the Carnot efficiency, and then print the corresponding Carnot efficiency to 3 decimal places. Also write the function.
35. Many mathematical models in engineering use the exponential function. The general form of the exponential decay function is:

$$y(t) = Ae^{-\tau t}$$

where A is the initial value at $t=0$, and τ is the time constant for the function. Write a script to study the effect of the time constant. To simplify the equation, set A equal to 1. Prompt the user for two different values for the time constant, and for beginning and ending values for the range of a t vector. Then, calculate two different y vectors using the above equation and the two time constants, and graph both exponential functions on the same graph within the range the user specified. Use a function to calculate y. Make one plot red. Be sure to label the graph and both axes. What happens to the decay rate as the time constant gets larger?

CHAPTER 4

Selection Statements

KEY TERMS

selection statements
branching statements
condition
action
temporary variable
error-checking
throwing an error
nesting statements
cascading if-else
"is" functions

CONTENTS

In the scripts and functions we've seen thus far, every statement was executed in sequence. This is not always desirable, and in this chapter we'll see how to make choices as to whether statements are executed or not, and how to choose between or among statements. The statements that accomplish this are called ***selection*** or ***branching*** statements.

The MATLAB® software has two basic statements that allow us to make choices: the **if** statement and the **switch** statement. The **if** statement has optional **else** and **elseif** clauses for branching. The **if** statement uses expressions that are logically **true** or **false**. These expressions use relational and logical operators. MATLAB also has "is" functions that test whether an attribute is **true** or not; these can be used with the selection statements.

4.1 THE IF STATEMENT

The **if** statement chooses whether another statement, or group of statements, is executed or not. The general form of the **if** statement is:

```
if condition
    action
end
```

A ***condition*** is a relational expression that is conceptually, or logically, **true** or **false**. The ***action*** is a statement, or a group of statements, that will be executed if the condition is **true**. When the **if** statement is executed, first the condition is

MATLAB®. http://dx.doi.org/10.1016/B978-0-12-804525-1.00004-0

evaluated. If the value of the condition is **true**, the action will be executed; if not, the action will not be executed. The action can be any number of statements until the reserved word **end**; the action is naturally bracketed by the reserved words **if** and **end**. (Note that this is different from the **end** that is used as an index into a vector or matrix.) The action is usually indented to make it easier to see.

For example, the following **if** statement checks to see whether the value of a variable is negative. If it is, the value is changed to a zero; otherwise, nothing is changed.

```
if num < 0
      num = 0
end
```

If statements can be entered in the Command Window, although they generally make more sense in scripts or functions. In the Command Window, the **if** line would be entered, followed by the Enter key, the action, the Enter key, and finally **end** and Enter. The results will follow immediately. For example, the preceding **if** statement is shown twice here.

```
>> num = -4;
>> if num < 0
      num = 0
   end
num =
      0

>> num = 5;
>> if num < 0
      num = 0
  end
>>
```

Note that the output from the assignment is not suppressed, so the result of the action will be shown if the action is executed. The first time, the value of the variable is negative so the action is executed and the variable is modified, but, in the second case, the variable is positive so the action is skipped.

This may be used, for example, to make sure that the square root function is not used on a negative number. The following script prompts the user for a number and prints the square root. If the user enters a negative number the **if** statement changes it to zero before taking the square root.

sqrtifexamp.m

```
% Prompt the user for a number and print its sqrt

num = input('Please enter a number: ');

% If the user entered a negative number, change it
if num < 0
    num = 0;
end
fprintf('The sqrt of %.1f is %.1f\n',num,sqrt(num))
```

Here are two examples of running this script:

```
>> sqrtifexamp
Please enter a number: -4.2
The sqrt of 0.0 is 0.0

>> sqrtifexamp
Please enter a number: 1.44
The sqrt of 1.4 is 1.2
```

Note that in the script the output from the assignment statement is suppressed. In this case, the action of the **if** statement was a single assignment statement. The action can be any number of valid statements. For example, we may wish to print a note to the user to say that the number entered was being changed. Also, instead of changing it to zero we will use the absolute value of the negative number entered by the user.

sqrtifexampii.m

```
% Prompt the user for a number and print its sqrt

num = input('Please enter a number: ');

% If the user entered a negative number, tell
% the user and change it
if num < 0
    disp('OK, we''ll use the absolute value')
    num = abs(num);
end
fprintf('The sqrt of %.1f is %.1f\n',num,sqrt(num))
```

```
>> sqrtifexampii
Please enter a number: -25
OK, we'll use the absolute value
The sqrt of 25.0 is 5.0
```

Note that as seen in this example, two single quotes in the **disp** statement are used to print one single quote.

PRACTICE 4.1

Write an **if** statement that would print "Hey, you get overtime!" if the value of a variable *hours* is greater than 40. Test the **if** statement for values of *hours* less than, equal to, and greater than 40. Will it be easier to do this in the Command Window or in a script?

QUICK QUESTION!

Assume that we want to create a vector of increasing integer values from *mymin* to *mymax*. We will write a function *createvec* that receives two input arguments, *mymin* and *mymax*, and returns a vector with values from *mymin* to *mymax* in steps of one. First, we would make sure that the value of *mymin* is less than the value of *mymax*. If not, we would need to exchange their values before creating the vector. How would we accomplish this?

Answer: To exchange values, a third variable, a temporary variable, is required. For example, let's say that we have two variables, *a* and *b*, storing the values:

```
a = 3;
b = 5;
```

To exchange values, we could *not* just assign the value of *b* to *a*, as follows:

```
a = b;
```

If that were done, then the value of *a* (the 3), is lost! Instead, we need to assign the value of *a* first to a ***temporary variable*** so that the value is not lost. The algorithm would be:

- assign the value of *a* to *temp*
- assign the value of *b* to *a*
- assign the value of *temp* to *b*

```
>> temp = a;
>> a = b
a =
     5
>> b = temp
b =
     3
```

Now, for the function. An **if** statement is used to determine whether or not the exchange is necessary.

createvec.m

```
function outvec = createvec(mymin, mymax)
% createvec creates a vector that iterates from a
%  specified minimum to a maximum
% Format of call: createvec(minimum, maximum)
% Returns a vector

%If the "minimum" isn't smaller than the "maximum",
% exchange the values using a temporary variable
if mymin > mymax
    temp = mymin;
    mymin = mymax;
    mymax = temp;
end

% Use the colon operator to create the vector
outvec = mymin:mymax;
end
```

Examples of calling the function are:

```
>> createvec(4,6)
ans =
     4    5    6

>> createvec(7,3)
ans =
     3    4    5    6    7
```

4.1.1 Representing Logical True and False

It has been stated that conceptually true expressions have the **logical** value of 1, and expressions that are conceptually false have the **logical** value of 0. Representing the concepts of **logical true** and **false** in MATLAB is slightly different: the concept of false is represented by the value of 0, but the concept of true can be represented by *any nonzero value* (not just 1). This can lead to some strange **logical** expressions. For example:

```
>> all(1:3)
ans =
     1
```

Also, consider the following **if** statement:

```
>> if 5
       disp('Yes, this is true!')
   end
Yes, this is true!
```

As 5 is a nonzero value, the condition is **true**. Therefore, when this **logical** expression is evaluated, it will be **true**, so the **disp** function will be executed and "Yes, this is true" is displayed. Of course, this is a pretty bizarre **if** statement, one that hopefully would never be encountered!

However, a simple mistake in an expression can lead to a similar result. For example, let's say that the user is prompted for a choice of 'Y' or 'N' for a yes/no question.

```
letter = input('Choice (Y/N): ','s');
```

In a script we might want to execute a particular action if the user responded with 'Y.' Most scripts would allow the user to enter either lowercase or uppercase; for example, either 'y' or 'Y' to indicate "yes." The proper expression that would return **true** if the value of *letter* was 'y' or 'Y' would be

```
letter == 'y' || letter == 'Y'
```

However, if by mistake this was written as:

```
letter == 'y' || 'Y'          %Note: incorrect!!
```

this expression would ALWAYS be **true**, regardless of the value of the variable *letter*. This is because `'Y'` is a nonzero value, so it is a **true** expression. The first part of the expression may be **false**, but as the second expression is **true** the entire expression would be **true**, regardless of the value of the variable *letter*.

4.2 THE IF-ELSE STATEMENT

The **if** statement chooses whether or not an action is executed. Choosing between two actions, or choosing from among several actions, is accomplished using **if-else**, nested **if-else**, and **switch** statements.

The **if-else** statement is used to choose between two statements, or sets of statements. The general form is:

```
if condition
    action1
else
    action2
end
```

First, the condition is evaluated. If it is **true**, then the set of statements designated as "action1" is executed, and that is the end of the **if-else** statement. If, instead, the condition is **false**, the second set of statements designated as "action2" is executed, and that is the end of the **if-else** statement. The first set of statements ("action1") is called the action of the **if** clause; it is what will be executed if the expression is **true**. The second set of statements ("action2") is called the action of the **else** clause; it is what will be executed if the expression is **false**. One of these actions, and only one, will be executed—which one depends on the value of the condition.

For example, to determine and print whether or not a random number in the range from 0 to 1 is less than 0.5, an **if-else** statement could be used:

```
if rand < 0.5
   disp('It was less than .5!')
else
   disp('It was not less than .5!')
end
```

PRACTICE 4.2

Write a script *printsindegorrad* that will:

- prompt the user for an angle
- prompt the user for (r)adians or (d)egrees, with radians as the default
- if the user enters 'd,' the **sind** function will be used to get the sine of the angle in degrees; otherwise, the **sin** function will be used. Which sine function to use will be based solely on whether the user entered a 'd' or not ('d' means degrees, so **sind** is used; otherwise, for any other character the default of radians is assumed so **sin** is used)
- print the result.

Here are examples of running the script:

```
>> printsindegorrad
Enter the angle: 45
(r)adians (the default) or (d)egrees: d
The sin is 0.71

>> printsindegorrad
Enter the angle: pi
(r)adians (the default) or (d)egrees: r
The sin is 0.00
```

One application of an **if-else** statement is to check for errors in the inputs to a script (this is called ***error-checking***). For example, an earlier script prompted the user for a radius, and then used that to calculate the area of a circle. However, it did not check to make sure that the radius was valid (e.g., a positive number). Here is a modified script that checks the radius:

checkradius.m

```
% This script calculates the area of a circle
% It error-checks the user's radius
radius = input('Please enter the radius: ');
if radius <= 0
    fprintf('Sorry; %.2f is not a valid radius\n',radius)
else
    area = calcarea(radius);
    fprintf('For a circle with a radius of %.2f,',radius)
    fprintf(' the area is %.2f\n',area)
end
```

Examples of running this script when the user enters invalid and then valid radii are shown as follows:

```
>> checkradius
Please enter the radius: -4
Sorry; -4.00 is not a valid radius

>> checkradius
Please enter the radius: 5.5
For a circle with a radius of 5.50, the area is 95.03
```

The **if-else** statement in this example chooses between two actions: printing an error message, or using the radius to calculate the area and then printing out the result. Note that the action of the **if** clause is a single statement, whereas the action of the **else** clause is a group of three statements.

MATLAB also has an **error** function that can be used to display an error message; the terminology is that this is ***throwing an error***. In the previous script, the if clause could be modified to use the **error** function rather than **fprintf**; the result will be displayed in red as with the error messages generated by MATLAB.

```
>> if radius <= 0
    error('Sorry; %.2f is not a valid radius\n', radius)
end

Sorry; -4.00 is not a valid radius
```

4.3 NESTED IF-ELSE STATEMENTS

The **if-else** statement is used to choose between two actions. To choose from more than two actions the **if-else** statements can be ***nested***, meaning one statement inside of another. For example, consider implementing the following continuous mathematical function $y=f(x)$:

```
y = 1    if  x < -1
y = x^2  if  -1 ≤ x ≤ 2
y = 4    if  x > 2
```

The value of *y* is based on the value of *x*, which could be in one of three possible ranges. Choosing which range could be accomplished with three separate **if** statements, is as follows:

```
if x < -1
   y = 1;
end
if x >= -1 && x <=2
   y = x^2;
end
if x > 2
   y = 4;
end
```

Note that the && in the expression of the second if statement is necessary. Writing the expression as $-1<=x<=2$ would be incorrect; recall from Chapter 1 that that expression would always be **true**, regardless of the value of the variable *x*.

As the three possibilities are mutually exclusive, the value of *y* can be determined by using three separate **if** statements. However, this is not a very efficient code: all three **logical** expressions must be evaluated, regardless of the range in which *x* falls. For example, if *x* is less than -1, the first expression is **true** and 1 would be assigned to *y*. However, the two expressions in the next two **if** statements are still evaluated. Instead of writing it this way, the statements can be nested so that the entire **if-else** statement ends when an expression is found to be **true**:

```
if x < -1
   y = 1;
else
   % If we are here, x must be >= -1
   % Use an if-else statement to choose
   %  between the two remaining ranges
   if  x <= 2
       y = x^2;
   else
       % No need to check
       % If we are here, x must be > 2
       y = 4;
   end
end
```

By using a nested **if-else** to choose from among the three possibilities, not all conditions must be tested as they were in the previous example. In this case, if x is less than -1, the statement to assign 1 to y is executed, and the **if-else** statement is completed so no other conditions are tested. If, however, x is not less than -1, then the **else** clause is executed. If the **else** clause is executed, then we already know that x is greater than or equal to -1 so that part does not need to be tested.

Instead, there are only two remaining possibilities: either x is less than or equal to 2, or it is greater than 2. An **if-else** statement is used to choose between those two possibilities. So, the action of the **else** clause was another **if-else** statement. Although it is long, all of the above code is one **if-else** statement, a nested **if-else** statement. The actions are indented to show the structure of the statement. Nesting **if-else** statements in this way can be used to choose from among 3, 4, 5, 6, ... the possibilities are practically endless!

This is actually an example of a particular kind of nested **if-else** called a *cascading* **if-else** statement. This is a type of nested **if-else** statement in which the conditions and actions cascade in a stair-like pattern.

Not all nested **if-else** statements are cascading. For example, consider the following (which assumes that a variable x has been initialized):

```
if x >= 0
   if x < 4
       disp('a')
   else
       disp('b')
   end
else
   disp('c')
end
```

4.3.1 The elseif Clause

THE PROGRAMMING CONCEPT

In some programming languages, choosing from multiple options means using nested **if-else** statements. However, MATLAB has another method of accomplishing this using the **elseif** clause.

THE EFFICIENT METHOD

To choose from among more than two actions, the **elseif** clause is used. For example, if there are *n* choices (where *n*>3 in this example), the following general form would be used:

Continued

THE EFFICIENT METHOD—CONT'D

```
if condition1
    action1
elseif condition2
    action2
elseif condition3
      action3
% etc: there can be many of these
else
    actionn     % the nth action
end
```

The actions of the **if**, **elseif**, and **else** clauses are naturally bracketed by the reserved words **if**, **elseif**, **else**, and **end**.

For example, the previous example could be written using the **elseif** clause, rather than nesting **if-else** statements:

```
if x < -1
   y = 1;
elseif x <= 2
   y = x^2;
else
   y = 4;
end
```

Note that in this example we only need one **end**. So, there are three ways of accomplishing the original task: using three separate **if** statements, using nested **if-else** statements, and using an **if** statement with **elseif** clauses, which is the simplest.

This could be implemented in a function that receives a value of *x* and returns the corresponding value of *y*:

calcy.m

```
function y = calcy(x)
% calcy calculates y as a function of x
% Format of call: calcy(x)
% y = 1          if    x < -1
% y = x^2        if    -1 <= x <= 2
% y = 4          if    x > 2

if x < -1
     y = 1;
elseif x <= 2
     y = x^2;
else
     y = 4;
end
end
```

```
>> x = 1.1;
>> y = calcy(x)
y =
  1.2100
```

QUICK QUESTION!

How could you write a function to determine whether an input argument is a scalar, a vector, or a matrix?

Answer: To do this, the **size** function can be used to find the dimensions of the input argument. If both the number of rows and columns is equal to 1, then the input argument is scalar. If, however, only one dimension is 1, the input argument is a vector (either a row or column vector). If neither dimension is 1, the input argument is a matrix. These three options can be tested using a nested **if-else** statement. In this example, the word 'scalar,' 'vector,' or 'matrix' is returned from the function.

findargtype.m

```
function outtype = findargtype(inputarg)
% findargtype determines whether the input
%    argument is a scalar, vector, or matrix
% Format of call: findargtype(inputArgument)
% Returns a string

[r c] = size(inputarg);
if r ==1 && c ==1
    outtype = 'scalar';
elseif r ==1 || c ==1
     outtype = 'vector';
else
     outtype = 'matrix';
end
end
```

Note that there is no need to check for the last case: if the input argument isn't a scalar or a vector, it must be a matrix! Examples of calling this function are:

```
>> findargtype(33)
ans =
scalar

>> disp(findargtype(2:5))
vector

>> findargtype(zeros(2,3))
ans =
matrix
```

PRACTICE 4.3

Modify the function *findargtype* to return either 'scalar,' 'row vector,' 'column vector,' or 'matrix,' depending on the input argument.

PRACTICE 4.4

Modify the original function *findargtype* to use three separate **if** statements instead of a nested **if-else** statement.

Another example demonstrates choosing from more than just a few options. The following function receives an integer quiz grade, which should be in the range from 0 to 10. The function then returns a corresponding letter grade, according to the following scheme: a 9 or 10 is an 'A,' an 8 is a 'B,' a 7 is a 'C,' a 6 is a 'D,' and anything below that is an 'F.' As the possibilities are mutually exclusive, we could implement the grading scheme using separate **if** statements. However, it is more efficient to have one **if-else** statement with multiple **elseif** clauses. Also, the function returns the letter 'X' if the quiz grade is not valid. The function assumes that the input is an integer.

letgrade.m

```
function grade = letgrade(quiz)
% letgrade returns the letter grade corresponding
%   to the integer quiz grade argument
% Format of call: letgrade(integerQuiz)
% Returns a character

% First, error-check
if quiz < 0 || quiz > 10
    grade = 'X';

% If here, it is valid so figure out the
%  corresponding letter grade
elseif quiz == 9 || quiz == 10
     grade = 'A';
elseif quiz == 8
     grade = 'B';
elseif quiz == 7
     grade = 'C';
elseif quiz == 6
     grade = 'D';
else
     grade = 'F';
end
end
```

Three examples of calling this function are:

```
>> quiz =8;
>> lettergrade = letgrade(quiz)
lettergrade =
B

>> quiz =4;
>> letgrade(quiz)
ans =
F

>> lg = letgrade(22)
lg =
X
```

In the part of this **if** statement that chooses the appropriate letter grade to return, all of the **logical** expressions are testing the value of the variable *quiz* to see if it is equal to several possible values, in sequence (first 9 or 10, then 8, then 7, etc.) This part can be replaced by a **switch** statement.

4.4 THE SWITCH STATEMENT

A **switch** statement can often be used in place of a nested **if-else** or an **if** statement with many **elseif** clauses. **Switch** statements are used when an expression is tested to see whether it is *equal to* one of several possible values.

The general form of the **switch** statement is:

```
switch switch_expression
  case caseexp1
    action1
  case caseexp2
    action2
  case caseexp3
    action3
  % etc: there can be many of these
  otherwise
    actionn
end
```

The **switch** statement starts with the reserved word **switch**, and ends with the reserved word **end**. The *switch_expression* is compared, in sequence, to the **case** expressions (*caseexp1*, *caseexp2*, etc.). If the value of the *switch_expression* matches *caseexp1*, for example, then *action1* is executed and the **switch** statement ends. If the value matches *caseexp3*, then *action3* is executed, and in general if the value matches *caseexpi* where *i* can be any integer from 1 to *n*, then *actioni* is executed. If the value of the *switch_expression* does not match any of the **case** expressions, the action after the word **otherwise** is executed (the *n*th action, *actionn*) if there is an **otherwise** (if not, no action is executed). It is not necessary to have an **otherwise** clause, although it is frequently useful. The *switch_expression* must be either a scalar or a string.

For the previous example, the **switch** statement can be used as follows:

switchletgrade.m

```
function grade = switchletgrade(quiz)
% switchletgrade returns the letter grade corresponding
%   to the integer quiz grade argument using switch
% Format of call: switchletgrade(integerQuiz)
% Returns a character

% First, error-check
if quiz < 0 || quiz > 10
    grade = 'X';
else
    % If here, it is valid so figure out the
    %  corresponding letter grade using a switch
    switch quiz
        case 10
            grade = 'A';
        case 9
            grade = 'A';
        case 8
            grade = 'B';
        case 7
            grade = 'C';
        case 6
            grade = 'D';
        otherwise
            grade = 'F';
    end
end
end
```

Here are two examples of calling this function:

Note
that it is assumed that the user will enter an integer value. If the user does not, either an error message will be printed or an incorrect result will be returned. Methods for remedying this will be discussed in Chapter 5.

```
>> quiz = 22;
>> lg = switchletgrade(quiz)
lg =
X

>> switchletgrade(9)
ans =
A
```

As the same action of printing 'A' is desired for more than one grade, these can be combined as follows:

```
switch quiz
    case {10,9}
        grade = 'A';
    case 8
        grade = 'B';
      % etc.
```

The curly braces around the **case** expressions 10 and 9 are necessary.

In this example, we error-checked first using an **if-else** statement. Then, if the grade was in the valid range, a **switch** statement was used to find the corresponding letter grade.

Sometimes the **otherwise** clause is used for the error message rather than first using an **if-else** statement. For example, if the user is supposed to enter only a 1, 3, or 5, the script might be organized as follows:

switcherror.m

```
% Example of otherwise for error message

choice = input('Enter a 1, 3, or 5: ');

switch choice
    case 1
        disp('It''s a one!!')
    case 3
        disp('It''s a three!!')
    case 5
        disp('It''s a five!!')
    otherwise
        disp('Follow directions next time!!')
end
```

In this example, actions are taken if the user correctly enters one of the valid options. If the user does not, the **otherwise** clause handles printing an error message. Note the use of two single quotes within the string to print one quote.

```
>> switcherror
Enter a 1, 3, or 5: 4
Follow directions next time!!
```

Note that the order of the case expressions does not matter, except that this is the order in which they will be evaluated.

MATLAB has a built-in function called **menu** that will display a Figure Window with pushbuttons for the options. A script that uses this **menu** function would then use either an **if-else** statement or a **switch** statement to take an appropriate action based on the button pushed. As of Version R2015b, however, the **menu** function is no longer recommended. Alternates will be found in Chapter 13 when Graphical User Interfaces are covered; in that chapter we will see how to create our own groups of pushbuttons, radio buttons, and other graphical objects.

4.5 THE "IS" FUNCTIONS IN MATLAB

There are a lot of functions that are built into MATLAB that test whether or not something is **true**; these functions have names that begin with the word "is." For example, we have already seen the use of the **isequal** function to compare arrays for equality. As another example, the function called **isletter** returns **logical** 1 if the character argument is a letter of the alphabet, or 0 if it is not:

```
>> isletter('h')
ans =
     1
>> isletter('4')
ans =
     0
```

The **isletter** function will return **logical true** or **false** so it can be used in a condition in an if statement. For example, here is the code that would prompt the user for a character, and then print whether or not it is a letter:

```
mychar = input('Please enter a char: ','s');
if isletter(mychar)
   disp('Is a letter')
else
   disp('Not a letter')
end
```

When used in an if statement, it is not necessary to test the value to see whether the result from **isletter** is equal to 1 or 0; this is redundant. In other words, in the condition of the if statement,

```
isletter(mychar)
```

and

```
isletter(mychar) == 1
```

would produce the same results.

QUICK QUESTION!

How can we write our own function *myisletter* to accomplish the same result as **isletter**?

Answer: The function would compare the character's position within the character encoding.

myisletter.m

```
function outlog = myisletter(inchar)
% myisletter returns true if the input argument
% is a letter of the alphabet or false if not
% Format of call: myisletter(input Character)
% Returns logical 1 or 0

outlog = inchar >= 'a' && inchar <= 'z' ...
         || inchar >= 'A' && inchar <= 'Z';
end
```

Note that it is necessary to check for both lowercase and uppercase letters.

The function **isempty** returns **logical true** if a variable is empty, **logical false** if it has a value, or an error message if the variable does not exist. Therefore, it can be used to determine whether a variable has a value yet or not. For example,

```
>> clear
>> isempty(evec)
Undefined function or variable 'evec'.

>> evec = [];
>> isempty(evec)
ans =
     1

>> evec = [evec 5];
>> isempty(evec)
ans =
     0
```

The **isempty** function will also determine whether or not a string variable is empty. This can be used to determine whether the user entered a string in an **input** function. In the following example, when prompted, the user simply hit the Return key.

```
>> istr = input('Please enter a string: ','s');
Please enter a string:
>> isempty(istr)
ans =
     1
```

PRACTICE 4.5

Prompt the user for a string, and then print either the string that the user entered or an error message if the user did not enter anything.

The **isa** function can be used to determine whether the first argument is a particular type.

```
>> num = 11;
>> isa(num, 'int16')
ans =
     0
>> isa(num,'double')
ans =
     1
```

The function **iskeyword** will determine whether or not a string is the name of a keyword in MATLAB, and therefore something that cannot be used as an identifier name. By itself (with no arguments), it will return the list of all keywords. Note that the names of functions like "sin" are not keywords, so their values can be overwritten if used as an identifier name.

```
>> iskeyword('sin')
ans =
     0
>> iskeyword('switch')
ans =
     1

>> iskeyword
ans =
    'break'
    'case'
    'catch'

     % etc.
```

There are many other "is" functions; the complete list can be found in the Help browser.

■ Explore Other Interesting Features

There are many other "is" functions. As more concepts are covered in the book, more and more of these functions will be introduced. Others that you may want to explore now include **isvarname**, and functions that will tell you whether an argument is a particular type or not (**ischar**, **isfloat**, **isinteger**, **islogical**, **isnumeric**, **isstr**, and **isreal**).

There are "is" functions to determine the type of an array: **isvector**, **isrow**, and **iscolumn**.

The **try/catch** functions are a particular type of **if-else** used to find and avoid potential errors. They may be a bit complicated to understand at this point, but keep them in mind for the future! ■

SUMMARY

COMMON PITFALLS

- Using = instead of `==` for equality in conditions
- Putting a space in the keyword **elseif**
- Not using quotes when comparing a string variable to a string, such as

  ```
  letter == y
  ```

  instead of

  ```
  letter == 'y'
  ```

- Not spelling out an entire **logical** expression. An example is typing

  ```
  radius || height <= 0
  ```

  instead of

  ```
  radius <= 0 || height <= 0
  ```

  or typing

  ```
  letter == 'y' || 'Y'
  ```

  instead of

  ```
  letter == 'y' || letter == 'Y'
  ```

  Note that these are logically incorrect, but would not result in error messages. Note also that the expression "`letter == 'y' || 'Y'`" will ALWAYS be **true**, regardless of the value of the variable *letter*, as `'Y'` is a nonzero value and therefore a **true** expression.
- Writing conditions that are more complicated than necessary, such as

  ```
  if (x <5) == 1
  ```

  instead of just

  ```
  if x < 5
  ```

  (The "==1" is redundant.)
- Using an **if** statement instead of an **if-else** statement for error-checking; for example,

  ```
  % Wrong method
  if error occurs
       print error message
  end

  continue rest of code
  ```

instead of

```
% Correct method
if error occurs
     print error message
else
     continue rest of code
end
```

In the first example, the error message would be printed but then the program would continue anyway.

PROGRAMMING STYLE GUIDELINES

- Use indentation to show the structure of a script or function. In particular, the actions in an **if** statement should be indented.
- When the **else** clause is not needed, use an **if** statement rather than an **if-else** statement. The following is an example:

```
if unit == 'i'
   len = len * 2.54;
else
   len = len; % this does nothing so skip it!
end
```

  Instead, just use:

```
if unit == 'i'
   len = len * 2.54;
end
```

- Do not put unnecessary conditions on **else** or **elseif** clauses. For example, the following prints one thing if the value of a variable *number* is equal to 5, and something else if it is not.

```
if number == 5
   disp('It is a 5')
elseif number ~= 5
   disp('It is not a 5')
end
```

  The second condition, however, is not necessary. Either the value is 5 or not, so just the **else** would handle this:

```
if number == 5
   disp('It is a 5')
else
   disp('It is not a 5')
end
```

MATLAB Reserved Words		
if	else	case
switch	elseif	otherwise

MATLAB Functions and Commands		
error	isletter	isa
menu	isempty	iskeyword

Exercises

1. Write a script that tests whether the user can follow instructions. It prompts the user to enter an 'x.' If the user enters anything other than an 'x,' it prints an error message—otherwise, the script does nothing.
2. Write a function *nexthour* that receives one integer argument, which is an hour of the day, and returns the next hour. This assumes a 12-hour clock; so, for example, the next hour after 12 would be 1. Here are two examples of calling this function.

```
>> fprintf('The next hour will be %d.\n',nexthour(3))
The next hour will be 4.
>> fprintf('The next hour will be %d.\n',nexthour(12))
The next hour will be 1.
```

3. The speed of a sound wave is affected by the temperature of the air. At 0 °C, the speed of a sound wave is 331 m/sec. The speed increases by approximately 0.6 m/sec for every degree (in Celsius) above 0; this is a reasonably accurate approximation for 0–50°C. So, our equation for the speed in terms of a temperature C is:

```
speed = 331 + 0.6 * C
```

Write a script *soundtemp* that will prompt the user for a temperature in Celsius in the range from 0 to 50 inclusive, and will calculate and print the speed of sound at that temperature if the user enters a temperature in that range, or an error message if not. Here are some examples of using the script:

```
>> soundtemp
Enter a temp in the range 0 to 50: -5.7
Error in temperature
>> soundtemp
Enter a temp in the range 0 to 50: 10
For a temperature of 10.0, the speed is 337.0
>> help soundtemp
Calculates and prints the speed of sound given a
temperature entered by the user
```

4. When would you use just an **if** statement and not an **if-else**?

5. Come up with "trigger words" in a problem statement that would tell you when it would be appropriate to use **if**, **if-else**, or **switch** statements.
6. Write a statement that will store **logical true** in a variable named "isit" if the value of a variable "*x*" is in the range from 0 to 10, or **logical false** if not. Do this with just one assignment statement, with no **if** or **if-else** statement!
7. The Pythagorean theorem states that for a right triangle, the relationship between the length of the hypotenuse *c* and the lengths of the other sides *a* and *b* is given by:

   $c^2 = a^2 + b^2$

   Write a script that will prompt the user for the lengths *a* and *c*, call a function *findb* to calculate and return the length of *b*, and print the result. Note that any values of *a* or *c* that are less than or equal to zero would not make sense, so the script should print an error message if the user enters any invalid value. Here is the function *findb*:

   findb.m

```
function b = findb(a,c)
% Calculates b from a and c
b = sqrt(c^2 - a^2);
end
```

8. The area A of a rhombus is defined as $A = \frac{d_1 d_2}{2}$, where d_1 and d_2 are the lengths of the two diagonals. Write a script *rhomb* that first prompts the user for the lengths of the two diagonals. If either is a negative number or zero, the script prints an error message. Otherwise, if they are both positive, it calls a function *rhombarea* to return the area of the rhombus, and prints the result. Write the function, also! The lengths of the diagonals, which you can assume are in inches, are passed to the *rhombarea* function.
9. A data file "parttolerance.dat" stores on one line, a part number, and the minimum and maximum values for the valid range that the part could weigh. Write a script "parttol" that will read these values from the file, prompt the user for a weight, and print whether or not that weight is within range.
   For example, IF the file stores the following:

```
>> type parttolerance.dat

123     44.205     44.287
```

   Here might be examples of executing the script:

```
>> parttol
Enter the part weight: 44.33
The part 123 is not in range
>> parttol
Enter the part weight: 44.25
The part 123 is within range
```

10. Write a script that will prompt the user for a character. It will create an *x*-vector that has 50 numbers, equally spaced between -2π and 2π, and then a *y*-vector

which is cos(x). If the user entered the character 'r,' it will plot these vectors with red *s—otherwise, for any other character it will plot the points with green+s.

11. Simplify this statement:

```
if number > 100
   number = 100;
else
   number = number;
end
```

12. Simplify this statement:

```
if val >= 10
   disp('Hello')
elseif val < 10
   disp('Hi')
end
```

13. The continuity equation in fluid dynamics for steady fluid flow through a stream tube equates the product of the density, velocity, and area at two points that have varying cross-sectional areas. For incompressible flow, the densities are constant so the equation is $A_1V_1 = A_2V_2$. If the areas and V_1 are known, V_2 can be found as $\frac{A_1}{A_2}V_1$. Therefore, whether the velocity at the second point increases or decreases depend on the areas at the two points. Write a script that will prompt the user for the two areas in square feet, and will print whether the velocity at the second point will increase, decrease, or remain the same as at the first point.

14. Write a function *eqfn* that will calculate $f(x) = x^2 + \frac{1}{x}$ for all elements of *x*. Since division by 0 is not possible, if any element in *x* is zero, the function will instead return a flag of −99. Here are examples of using this function:

```
>> vec = [5  0  11  2];
>> eqfn(vec)
ans =
   -99
>> result = eqfn(4)
result =
   16.2500
>> eqfn(2:5)
ans =
    4.5000   9.3333   16.2500   25.2000
```

15. In chemistry, the pH of an aqueous solution is a measure of its acidity. The pH scale ranges from 0 to 14, inclusive. A solution with a pH of 7 is said to be *neutral*, a solution with a pH greater than 7 is *basic*, and a solution with a pH less than 7 is *acidic*. Write a script that will prompt the user for the pH of a solution, and will print whether it is neutral, basic, or acidic. If the user enters an invalid pH, an error message will be printed.

16. Write a function *flipvec* that will receive one input argument. If the input argument is a row vector, the function will reverse the order and return a new row vector. If the input argument is a column vector, the function will reverse the order and return a new column vector. If the input argument is a matrix or a scalar, the function will return the input argument unchanged.
17. In a script, the user is supposed to enter either a 'y' or 'n' in response to a prompt. The user's input is read into a character variable called "letter." The script will print "OK, continuing" if the user enters either a 'y' or 'Y' or it will print "OK, halting" if the user enters a 'n' or 'N' or "Error" if the user enters anything else. Put this statement in the script first:

    ```
    letter = input('Enter your answer: ', 's');
    ```

    Write the script using a single nested **if-else** statement (**elseif** clause is permitted).
18. Write the script from the previous exercise using a **switch** statement instead.
19. In aerodynamics, the Mach number is a critical quantity. It is defined as the ratio of the speed of an object (e.g., an aircraft) to the speed of sound. If the Mach number is less than 1, the flow is subsonic; if the Mach number is equal to 1, the flow is transonic; and if the Mach number is greater than 1, the flow is supersonic. Write a script that will prompt the user for the speed of an aircraft and the speed of sound at the aircraft's current altitude and will print whether the condition is subsonic, transonic, or supersonic.
20. Write a script that will generate one random integer and will print whether the random integer is an even or an odd number. (Hint: an even number is divisible by 2, whereas an odd number is not; so check the remainder after dividing by 2.)

    Global temperature changes have resulted in new patterns of storms in many parts of the world. Tracking wind speeds and a variety of categories of storms is important in understanding the ramifications of these temperature variations. Programs that work with storm data will use selection statements to determine the severity of storms and also to make decisions based on the data.
21. Whether a storm is a tropical depression, tropical storm, or hurricane is determined by the average sustained wind speed. In miles per hour, a storm is a tropical depression if the winds are less than 38 mph. It is a tropical storm if the winds are between 39 and 73 mph, and it is a hurricane if the wind speeds are >=74 mph. Write a script that will prompt the user for the wind speed of the storm, and will print which type of storm it is.
22. The Beaufort Wind Scale is used to characterize the strength of winds. The scale uses integer values and goes from a force of 0, which is no wind, up to 12, which is a hurricane. The following script first generates a random force value. Then, it prints a message regarding what type of wind that force represents, using a **switch** statement. You are to rewrite this **switch** statement as one nested **if-else** statement that accomplishes exactly the same thing. You may use **else** and/or **elseif** clauses.

```
ranforce = randi([0, 12]);
switch ranforce
    case 0
        disp('There is no wind')
    case {1,2,3,4,5,6}
        disp('There is a breeze')
    case {7,8,9}
        disp('This is a gale')
    case {10,11}
        disp('It is a storm')
    case 12
        disp('Hello, Hurricane!')
end
```

23. Rewrite the following **switch** statement as one nested **if-else** statement (**elseif** clauses may be used). Assume that there is a variable *letter* and that it has been initialized.

```
switch letter
      case 'x'
            disp('Hello')
      case {'y', 'Y'}
            disp('Yes')
      case 'Q'
            disp('Quit')
      otherwise
            disp('Error')
end
```

24. Rewrite the following nested **if-else** statement as a **switch** statement that accomplishes exactly the same thing. Assume that *num* is an integer variable that has been initialized, and that there are functions *f1*, *f2*, *f3*, and *f4*. Do not use any **if** or **if-else** statements in the actions in the **switch** statement, only calls to the four functions.

```
if num < -2 || num > 4
    f1(num)
else
    if num <= 2
        if num >= 0
            f2(num)
        else
            f3(num)
        end
    else
        f4(num)
    end
end
```

25. Write a script *areaMenu* that will print a list consisting of "cylinder," "circle," and "rectangle." It prompts the user to choose one, and then prompts the user

for the appropriate quantities (e.g., the radius of the circle) and then prints its area. If the user enters an invalid choice, the script simply prints an error message. The script should use a nested **if-else** statement to accomplish this. Here are two examples of running it (units are assumed to be inches).

```
>> areaMenu
Menu
1. Cylinder
2. Circle
3. Rectangle
Please choose one: 2
Enter the radius of the circle: 4.1
The area is 52.81

>> areaMenu
Menu
1. Cylinder
2. Circle
3. Rectangle
Please choose one: 3
Enter the length: 4
Enter the width: 6
The area is 24.00
```

26. Modify *the areaMenu* script to use a **switch** statement to decide which area to calculate.
27. Write a script that will prompt the user for a string and then print whether it was empty or not.
28. Simplify this statement:

```
if iskeyword('else') == 1
   disp('Cannot use as a variable name')
end
```

29. Store a value in a variable and then use **isa** to test to see whether or not it is the type **double**.
30. Write a function called "makemat" that will receive two row vectors as input arguments, and from them create and return a matrix with two rows. You may not assume that the length of the vectors is known. Also, the vectors may be of different lengths. If that is the case, add 0's to the end of one vector first to make it as long as the other. For example, a call to the function might be:

```
>>makemat(1:4, 2:7)
ans =
    1    2    3    4    0    0
    2    3    4    5    6    7
```

CHAPTER 5

Loop Statements and Vectorizing Code

KEY TERMS

looping statements
counted loops
conditional loops
action
vectorized code
iterate
loop or iterator variable
echo printing
running sum
running product
preallocate
nested loop
outer loop
inner loop
infinite loop
factorial
sentinel
counting
error-checking

CONTENTS

Consider the problem of calculating the area of a circle with a radius of 0.3 cm. A MATLAB® program certainly is not needed to do that; you'd use your calculator instead, and punch in $\pi * 0.3^2$. However, if a table of circle areas is desired, for radii ranging from 0.1 to 100 cm in steps of 0.05 (e.g., 0.1, 0.15, 0.2, etc.), it would be very tedious to use a calculator and write it all down. One of the great uses of programming languages and software packages such as MATLAB is the ability to repeat a process such as this.

This chapter will cover statements in MATLAB that allow other statement(s) to be repeated. The statements that do this are called *looping statements* or ***loops***. There are two basic kinds of loops in programming: ***counted loops*** and ***conditional loops***. A counted loop is a loop that repeats statements a specified number of times (so, ahead of time it is known how many times the statements are to be repeated). In a counted loop, for example, you might say "repeat these statements 10 times". A conditional loop also repeats statements, but ahead of time it is not known *how many* times the statements will need to be repeated. With a conditional loop, for example, you might say "repeat these statements until this condition becomes false". The statement(s) that are repeated in any loop are called the ***action*** of the loop.

There are two different loop statements in MATLAB: the **for** statement and the **while** statement. In practice, the **for** statement is used as the counted loop, and

MATLAB®. http://dx.doi.org/10.1016/B978-0-12-804525-1.00005-2

the **while** is usually used as the conditional loop. To keep it simple, that is how they will be presented here.

In many programming languages, looping through the elements in a vector or matrix is a very fundamental concept. In MATLAB, however, as it is written to work with vectors and matrices, looping through elements is usually not necessary. Instead, "vectorized code" is used, which means replacing the loops through matrices with the use of built-in functions and operators. Both methods will be described in this chapter. The earlier sections will focus on "the programming concepts", using loops. These will be contrasted with "the efficient methods", using ***vectorized code***. Loops are still relevant and necessary in MATLAB in other contexts, just not normally when working with vectors or matrices.

5.1 THE FOR LOOP

The **for** statement, or the **for** loop, is used when it is necessary to repeat statement(s) in a script or function, and when it *is* known ahead of time how many times the statements will be repeated. The statements that are repeated are called the action of the loop. For example, it may be known that the action of the loop will be repeated five times. The terminology used is that we ***iterate*** through the action of the loop five times.

The variable that is used to iterate through values is called a ***loop variable*** or an ***iterator variable***. For example, the variable might iterate through the integers 1 through 5 (e.g., 1, 2, 3, 4, and then 5). Although, in general, variable names should be mnemonic, it is common in many languages for an iterator variable to be given the name *i* (and if more than one iterator variable is needed, *i*, *j*, *k*, *l*, etc.) This is historical, and is because of the way integer variables were named in Fortran. However, in MATLAB both **i** and **j** are built-in functions that return the value $\sqrt{-1}$, so using either as a loop variable will override that value. If that is not an issue, then it is okay to use *i* as a loop variable.

The general form of the **for** loop is:

```
for loopvar = range
    action
end
```

where *loopvar* is the loop variable, "range" is the range of values through which the loop variable is to iterate, and the action of the loop consists of all statements up to the **end**. Just like with **if** statements, the action is indented to make it easier to see. The range can be specified using any vector, but normally the easiest way to specify the range of values is to use the colon operator.

As an example, we will print a column of numbers from 1 to 5.

THE PROGRAMMING CONCEPT

The loop could be entered in the Command Window, although, like **if** and **switch** statements, loops will make more sense in scripts and functions. In the Command Window, the results would appear after the **for** loop:

```
>> for i = 1:5
       fprintf('%d\n',i)
   end
1
2
3
4
5
```

What the **for** statement accomplished was to print the value of *i* and then the newline character for every value of *i*, from 1 through 5 in steps of 1. The first thing that happens is that *i* is initialized to have the value 1. Then, the action of the loop is executed, which is the **fprintf** statement that prints the value of *i* (1), and then the newline character to move the cursor down. Then, *i* is incremented to have the value of 2. Next, the action of the loop is executed, which prints 2 and the newline. Then, *i* is incremented to 3 and that is printed; then, *i* is incremented to 4 and that is printed; and then, finally, *i* is incremented to 5 and that is printed. The final value of *i* is 5; this value can be used once the loop has finished.

THE EFFICIENT METHOD

Of course, **disp** could also be used to print a column vector, to achieve the same result:

```
>> disp([1:5]')
     1
     2
     3
     4
     5
```

QUICK QUESTION!

How could you print this column of integers (using the programming method):

```
  0
 50
100
150
200
```

Answer: In a loop, you could print these values starting at 0, incrementing by 50 and ending at 200. Each is printed using a field width of 3.

```
>> for i = 0:50:200
       fprintf('%3d\n',i)
   end
```

5.1.1 For Loops that do not Use the Iterator Variable in the Action

In the previous example, the value of the loop variable was used in the action of the **for** loop: it was printed. It is not always necessary to actually use the value of the loop variable, however. Sometimes the variable is simply used to iterate, or repeat, an action a specified number of times. For example,

```
for i = 1:3
    fprintf('I will not chew gum\n')
end
```

produces the output:

```
I will not chew gum
I will not chew gum
I will not chew gum
```

The variable *i* is necessary to repeat the action three times, even though the value of *i* is not used in the action of the loop.

QUICK QUESTION!

What would be the result of the following **for** loop?

```
for i = 4:2:8
    fprintf('I will not chew gum\n')
end
```

Answer: Exactly the same output as above! It doesn't matter that the loop variable iterates through the values 4, then 6, then 8 instead of 1, 2, 3. As the loop variable is not used in the action, this is just another way of specifying that the action should be repeated three times. Of course, using 1:3 makes more sense!

PRACTICE 5.1

Write a **for** loop that will print a column of five *'s.

5.1.2 Input in a for Loop

The following script repeats the process of prompting the user for a number and ***echo printing*** the number (which means simply printing it back out). A **for** loop specifies how many times this is to occur. This is another example in which the loop variable is not used in the action, but instead, just specifies how many times to repeat the action.

forecho.m

```
% This script loops to repeat the action of
% prompting the user for a number and echo-printing it

for iv = 1:3
    inputnum = input('Enter a number: ');
    fprintf('You entered %.1f\n',inputnum)
end
```

```
>> forecho
Enter a number: 33
You entered 33.0
Enter a number: 1.1
You entered 1.1
Enter a number: 55
You entered 55.0
```

In this example, the loop variable *iv* iterates through the values 1–3, so the action is repeated three times. The action consists of prompting the user for a number and echo printing it with one decimal place.

5.1.3 Finding Sums and Products

A very common application of a **for** loop is to calculate sums and products. For example, instead of just echo printing the numbers that the user enters, we could calculate the sum of the numbers. In order to do this, we need to add each value to a ***running sum***. A running sum keeps changing, as we keep adding to it. First, the sum has to be initialized to 0.

As an example, we will write a script *sumnnums* that will sum the *n* numbers entered by the user; *n* is a random integer that is generated. In a script to calculate the sum, we need a loop or iterator variable *i*, and also a variable to store the running sum. In this case we will use a variable *runsum* as the running sum. Every time through the loop, the next value that the user enters is added to the value of *runsum*. This script will print the end result, which is the sum of all the numbers, stored in the variable *runsum*.

sumnnums.m

```
% sumnnums calculates the sum of the n numbers
% entered by the user

n = randi([3 10]);
runsum = 0;
for i = 1:n
    inputnum = input('Enter a number: ');
    runsum = runsum + inputnum;
end
fprintf('The sum is %.2f\n', runsum)
```

Here is an example in which 3 is generated to be the value of the variable *n*; the script calculates and prints the sum of the numbers the user enters, 4 + 3.2 + 1.1, or 8.3:

```
>> sumnnums
Enter a number: 4
Enter a number: 3.2
Enter a number: 1.1
The sum is 8.30
```

Another very common application of a **for** loop is to find a ***running product***. With a product, the running product must be initialized to 1 (as opposed to a running sum, which is initialized to 0).

PRACTICE 5.2

Write a script *prodnnums* that is similar to the *sumnnums* script but will calculate and print the product of the numbers entered by the user.

5.1.4 Preallocating Vectors

When numbers are entered by the user, it is often necessary to store them in a vector. There are two basic methods that could be used to accomplish this. One method is to start with an empty vector and extend the vector by adding each number to it as the numbers are entered by the user. Extending a vector, however, is very inefficient. What happens is that every time a vector is extended, a new "chunk" of memory must be found that is large enough for the new vector, and all of the values must be copied from the original location in memory to the new one. This can take a long time to execute.

A better method is to ***preallocate*** the vector to the correct size and then change the value of each element to be the numbers that the user enters. This method involves referring to each index in the result vector, and placing each number into the next element in the result vector. This method is far superior, if it is known ahead of time how many elements the vector will have. One common method is to use the **zeros** function to preallocate the vector to the correct length.

The following is a script that accomplishes this and prints the resulting vector. The script generates a random integer *n* and repeats the process *n* times. As it is known that the resulting vector will have *n* elements, the vector can be preallocated.

forgenvec.m

```
% forgenvec creates a vector of length n
% It prompts the user and puts n numbers into a vector

n = randi([4   8]);
numvec = zeros(1,n);
for iv = 1:n
    inputnum = input('Enter a number: ');
    numvec(iv) = inputnum;
end
fprintf('The vector is: \n')
disp(numvec)
```

Next is an example of executing this script.

```
>> forgenvec
Enter a number: 44
Enter a number: 2.3
Enter a number: 11
The vector is:
   44.0000    2.3000    11.0000
```

It is very important to notice that the loop variable iv is used as the index into the vector.

QUICK QUESTION!

If you need to just print the sum or average of the numbers that the user enters, would you need to store them in a vector variable?

Answer: No. You could just add each to a running sum as you read them in a loop.

QUICK QUESTION!

What if you wanted to calculate how many of the numbers that the user entered were greater than the average?

Answer: Yes, then you would need to store them in a vector because you would have to go back through them to count how many were greater than the average (or, alternatively, you could go back and ask the user to enter them again!!).

5.1.5 For Loop Example: Subplot

A function that is very useful with all types of plots is **subplot**, which creates a matrix of plots in the current Figure Window. Three arguments are passed to it in the form **subplot(r,c,n)**, where *r* and *c* are the dimensions of the matrix and *n* is the number of the particular plot within this matrix. The plots are numbered rowwise starting in the upper left corner. In many cases, it is useful to create a **subplot** in a <u>for</u> loop so the loop variable can iterate through the integers 1 through *n*.

For example, if it is desired to have three plots next to each other in one Figure Window, the function would be called as **subplot(1,3,n)**. The matrix dimensions in the Figure Window would be *1x3* in this case, and from left to right the individual plots would be numbered 1, 2, and then 3 (these would be the values of *n*). The first two arguments would always be 1 and 3, as they specify the dimensions of the matrix within the Figure Window.

When the **subplot** function is called in a loop, the first two arguments will always be the same as they give the dimensions of the matrix. The third argument will iterate through the numbers assigned to the elements of the matrix. When the **subplot** function is called, it makes the specified element the "active" plot; then, any plot function can be used, complete with formatting such as axis labeling and titles within that element. Note that the **subplot** function just specifies the dimensions of the matrix in the Figure Window, and which is the "active" element; **subplot** itself does not plot anything.

For example, the following **subplot** shows the difference, in one Figure Window, between using 20 points and 40 points to plot **sin(x)** between 0 and $2 * \pi$. The **subplot** function creates a *1 x 2* row vector of plots in the Figure Window, so that the two plots are shown side by side. The loop variable *i* iterates through the values 1 and then 2.

The first time through the loop, when *i* has the value 1, 20 * 1 or 20 points are used, and the value of the third argument to the **subplot** function is 1. The second time through the loop, 40 points are used and the third argument to **subplot** is 2. The resulting Figure Window with both plots is shown in Fig 5.1.

subplotex.m

```
% Demonstrates subplot using a for loop
for i = 1:2
    x = linspace(0,2*pi,20*i);
    y = sin(x);
    subplot(1,2,i)
    plot(x,y,'ko')
    xlabel('x')
    ylabel('sin(x)')
    title('sin plot')
end
```

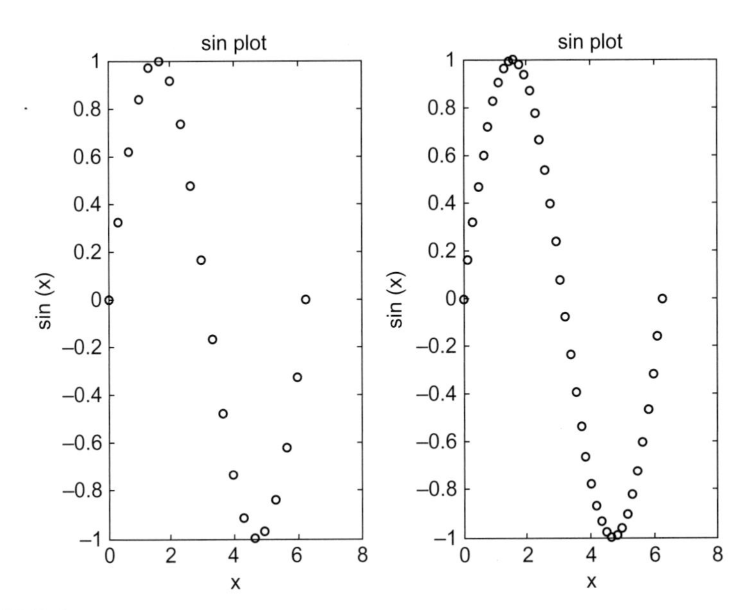

FIGURE 5.1
Subplot to demonstrate a plot using 20 points and 40 points.

Note that once string manipulating functions have been covered in Chapter 7, it will be possible to have customized titles (e.g., showing the number of points).

5.2 NESTED FOR LOOPS

The action of a loop can be any valid statement(s). When the action of a loop is another loop, this is called a *nested loop*.

The general form of a nested **for** loop is as follows:

```
for loopvarone = rangeone              ← outer loop

    % actionone includes the inner loop

    for loopvartwo = rangetwo          ← inner loop
        actiontwo
    end
end
```

The first **for** loop is called the ***outer loop***; the second **for** loop is called the ***inner loop***. The action of the outer loop consists (in part; there could be other statements) of the entire inner loop.

As an example, a nested **for** loop will be demonstrated in a script that will print a box of stars (*). Variables in the script will specify how many rows and columns to print. For example, if *rows* has the value 3 and *columns* has the value 5, a *3 x 5* box would be printed. As lines of output are controlled by printing the newline character, the basic algorithm is as follows.

- For every row of output:
  - Print the required number of stars
  - Move the cursor down to the next line (print '\n')

printstars.m

```
% Prints a box of stars
% How many will be specified by two variables
%  for the number of rows and columns

rows = 3;
columns = 5;
% loop over the rows
for i=1:rows
    %  for every row loop to print *'s and then one \n
    for j=1:columns
        fprintf('*')
    end
    fprintf('\n')
end
```

Executing the script displays the output:

```
>> printstars
*****
*****
*****
```

The variable *rows* specifies the number of rows to print, and the variable *columns* specifies how many stars to print in each row. There are two loop variables: *i* is the loop variable over the rows and *j* is the loop variable over the columns. As the number of rows is known and the number of columns is known (given by the variables *rows* and *columns*), **for** loops are used. There is one **for** loop to loop over the rows and another to print the required number of stars for every row.

The values of the loop variables are not used within the loops, but are used simply to iterate the correct number of times. The first **for** loop specifies that the action will be repeated "rows" times. The action of this loop is to print stars and then the newline character. Specifically, the action is to loop to print *columns* stars (e.g., five stars) across on one line. Then, the newline character is printed after all five stars to move the cursor down to the next line.

In this case, the outer loop is over the rows and the inner loop is over the columns. The outer loop must be over the rows because the script is printing a certain number of rows of output. For each row, a loop is necessary to print the required number of stars; this is the inner **for** loop.

When this script is executed, first the outer loop variable *i* is initialized to 1. Then, the action is executed. The action consists of the inner loop and then printing the newline character. So, while the outer loop variable has the value 1, the inner loop variable *j* iterates through all of its values. As the value of *columns* is 5, the inner loop will print a single star five times. Then, the newline character is printed and then the outer loop variable *i* is incremented to 2. The action of the outer loop is then executed again, meaning the inner loop will print five stars, and then the newline character will be printed. This continues, and in all, the action of the outer loop will be executed *rows* times.

Notice that the action of the outer loop consists of two statements (the **for** loop and an **fprintf** statement). The action of the inner loop, however, is only a single **fprintf** statement.

The **fprintf** statement to print the newline character must be separate from the other **fprintf** statement that prints the star character. If we simply had

```
fprintf('*\n')
```

as the action of the inner loop, this would print a long column of 15 stars, not a *3 x 5* box.

QUICK QUESTION!

How could this script be modified to print a triangle of stars instead of a box such as the following:

```
*
**
***
```

Answer: In this case, the number of stars to print in each row is the same as the row number (e.g., one star is printed in row 1, two stars in row 2, and so on). The inner **for** loop does not loop to columns, but to the value of the row loop variable (so we do not need the variable *columns*):

printtristars.m

```
% Prints a triangle of stars
% How many will be specified by a variable
%  for the number of rows
rows = 3;
for i=1:rows
    % inner loop just iterates to the value of i
    for j=1:i
        fprintf('*')
    end
    fprintf('\n')
end
```

In the previous examples, the loop variables were just used to specify the number of times the action is to be repeated. In the next example, the actual values of the loop variables will be printed.

printloopvars.m

```
% Displays the loop variables
for i = 1:3
    for j = 1:2
        fprintf('i=%d, j=%d\n',i,j)
    end
    fprintf('\n')
end
```

Executing this script would print the values of both *i* and *j* on one line every time the action of the inner loop is executed. The action of the outer loop consists of the inner loop and printing a newline character, so there is a separation between the actions of the outer loop:

```
>> printloopvars
i=1, j=1
i=1, j=2

i=2, j=1
i=2, j=2

i=3, j=1
i=3, j=2
```

Now, instead of just printing the loop variables, we can use them to produce a multiplication table, by multiplying the values of the loop variables.

The following function *multtable* calculates and returns a matrix which is a multiplication table. Two arguments are passed to the function, which are the number of rows and columns for this matrix.

multtable.m

```
function outmat = multtable(rows, columns)
% multtable returns a matrix which is a
% multiplication table
% Format: multtable(nRows, nColumns)

% Preallocate the matrix
outmat = zeros(rows,columns);
for i = 1:rows
    for j = 1:columns
        outmat(i,j) = i * j;
    end
end
end
```

In the following example of calling this function, the resulting matrix has three rows and five columns:

```
>> multtable(3,5)
ans =
     1     2     3     4     5
     2     4     6     8    10
     3     6     9    12    15
```

Note that this is a function that returns a matrix. It preallocates the matrix to zeros, and then replaces each element. As the number of rows and columns are known, **for** loops are used. The outer loop loops over the rows and the inner loop loops over the columns. The action of the nested loop calculates i * j for all values of *i* and *j*. *Just like with vectors, it is again important to notice that the loop variables are used as the indices into the matrix.*

First, when *i* has the value 1, *j* iterates through the values 1 through 5, so first we are calculating 1*1, then 1*2, then 1*3, then 1*4, and finally, 1*5. These are the values in the first row (first in element (1,1), then (1,2), then (1,3), then (1,4), and finally (1,5)). Then, when *i* has the value 2, the elements in the second row of the output matrix are calculated, as *j* again iterates through the values from 1 through 5. Finally, when *i* has the value 3, the values in the third row are calculated (3*1, 3*2, 3*3, 3*4, and 3*5).

This function could be used in a script that prompts the user for the number of rows and columns, calls this function to return a multiplication table, and writes the resulting matrix to a file:

createmulttab.m

```
% Prompt the user for rows and columns and
% create a multiplication table to store in
% a file "mymulttable.dat"

num_rows = input('Enter the number of rows: ');
num_cols = input('Enter the number of columns: ');
multmatrix = multtable(num_rows, num_cols);
save mymulttable.dat multmatrix -ascii
```

The following is an example of running this script, and then loading from the file into a matrix in order to verify that the file was created:

```
>> createmulttab
Enter the number of rows: 6
Enter the number of columns: 4

>> load mymulttable.dat
```

```
>> mymulttable
mymulttable =
        1        2        3        4
        2        4        6        8
        3        6        9       12
        4        8       12       16
        5       10       15       20
        6       12       18       24
```

PRACTICE 5.3

For each of the following (they are separate), determine what would be printed. Then, check your answers by trying them in MATLAB.

```
mat = [7   11   3;   3:5];
[r, c] = size(mat);
for i = 1:r
    fprintf('The sum is %d\n', sum(mat(i,:)))
end
-----------------------------------------
for i = 1:2
    fprintf('%d: ', i)
    for j = 1:4
        fprintf('%d ', j)
    end
    fprintf('\n')
end
```

5.2.1 Combining Nested for Loops and if Statements

The statements inside of a nested loop can be any valid statements, including any selection statement. For example, there could be an **if** or **if-else** statement as the action, or part of the action, in a loop.

As an example, assume there is a file called "datavals.dat" containing results recorded from an experiment. However, some were recorded erroneously. The numbers are all supposed to be positive. The following script reads from this file into a matrix. It prints the sum from each row of only the positive numbers. We will assume that the file contains integers, but will not assume how many lines are in the file nor how many numbers per line (although we will assume that there are the same number of integers on every line).

sumonlypos.m

```
% Sums only positive numbers from file
% Reads from the file into a matrix and then
%   calculates and prints the sum of only the
%   positive numbers from each row

load datavals.dat
[r c] = size(datavals);

for row = 1:r
    runsum = 0;
    for col = 1:c
        if datavals(row,col) >= 0
            runsum = runsum+datavals(row,col);
        end
    end
    fprintf('The sum for row %d is %d\n',row,runsum)
end
```

For example, *if* the file contains:

```
33  -11   2
 4    5   9
22    5  -7
 2   11   3
```

the output from the program would look like this:

```
>> sumonlypos
The sum for row 1 is 35
The sum for row 2 is 18
The sum for row 3 is 27
The sum for row 4 is 16
```

The file is loaded and the data are stored in a matrix variable. The script finds the dimensions of the matrix and then loops through all of the elements in the matrix by using a nested loop; the outer loop iterates through the rows and the inner loop iterates through the columns. This is important; as an action is desired for every row, the outer loop has to be over the rows. For each element an **if-else** statement determines whether the element is positive or not. It only adds the positive values to the row sum. As the sum is found for each row, the *runsum* variable is initialized to 0 for every row, meaning inside of the outer loop.

QUICK QUESTION!

Would it matter if the order of the loops was reversed in this example, so that the outer loop iterates over the columns and the inner loop over the rows?

Answer: Yes, as we want a sum for every row the outer loop must be over the rows.

QUICK QUESTION!

What would you have to change in order to calculate and print the sum of only the positive numbers from each column instead of each row?

Answer: You would reverse the two loops, and change the sentence to say "The sum of column...". That is all that would change. The elements in the matrix would still be referenced as datavals(row,col). The row index is always given first, then the column index – regardless of the order of the loops.

PRACTICE 5.4

Write a function *mymatmin* that finds the minimum value in each column of a matrix argument and returns a vector of the column minimums. An example of calling the function follows:

```
>> mat = randi(20,3,4)
mat =
    15    19    17     5
     6    14    13    13
     9     5     3    13

>> mymatmin(mat)
ans =
     6     5     3     5
```

QUICK QUESTION!

Would the function *mymatmin* in Practice 5.4 also work for a vector argument?

Answer: Yes, it should, as a vector is just a subset of a matrix. In this case, one of the loop actions would be executed only one time (for the rows if it is a row vector or for the columns if it is a column vector).

5.3 WHILE LOOPS

The **while** statement is used as the conditional loop in MATLAB; it is used to repeat an action when ahead of time it is *not known how many* times the action will be repeated. The general form of the **while** statement is:

```
while condition
    action
end
```

The action, which consists of any number of statement(s), is executed as long as the condition is true.

The way it works is that first the condition is evaluated. If it is logically **true**, the action is executed. So, to begin with, the **while** statement is just like an **if** statement. However, at that point, the condition is evaluated again. If it is still **true**, the action is executed again. Then, the condition is evaluated again. If it is still **true**, the action is executed again. Then, the condition is....eventually, this has to stop! Eventually, something in the action has to change something in the condition so it becomes **false**. The condition must eventually become **false** to avoid an ***infinite loop***. (If this happens, Ctrl-C will exit the loop.)

As an example of a conditional loop, we will write a function that will find the first ***factorial*** that is greater than the input argument *high*. For an integer n, the factorial of n, written as n!, is defined as $n! = 1 \times 2 \times 3 \times 4 \times \ldots \times n$. To calculate a factorial, a **for** loop would be used. However, in this case we do not know the value of n, so we have to keep calculating the next factorial until a level is reached, which means using a **while** loop.

The basic algorithm is to have two variables: one that iterates through the values 1,2, 3, and so on; and one that stores the factorial of the iterator at each step. We start with 1 and 1 factorial, which is 1. Then, we check the factorial. If it is not greater than *high*, the iterator variable will then increment to 2 and find its factorial (2). If this is not greater than *high*, the iterator will then increment to 3 and the function will find its factorial (6). This continues until we get to the first factorial that is greater than *high*.

So, the process of incrementing a variable and finding its factorial is repeated until we get to the first value greater than **high**. This is implemented using a **while** loop:

factgthigh.m

```
function facgt = factgthigh(high)
% factgthigh returns the first factorial > input
% Format: factgthigh(inputInteger)

i=0;
fac=1;
while fac <= high
     i=i+1;
     fac = fac * i;
end
facgt = fac;
end
```

An example of calling the function, passing 5000 for the value of the input argument *high*, follows:

```
>> factgthigh(5000)
ans =
          5040
```

The iterator variable *i* is initialized to 0, and the running product variable *fac*, which will store the factorial of each value of *i*, is initialized to 1. The first time the **while** loop is executed, the condition is **true**: 1 is less than or equal to 5000. So, the action of the loop is executed, which is to increment *i* to 1 and *fac* becomes 1 (1×1).

After the execution of the action of the loop, the condition is evaluated again. As it will still be **true**, the action is executed: *i* is incremented to 2 and *fac* will get the value 2 (1×2). The value 2 is still <= 5000, so the action will be executed again: *i* will be incremented to 3 and *fac* will get the value 6 (2×3). This continues, until the first value of *fac* is found that is greater than 5000. As soon as *fac* gets to this value, the condition will be **false** and the **while** loop will end. At that point the factorial is assigned to the output argument, which returns the value.

The reason that *i* is initialized to 0 rather than 1 is that the first time the loop action is executed, *i* becomes 1 and *fac* becomes 1, so we have 1 and 1!, which is 1.

5.3.1 Multiple Conditions in a while Loop

In the *factgthigh* function, the condition in the **while** loop consisted of one expression, which tested whether or not the variable *fac* was less than or equal to the variable *high*. In many cases, however, the condition will be more complicated than that and could use either the **or** operator || or the **and** operator &&. For example, it may be that it is desired to stay in a **while** loop as long as a variable *x* is in a particular range:

```
while x >= 0 && x <= 100
```

As another example, continuing the action of a loop may be desired as long as at least one of two variables is in a specified range:

```
while x < 50 || y < 100
```

5.3.2 Reading from a File Using a while Loop

The following example illustrates reading from a data file using a **while** loop. Data from an experiment has been recorded in a file called "experd.dat". The file has some weights followed by a −99 and then more weights, all on the same

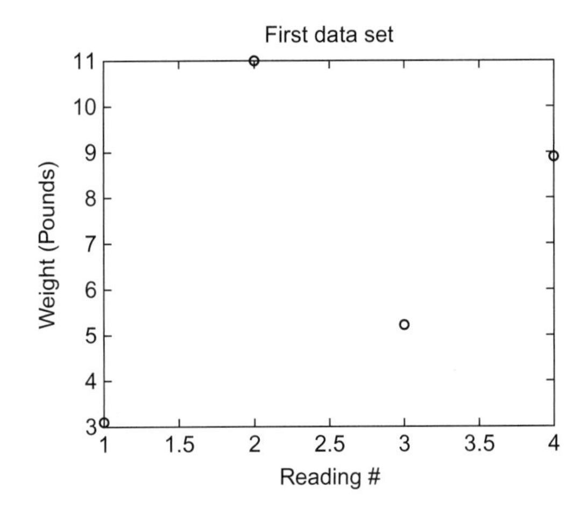

FIGURE 5.2
Plot of some (but not all) data from a file.

line. The only data values that we are interested in, however, are those before −99. The −99 is an example of a ***sentinel***, which is a marker in between data sets.

The algorithm for the script is as follows:

- Read the data from the file into a vector.
- Create a new vector variable *newvec* that only has the data values up to but not including the −99.
- Plot the new vector values, using black circles.

For example, *if* the file contains the following:

```
3.1  11  5.2  8.9  -99  4.4 62
```

The plot produced would look like Fig 5.2.

For simplicity, we will assume that the file format is as specified. Using **load** will create a vector with the name *experd*, which contains the values from the file.

THE PROGRAMMING CONCEPT

Using the programming method, we would loop through the vector until the −99 is found, creating the new vector by storing each element from *experd* in the vector *newvec*.

Continued

THE PROGRAMMING CONCEPT—CONT'D

findvalwhile.m

```
% Reads data from a file, but only plots the numbers
% up to a flag of -99. Uses a while loop.

load experd.dat

i = 1;
while experd(i) ~= -99
     newvec(i) = experd(i);
     i = i + 1;
end

plot(newvec,'ko')
xlabel('Reading #')
ylabel('Weight(pounds)')
title('First Data Set')
```

Note that this extends the vector *newvec* every time the action of the loop is executed.

THE EFFICIENT METHOD

Using the **find** function, we can locate the index of the element that stores the −99. Then, the new vector comprises all of the original vector from the first element to the index *before* the index of the element that stores the −99.

findval.m

```
% Reads data from a file, but only plots the numbers
% up to a flag of -99.Uses find and the colon operator

load experd.dat

Where = find(experd == -99);
newvec = experd(1:where-1);

plot(newvec,'ko')
xlabel('Reading #')
ylabel('Weight(pounds)')
title('First Data Set')
```

5.3.3 Input in a while Loop

Sometimes a **while** loop is used to process input from the user as long as the user is entering data in a correct format. The following script repeats the process of prompting the user, reading in a positive number, and echo printing it, as long as the user correctly enters positive numbers when prompted. As soon as the user types in a negative number, the script will print "OK" and end.

whileposnum.m

```
% Prompts the user and echo prints the numbers entered
% until the user enters a negative number

inputnum=input('Enter a positive number: ');
while inputnum >= 0
    fprintf('You entered a %d.\n\n',inputnum)
    inputnum = input('Enter a positive number: ');
end
fprintf('OK!\n')
```

When the script is executed, the input/output might look like this:

```
>> whileposnum
Enter a positive number: 6
You entered a 6.

Enter a positive number: -2
OK!
```

Note that the prompt is repeated in the script: once before the loop, and then again at the end of the action. This is done so that every time the condition is evaluated, there is a new value of *inputnum* to check. If the user enters a negative number the first time, no values would be echo printed:

Note
This example illustrates a very important feature of ***while*** *loops: it is possible that the action will not be executed at all, if the value of the condition is false the first time it is evaluated.*

```
>> whileposnum
Enter a positive number: -33
OK!
```

As we have seen previously, MATLAB will give an error message if a character is entered rather than a number.

```
>> whileposnum
Enter a positive number: a
Error using input
Undefined function or variable 'a'.

Error in whileposnum (line 4)
inputnum=input('Enter a positive number: ');

Enter a positive number: -4
OK!
```

However, if the character is actually the name of a variable, it will use the value of that variable as the input. For example:

```
>> a = 5;
>> whileposnum
Enter a positive number: a
You entered a 5.

Enter a positive number: -4
OK!
```

EXTENDING A VECTOR

If it is desired to store all of the positive numbers that the user enters, we would store them one at a time in a vector. However, as we do not know ahead of time how many elements we will need, we cannot preallocate to the correct size. The two methods of extending a vector one element at a time are shown here. We can start with an empty vector and concatenate each value to the vector, or we can increment an index.

```
numvec = [];
inputnum=input('Enter a positive number: ');
while inputnum >= 0
    numvec = [numvec inputnum];
    inputnum = input('Enter a positive number: ');
end

% OR:

i = 0;
inputnum=input('Enter a positive number: ');
while inputnum >= 0
    i = i+1;
    numvec(i) = inputnum;
    inputnum = input('Enter a positive number: ');
end
```

Keep in mind that this is inefficient and should be avoided if the array can be preallocated.

5.3.4 Counting in a While Loop

Although **while** loops are used when the number of times the action will be repeated is not known ahead of time, it is often useful to know how many times the action was, in fact, repeated. In that case, it is necessary to ***count*** the number of times the action is executed. The following variation on the previous script counts the number of positive numbers that the user successfully enters.

countposnum.m

```
% Prompts the user for positive numbers and echo prints as
% long as the user enters positive numbers

% Counts the positive numbers entered by the user
counter=0;
inputnum=input('Enter a positive number: ');
while inputnum >= 0
    fprintf('You entered a %d.\n\n',inputnum)
    counter = counter+1;
    inputnum = input('Enter a positive number: ');
end
fprintf('Thanks, you entered %d positive numbers.\n',counter)
```

The script initializes a variable *counter* to 0. Then, in the **while** loop action, every time the user successfully enters a number, the script increments the counter variable. At the end of the script it prints the number of positive numbers that were entered.

```
>> countposnum
Enter a positive number: 4
You entered a 4.

Enter a positive number: 11
You entered a 11.

Enter a positive number: -4
Thanks, you entered 2 positive numbers.
```

PRACTICE 5.5

Write a script *avenegnum* that will repeat the process of prompting the user for negative numbers, until the user enters a zero or positive number, as just shown. Instead of echo printing them, however, the script will print the average (of just the negative numbers). If no negative numbers are entered, the script will print an error message instead of the average. Use the programming method. Examples of executing this script follow:

```
>> avenegnum
Enter a negative number: 5
No negative numbers to average.

>> avenegnum
Enter a negative number: -8
Enter a negative number: -3
Enter a negative number: -4
Enter a negative number: 6
The average was -5.00
```

5.3.5 Error-Checking User Input in a while loop

In most applications, when the user is prompted to enter something, there is a valid range of values. If the user enters an incorrect value, rather than having the program carry on with an incorrect value, or just printing an error message, the program should repeat the prompt. The program should keep prompting the user, reading the value, and checking it until the user enters a value that is in the correct range. This is a very common application of a conditional loop: looping until the user correctly enters a value in a program. This is called *error-checking*.

For example, the following script prompts the user to enter a positive number, and loops to print an error message and repeat the prompt until the user finally enters a positive number.

readonenum.m

```
% Loop until the user enters a positive number

inputnum=input('Enter a positive number: ');
while inputnum < 0
    inputnum = input('Invalid! Enter a positive number: ');
end
fprintf('Thanks, you entered a %.1f \n',inputnum)
```

An example of running this script follows:

Note

MATLAB itself catches the character input and prints an error message, and repeats the prompt when the *c* was entered.

```
>> readonenum
Enter a positive number: -5
Invalid! Enter a positive number: -2.2
Invalid! Enter a positive number: c
Error using input
Undefined function or variable 'c'.
Error in readonenum (line 5)
 inputnum = input('Invalid! Enter a positive number: ');
Invalid! Enter a positive number: 44
Thanks, you entered a 44.0
```

QUICK QUESTION!

How could we vary the previous example so that the script asks the user to enter positive numbers *n* times, where *n* is an integer defined to be 3?

Answer: Every time the user enters a value, the script checks and in a **while** loop keeps telling the user that it's invalid until a valid positive number is entered. By putting the error-check in a **for** loop that repeats n times, the user is forced eventually to enter three positive numbers, as shown in the following.

readnnums.m

```
% Loop until the user enters n positive numbers
n=3;
fprintf('Please enter %d positive numbers\n\n',n)
for i=1:n
    inputnum=input('Enter a positive number: ');
    while inputnum < 0
        inputnum = input('Invalid! Enter a positive number: ');
    end
    fprintf('Thanks, you entered a %.1f \n',inputnum)
end
```

QUICK QUESTION!—CONT'D

```
>> readnnums
Please enter 3 positive numbers

Enter a positive number: 5.2
Thanks, you entered a 5.2
Enter a positive number: 6
Thanks, you entered a 6.0
Enter a positive number: -7.7
Invalid! Enter a positive number: 5
Thanks, you entered a 5.0
```

5.3.5.1 Error-Checking for Integers

As MATLAB uses the type **double** by default for all values, to check to make sure that the user has entered an integer, the program has to convert the input value to an integer type (e.g., **int32**) and then check to see whether that is equal to the original input. The following examples illustrate the concept.

If the value of the variable *num* is a real number, converting it to the type **int32** will round it, so the result is not the same as the original value.

```
>> num = 3.3;
>> inum = int32(num)
inum =
           3
>> num == inum
ans =
    0
```

If, however, the value of the variable *num* is an integer, converting it to an integer type will not change the value.

```
>> num = 4;
>> inum = int32(num)
inum =
           4
>> num == inum
ans =
     1
```

The following script uses this idea to error-check for integer data; it loops until the user correctly enters an integer.

Note
This assumes that the user enters something. Use the **isempty** function to be sure.

readoneint.m

```
% Error-check until the user enters an integer
inputnum = input('Enter an integer: ');
num2 = int32(inputnum);
while num2 ~= inputnum
    inputnum = input('Invalid! Enter an integer: ');
    num2 = int32(inputnum);
end
fprintf('Thanks, you entered a %d \n',inputnum)
```

Examples of running this script are:

```
>> readoneint
Enter an integer: 9.5
Invalid! Enter an integer: 3.6
Invalid! Enter an integer: -11
Thanks, you entered a -11

>> readoneint
Enter an integer: 5
Thanks, you entered a 5
```

Putting these ideas together, the following script loops until the user correctly enters a positive integer. There are two parts to the condition, the value must be positive and must be an integer.

readoneposint.m

```
% Error checks until the user enters a positive integer
inputnum = input('Enter a positive integer: ');
num2 = int32(inputnum);
while num2 ~= inputnum || num2 < 0
    inputnum = input('Invalid! Enter a positive integer: ');
    num2 = int32(inputnum);
end
fprintf('Thanks, you entered a %d \n',inputnum)
```

```
>> readoneposint
Enter a positive integer: 5.5
Invalid! Enter a positive integer: -4
Invalid! Enter a positive integer: 11
Thanks, you entered a 11
```

PRACTICE 5.6

Modify the script *readoneposint* to read *n* positive integers, instead of just one.

5.4 LOOPS WITH VECTORS AND MATRICES; VECTORIZING

In most programming languages when performing an operation on a vector, a **for** loop is used to loop through the entire vector, using the loop variable as the index into the vector. In general, in MATLAB, assuming there is a vector variable *vec*, the indices range from 1 to the length of the vector, and the **for** statement loops through all of the elements performing the same operation on each one:

```
for i = 1:length(vec)
    % do something with vec(i)
end
```

In fact, this is one reason to store values in a vector. Typically, values in a vector represent "the same thing", so, typically in a program the same operation would be performed on every element.

Similarly, for an operation on a matrix, a nested loop would be required, and the loop variables over the rows and columns are used as the subscripts into the matrix. In general, assuming a matrix variable *mat*, we use **size** to return separately the number of rows and columns, and use these variables in the **for** loops. If an action is desired for every row in the matrix, the nested for loop would look like this:

```
[r, c] = size(mat);
for row = 1:r
   for col = 1:c
      % do something with mat(row,col)
   end
end
```

If, instead, an action is desired for every column in the matrix, the outer loop would be over the columns. (Note, however, that the reference to a matrix element always refers to the row index first and then the column index.)

```
[r, c] = size(mat);
for col = 1:c
   for row = 1:r
      % do something with mat(row,col)
   end
end
```

Typically, this is not necessary in MATLAB! Although **for** loops are very useful for many other applications in MATLAB, they are not typically used for operations on vectors or matrices; instead, the efficient method is to use built-in functions and/or operators. This is called vectorized code. The use of loops and selection statements with vectors and matrices is a basic programming concept with many other languages, and so both "the programming concept" and "the efficient method" are highlighted in this section and, to some extent, throughout the rest of this book.

5.4.1 Vectorizing Sums and Products

For example, let's say that we want to perform a scalar multiplication, in this case multiplying every element of a vector *v* by 3, and store the result back in *v*, where *v* is initialized as follows:

```
>> v= [3  7  2  1];
```

THE PROGRAMMING CONCEPT

To accomplish this, we can loop through all of the elements in the vector and multiply each element by 3. In the following, the output is suppressed in the loop, and then the resulting vector is shown:

```
>> for i = 1:length(v)
       v(i) = v(i) *3;
   end
>> v
v =
     9    21     6     3
```

THE EFFICIENT METHOD

```
>> v = v * 3
```

How could we calculate the factorial of n, n! = 1 * 2 * 3 * 4 *... * n?

THE PROGRAMMING CONCEPT

The basic algorithm is to initialize a running product to 1 and multiply the running product by every integer from 1 to n. This is implemented in a function:

myfact.m

```
function runprod = myfact(n)
% myfact returns n!
% Format of call: myfact(n)

runprod = 1;
for i = 1:n
    runprod = runprod * i;
end
end
```

THE PROGRAMMING CONCEPT—CONT'D

Any positive integer argument could be passed to this function, and it will calculate the factorial of that number. For example, if 5 is passed, the function will calculate and return 1*2*3*4*5, or 120:

```
>> myfact(5)
ans =
   120
```

THE EFFICIENT METHOD

MATLAB has a built-in function, **factorial**, that will find the factorial of an integer n. The **prod** function could also be used to find the product of the vector 1:5.

```
>> factorial(5)
ans =
   120
>> prod(1:5)
ans =
   120
```

QUICK QUESTION!

MATLAB has a **cumsum** function that will return a vector of all of the running sums of an input vector. However, many other languages do not, so how could we write our own?

Answer: Essentially, there are two programming methods that could be used to simulate the **cumsum** function. One method is to start with an empty vector and extend the vector by adding each running sum to it as the running sums are calculated. A better method is to preallocate the vector to the correct size and then change the value of each element to be successive running sums.

myveccumsum.m

```
function outvec = myveccumsum(vec)
% myveccumsum imitates cumsum for a vector
% It preallocates the output vector
% Format: myveccumsum(vector)

outvec = zeros(size(vec));
runsum = 0;
for i = 1:length(vec)
    runsum = runsum + vec(i);
    outvec(i) = runsum;
end
end
```

An example of calling the function follows:

```
>> myveccumsum([5   9   4])
ans =
     5  14  18
```

PRACTICE 5.7

Write a function that imitates the **cumprod** function. Use the method of preallocating the output vector.

QUICK QUESTION!

How would we sum each individual column of a matrix?

Answer: The programming method would require a nested loop in which the outer loop is over the columns. The function will sum each column and return a row vector containing the results.

matcolsum.m

```
function outsum = matcolsum(mat)
% matcolsum finds the sum of every column in a matrix
% Returns a vector of the column sums
% Format: matcolsum(matrix)

[row, col] = size(mat);

% Preallocate the vector to the number of columns
outsum = zeros(1,col);

% Every column is being summed so the outer loop
% has to be over the columns
for i = 1:col
    % Initialize the running sum to 0 for every column
    runsum = 0;
    for j = 1:row
        runsum = runsum+mat(j,i);
    end
    outsum(i) = runsum;
end
end
```

Note that the output argument will be a row vector containing the same number of columns as the input argument matrix. Also, as the function is calculating a sum for each column, the runsum variable must be initialized to 0 for every column, so it is initialized inside of the outer loop.

```
>> mat = [3:5; 2 5 7]
mat =
     3     4     5
     2     5     7
>> matcolsum(mat)
ans =
     5     9    12
```

Of course, the built-in **sum** function in MATLAB would accomplish the same thing, as we have already seen.

PRACTICE 5.8

Modify the function *matcolsum*. Create a function *matrowsum* to calculate and return a vector of all of the row sums instead of column sums. For example, calling it and passing the *mat* variable above would result in the following:

```
>> matrowsum(mat)
ans =
    12      14
```

5.4.2 Vectorizing Loops with Selection Statements

In many applications, it is useful to determine whether numbers in a matrix are positive, zero, or negative.

THE PROGRAMMING CONCEPT

A function *signum* follows that will accomplish this:

signum.m

```
function outmat = signum(mat)
% signum imitates the sign function
% Format: signum(matrix)

[r, c] = size(mat);
for i = 1:r
    for j = 1:c
        if mat(i,j) > 0
            outmat(i,j) = 1;
        elseif mat(i,j) == 0
            outmat(i,j) = 0;
        else
            outmat(i,j) = -1;
        end
    end
end
end
```

Here is an example of using this function:

```
>> mat = [0  4  -3;  -1  0  2]
mat =
     0     4    -3
    -1     0     2
```

Continued

THE PROGRAMMING CONCEPT—CONT'D

```
>> signum(mat)
ans =
     0     1    -1
    -1     0     1
```

THE EFFICIENT METHOD

Close inspection reveals that the function accomplishes the same task as the built-in sign function!

```
>> sign(mat)
ans =
     0     1    -1
    -1     0     1
```

Another example of a common application on a vector is to find the minimum and/or maximum value in the vector.

THE PROGRAMMING CONCEPT

For instance, the algorithm to find the minimum value in a vector is as follows:

- The working minimum (the minimum that has been found so far) is the first element in the vector to begin with.
- Loop through the rest of the vector (from the second element to the end).
  - If any element is less than the working minimum, then that element is the new working minimum.

The following function implements this algorithm and returns the minimum value found in the vector.

myminvec.m

```
function outmin = myminvec(vec)
% myminvec returns the minimum value in a vector
% Format: myminvec(vector)

outmin = vec(1);
for i = 2:length(vec)
    if vec(i) < outmin
        outmin = vec(i);
    end
end
end
```

THE PROGRAMMING CONCEPT—CONT'D

```
>> vec = [3  8  99  -1];
>> myminvec(vec)
ans =
    -1

>> vec = [3  8  99  11];
>> myminvec(vec)
ans =
     3
```

Note

An **if** statement is used in the loop rather than an **if-else** statement. If the value of the next element in the vector is less than *outmin*, then the value of *outmin* is changed; otherwise, no action is necessary.

THE EFFICIENT METHOD

Use the **min** function:

```
>> vec = [5  9  4];
>> min(vec)
ans =
     4
```

QUICK QUESTION!

Determine what the following function accomplishes:

xxx.m

```
function logresult = xxx(vec)
% QQ for you - what does this do?

logresult = false;
i = 1;
while i <= length(vec) && logresult == false
    if vec(i) ~= 0
       logresult = true;
    end
    i = i+1;
end
end
```

Answer: The output produced by this function is the same as the **any** function for a vector. It initializes the output argument to **false**. It then loops through the vector and, if any element is nonzero, changes the output argument to **true**. It loops until either a nonzero value is found or it has gone through all elements.

QUICK QUESTION!

Determine what the following function accomplishes.

yyy.m

```
function logresult = yyy(mat)
% QQ for you - what does this do?

count = 0;
[r, c] = size(mat);
for i = 1:r
    for j = 1:c
        if mat(i,j) ~= 0
            count = count+1;
        end
     end
end

logresult = count == numel(mat);
end
```

Answer: The output produced by this function is the same as the **all** function.

As another example, we will write a function that will receive a vector and an integer as input arguments and will return a logical vector that stores **logical true** only for elements of the vector that are greater than the integer and **false** for the other elements.Note that as the vector was preallocated to **false**, the **else** clause is not necessary.

THE PROGRAMMING CONCEPT

The function receives two input arguments: the vector, and an integer *n* with which to compare. It loops through every element in the input vector, and stores in the result vector either **true** or **false** depending on whether vec(i) > n is **true** or **false**.

testvecgtn.m

```
function outvec = testvecgtn(vec,n)
% testvecgtn tests whether elements in vector
%     are greater than n or not
% Format: testvecgtn(vector, n)

% Preallocate the vector to logical false
outvec = false(size(vec));
for i = 1:length(vec)
    % If an element is > n, change to true
    if vec(i) > n
        outvec(i) = true;
    end
end
end
```

THE PROGRAMMING CONCEPT—CONT'D

```
>> ov = testvecgtn([44 2 11 -3 5 8], 6)
ov =
    1    0    1    0    0    1
>> class(ov)
ans =
logical
```

THE EFFICIENT METHOD

As we have seen, the relational operator > will automatically create a **logical** vector.

testvecgtnii.m

```
function outvec = testvecgtnii(vec,n)
% testvecgtnii tests whether elements in vector
%     are greater than n or not with no loop
% Format: testvecgtnii(vector, n)

outvec = vec > n;
end
```

PRACTICE 5.9

Call the function *testvecgtnii*, passing a vector and a value for n. Use MATLAB code to count how many values in the vector were greater than n.

5.4.3 TIPS FOR WRITING EFFICIENT CODE

To be able to write efficient code in MATLAB, including vectorizing, there are several important features to keep in mind:

- Scalar and array operations
- Logical vectors
- Built-in functions
- Preallocation of vectors

There are many functions in MATLAB that can be utilized instead of code that uses loops and selection statements. These functions have been demonstrated already but it is worth repeating them to emphasize their utility:

- **sum** and **prod**: find the sum or product of every element in a vector or column in a matrix

- **cumsum** and **cumprod**: return a vector or matrix of the cumulative (running) sums or products
- **min** and **max**: find the minimum value in a vector or in every column of a matrix
- **any**, **all**, **find**: work with logical expressions
- "is" functions, such as **isletter** and **isequal**: return **logical** values

In almost all cases, code that is faster to write by the programmer is also faster for MATLAB to execute. So, "efficient code" means that it is both efficient for the programmer and for MATLAB.

PRACTICE 5.10

Vectorize the following (rewrite the code efficiently):

```
i = 0;
for inc = 0: 0.5: 3
    i = i+1;
    myvec(i) = sqrt(inc);
end
---------------------------------------
[r c] = size(mat);
newmat = zeros(r,c);
for i = 1:r
    for j = 1:c
        newmat(i,j) = sign(mat(i,j));
    end
end
```

MATLAB has a built-in function **checkcode** that can detect potential problems within scripts and functions. Consider, for example, the following script that extends a vector within a loop:

badcode.m

```
for j = 1:4
    vec(j) = j
end
```

The function **checkcode** will flag this, as well as the good programming practice of suppressing output within scripts:

```
>> checkcode('badcode')
L 2 (C 5-7): The variable 'vec' appears to change size on every loop
iteration (within a script). Consider preallocating for speed.
L 2 (C 12): Terminate statement with semicolon to suppress output (within
a script).
```

The same information is shown in Code Analyzer Reports, which can be produced within MATLAB for one file (script or function) or for all code files within a folder. Clicking on the down arrow for the Current Folder, and then choosing Reports and then Code Analyzer Report will check the code for all files within the Current Folder. When viewing a file within the Editor, click on the down arrow and then Show Code Analyzer Report for a report on just that one file.

5.5 TIMING

MATLAB has built-in functions that determine how long it takes code to execute. One set of related functions is **tic/toc**. These functions are placed around code, and will print the time it took for the code to execute. Essentially, the function **tic** turns a timer on, and then **toc** evaluates the timer and prints the result. Here is a script that illustrates these functions.

fortictoc.m

```
tic
mysum = 0;
for i = 1:20000000
    mysum = mysum+i;
end
toc
```

```
>> fortictoc
Elapsed time is 0.087294 seconds.
```

Here is an example of a script that demonstrates how much preallocating a vector speeds up the code.

Note

When using timing functions such as **tic/toc**, be aware that other processes running in the background (e.g., any web browser) will affect the speed of your code.

tictocprealloc.m

```
% This shows the timing difference between
% preallocating a vector vs. not

Clear
disp('No preallocation')
tic
for i = 1:10000
    x(i) = sqrt(i);
end
toc

disp('Preallocation')
tic
y = zeros(1,10000)
for i = 1:10000
    y(i) = sqrt(i);
end
toc
```

```
>> tictocprealloc
No preallocation
Elapsed time is 0.005070 seconds.
Preallocation
Elapsed time is 0.000273 seconds.
```

QUICK QUESTION!

Preallocation can speed up code, but to preallocate it is necessary to know the desired size. What if you do not know the eventual size of a vector (or matrix)? Does that mean that you have to extend it rather than preallocating?

Answer: If you know the maximum size that it could possibly be, you can preallocate to a size that is larger than necessary and then delete the "unused" elements. To do that, you would have to count the number of elements that are actually used. For example, if you have a vector *vec* that has been preallocated and a variable *count* that stores the number of elements that were actually used, this will trim the unnecessary elements:

```
vec = vec(1:count)
```

MATLAB also has a Profiler that will generate detailed reports on execution time of codes. In newer versions of MATLAB, from the Editor, click on Run and Time; this will bring up a report in the Profile Viewer. Choose the function name to see a very detailed report, including a Code Analyzer Report. From the Command Window, this can be accessed using **profile on** and **profile off**, and **profile viewer**.

```
>> profile on
>> tictocprealloc
No preallocation
Elapsed time is 0.047721 seconds.
Preallocation
Elapsed time is 0.040621 seconds.
>> profile viewer
>> profile off
```

■ Explore Other Interesting Features

Explore what happens when you use a matrix rather than a vector to specify the range in a **for** loop. For example,

```
for i = mat
    disp(i)
end
```

Take a guess before you investigate!

Try the **pause** function in loops.

Investigate the **vectorize** function.

The **tic** and **toc** functions are in the **timefun** help topic. Type **help timefun** to investigate some of the other timing functions. ■

SUMMARY

COMMON PITFALLS

- Forgetting to initialize a running sum or count variable to 0
- Forgetting to initialize a running product variable to 1
- In cases where loops are necessary, not realizing that if an action is required for every row in a matrix, the outer loop must be over the rows (and if an action is required for every column, the outer loop must be over the columns)
- Not realizing that it is possible that the action of a **while** loop will never be executed
- Not error-checking input into a program
- Forgetting to vectorize code whenever possible. If it is not necessary to use loops in MATLAB, don't!
- Forgetting that **subplot** numbers the plots rowwise rather than columnwise.
- Not realizing that the **subplot** function just creates a matrix within the Figure Window. Each part of this matrix must then be filled with a plot, using any type of plot function.

PROGRAMMING STYLE GUIDELINES

- Use loops for repetition only when necessary
  - **for** statements as counted loops
  - **while** statements as conditional loops
- Do not use *i* or *j* for iterator variable names if the use of the built-in constants **i** and **j** is desired.
- Indent the action of loops.
- If the loop variable is just being used to specify how many times the action of the loop is to be executed, use the colon operator 1:n where *n* is the number of times the action is to be executed.
- Preallocate vectors and matrices whenever possible (when the size is known ahead of time).
- When data are read in a loop, only store them in an array if it will be necessary to access the individual data values again.

MATLAB Reserved Words
for
while
end

MATLAB Functions and Commands	
subplot	profile
factorial	
checkcode	
tic/toc	

Exercises

1. Write a **for** loop that will print the column of real numbers from 2.7 to 3.5 in steps of 0.2.
2. In the Command Window, write a **for** loop that will iterate through the integers from 32 to 255. For each, show the corresponding character from the character encoding. Play with this! Try printing characters beyond the standard ASCII, in small groups. For example, print the characters that correspond to integers from 300 to 340.
3. Prompt the user for an integer *n* and print "I love this stuff!" n times.
4. When would it matter if a **for** loop contained `for i = 1:4` vs. `for i = [3 5 2 6]`, and when would it not matter?
5. Write a function *sumsteps2* that calculates and returns the sum of 1 to *n* in steps of 2, where *n* is an argument passed to the function. For example, if 11 is passed, it will return 1+3+5+7+9+11. Do this using a **for** loop. Calling the function will look like this:

```
>> sumsteps2(11)
ans =
    36
```

6. Write a function *prodby2* that will receive a value of a positive integer *n* and will calculate and return the product of the odd integers from 1 to *n* (or from 1 to *n-1* if *n* is even). Use a **for** loop.
7. Write a script that will:
   - generate a random integer in the inclusive range from 2 to 5
   - loop that many times to
     - prompt the user for a number
     - print the sum of the numbers entered so far with one decimal place
8. Write a script that will load data from a file into a matrix. Create the data file first, and make sure that there is the same number of values on every line in the file so that it can be loaded into a matrix. Using a **for** loop, it will then create a subplot

for every row in the matrix, and will plot the numbers from each row element in the Figure Window.

9. Write the code that will prompt the user for 4 numbers, and store them in a vector. Make sure that you preallocate the vector!
10. Write a **for** loop that will print the elements from a vector variable in sentence format, regardless of the length of the vector. For example, if this is the vector:

```
>> vec = [5.5  11  3.45];
```

this would be the result:

```
Element 1 is 5.50.
Element 2 is 11.00.
Element 3 is 3.45.
```

The **for** loop should work regardless of how many elements are in the vector.

11. Execute this script and be amazed by the results! You can try more points to get a clearer picture, but it may take a while to run.

```
clear
clf
x = rand;
y = rand;
plot(x,y)
hold on
for it = 1:10000
   choic = round(rand*2);
   if choic == 0
      x = x/2;
      y = y/2;
   elseif choic == 1
      x = (x+1)/2;
      y = y/2;
   else
      x = (x+0.5)/2;
      y = (y+1)/2;
   end
   plot(x,y)
   hold on
end
```

12. A machine cuts N pieces of a pipe. After each cut, each piece of pipe is weighed and its length is measured; these 2 values are then stored in a file called *pipe.dat* (first the weight and then the length on each line of the file). Ignoring units, the weight is supposed to be between 2.1 and 2.3, inclusive, and the length is supposed to be between 10.3 and 10.4, inclusive. The following is just the beginning of what will be a long script to work with these data. For now, the script will just count how many rejects there are. A reject is any piece of pipe that has an invalid

weight and/or length. For a simple example, if N is 3 (meaning three lines in the file) and the file stores:

```
2.14 10.30
2.32 10.36
2.20 10.35
```

There is only one reject, the second one, as it weighs too much. The script would print:
There were 1 rejects.

13. Come up with "trigger" words in a problem statement that would tell you when it's appropriate to use **for** loops and/or nested **for** loops.
14. With a matrix, when would:
    - your outer loop be over the rows
    - your outer loop be over the columns
    - not matter which is the outer and which is the inner loop?
15. Write a function *myones* that will receive two input arguments *n* and *m* and will return an *nxm* matrix of all ones. Do NOT use any built-in functions (so, yes, the code will be inefficient). Here is an example of calling the function:
16. Write a script that will print the following multiplication table:

```
1
2  4
3  6  9
4  8  12  16
5  10  15  20  25
```

17. Write a function that will receive a matrix as an input argument, and will calculate and return the overall average of all numbers in the matrix. Use loops, not built-in functions, to calculate the average.
18. Write an algorithm for an ATM program. Think about where there would be selection statements, menus, loops (counted vs. conditional), etc. – but – don't write MATLAB code, just an algorithm (pseudo-code).
19. Trace this to figure out what the result will be, and then type it into MATLAB to verify the results.

```
count = 0;
number = 8;
while number > 3
     fprintf('number is %d\n', number)
     number = number - 2;
     count = count+1;
end
fprintf('count is %d\n', count)
```

20. The inverse of the mathematical constant *e* can be approximated as follows:

$$\frac{1}{e} \approx \left(1 - \frac{1}{n}\right)^{n}$$

Write a script that will loop through values of n until the difference between the approximation and the actual value is less than 0.0001. The script should then print out the built-in value of e^{-1} and the approximation to 4 decimal places, and also print the value of n required for such accuracy.

21. Write a script that will generate random integers in the range from 0 to 50, and print them, until one is finally generated that is greater than 25. The script should print how many attempts it took.
22. Write a script that will prompt the user for a keyword in MATLAB, error-checking until a keyword is entered.
23. A blizzard is a massive snowstorm. Definitions vary, but for our purposes we will assume that a blizzard is characterized by both winds of 30 mph or higher and blowing snow that leads to visibility of 0.5 miles or less, sustained for at least four hours. Data from a storm one day has been stored in a file *stormtrack.dat*. There are 24 lines in the file, one for each hour of the day. Each line in the file has the wind speed and visibility at a location. Create a sample data file. Read this data from the file and determine whether blizzard conditions were met during this day or not.
24. Given the following loop:

```
while x < 10
     action
end
```

- For what values of the variable x would the action of the loop be skipped entirely?
- If the variable x is initialized to have the value of 5 before the loop, what would the action have to include in order for this to not be an infinite loop?

25. Write a script called *prtemps* that will prompt the user for a maximum Celsius value in the range from −16 to 20; error-check to make sure it is in that range. Then, print a table showing degrees Fahrenheit and degrees Celsius until this maximum is reached. The first value that exceeds the maximum should not be printed. The table should start at 0 degrees Fahrenheit, and increment by 5 degrees Fahrenheit until the max (in Celsius) is reached. Both temperatures should be printed with a field width of 6 and one decimal place. The formula is $C = 5/9\ (F-32)$.
26. Vectorize this code! Write *one* assignment statement that will accomplish exactly the same thing as the given code (assume that the variable *vec* has been initialized):

```
result = 0;
for i = 1:length(vec)
    result = result + vec(i);
end
```

27. Vectorize this code! Write *one* assignment statement that will accomplish exactly the same thing as the given code (assume that the variable *vec* has been initialized):

```
newv = zeros(size(vec));
myprod = 1;
for i = 1:length(vec)
   myprod = myprod * vec(i);
   newv(i) = myprod;
end
newv % Note: this is just to display the value
```

28. The following code was written by somebody who does not know how to use MATLAB efficiently. Rewrite this as a single statement that will accomplish exactly the same thing for a matrix variable *mat* (e.g., vectorize this code):

```
[r c] = size(mat);
for i = 1:r
    for j = 1:c
        mat(i,j) = mat(i,j) * 2;
   end
end
```

29. Vectorize the following code. Write one assignment statement that would accomplish the same thing. Assume that *mat* is a matrix variable that has been initialized.

```
[r,c] = size(mat);
val = mat(1,1);
for i = 1:r
    for j = 1:c
        if mat(i,j) < val
            val = mat(i,j);
        end
    end
end
val % just for display
```

30. Vectorize the following code. Write statement(s) that accomplish the same thing, eliminating the loop. Assume that there is a vector v that has a negative number in it, e.g:

```
>> v = [4 11 22 5 33 -8 3 99 52];
```

```
newv = [];
i = 1;
while v(i) >= 0
   newv(i) = v(i);
   i = i+1;
end
newv % Note: just to display
```

31. Give some examples of when you would need to use a counted loop in MATLAB, and when you would not.
32. For each of the following, decide whether you would use a **for** loop, a **while** loop, a nested loop (and if so what kind, e.g., a **for** loop inside of another **for** loop, a **while** loop inside of a **for** loop, etc.), or no loop at all. DO NOT WRITE THE ACTUAL CODE.
    - sum the integers 1 through 50:
    - add 3 to all numbers in a vector:
    - prompt the user for a string and keep doing this until the string that the user enters is a keyword in MATLAB:
    - find the minimum in every column of a matrix:
    - prompt the user for 5 numbers and find their sum:
    - prompt the user for 10 numbers, find the average and also find how many of the numbers were greater than the average:
    - generate a random integer n in the range from 10 to 20. Prompt the user for n positive numbers, error-checking to make sure you get n positive numbers (and just echo print each one):
    - prompt the user for positive numbers until the user enters a negative number. Calculate and print the average of the positive numbers, or an error message if none are entered:
33. Write a script that will prompt the user for a quiz grade and error-check until the user enters a valid quiz grade. The script will then echo print the grade. For this case, valid grades are in the range from 0 to 10 in steps of 0.5. Do this by creating a vector of valid grades and then use **any** or **all** in the condition in the **while** loop.
34. Which is faster: using **false** or using **logical**(0) to preallocate a matrix to all **logical** zeros? Write a script to test this.
35. Which is faster: using a **switch** statement or using a nested **if-else**? Write a script to test this.
36. Write a script *beautyofmath* that produces the following output. The script should iterate from 1 to 9 to produce the expressions on the left, perform the specified operation to get the results shown on the right and print exactly in the format shown here.

```
>> beautyofmath
1 x 8 + 1 = 9
12 x 8 + 2 = 98
123 x 8 + 3 = 987
1234 x 8 + 4 = 9876
12345 x 8 + 5 = 98765
123456 x 8 + 6 = 987654
1234567 x 8 + 7 = 9876543
12345678 x 8 + 8 = 98765432
123456789 x 8 + 9 = 987654321
```

37. The Wind Chill Factor (WCF) measures how cold it feels with a given air temperature T (in degrees Fahrenheit) and wind speed V (in miles per hour). One formula for WCF is

$$WCF = 35.7 + 0.6\,T - 35.7\left(V^{0.16}\right) + 0.43\,T\left(V^{0.16}\right)$$

Write a function to receive the temperature and wind speed as input arguments, and return the WCF. Using loops, print a table showing wind chill factors for temperatures ranging from −20 to 55 in steps of 5, and wind speeds ranging from 0 to 55 in steps of 5. Call the function to calculate each wind chill factor.
38. Instead of printing the WCFs in the previous problem, create a matrix of WCFs and write them to a file. Use the programming method, using nested loops.
39. Write a script that will prompt the user for N integers, and then write the positive numbers (>=0) to an ASCII file called *pos.dat* and the negative numbers to an ASCII file called *neg.dat*. Error-check to make sure that the user enters N integers.
40. Write a script to add two 30-digit numbers and print the result. This is not as easy as it might sound at first, because integer types may not be able to store a value this large. One way to handle large integers is to store them in vectors, where each element in the vector stores a digit of the integer. Your script should initialize two 30-digit integers, storing each in a vector, and then add these integers, also storing the result in a vector. Create the original numbers using the **randi** function. Hint: add 2 numbers on paper first, and pay attention to what you do!
41. Write a "Guess My Number Game" program. The program generates a random integer in a specified range, and the user (the player) has to guess the number. The program allows the user to play as many times as he/she would like; at the conclusion of each game, the program asks whether the player wants to play again.
The basic algorithm is:
    1. The program starts by printing instructions on the screen.
    2. For every game:
        - the program generates a new random integer in the range from MIN to MAX. Treat MIN and MAX like constants; start by initializing them to 1 and 100
        - loop to prompt the player for a guess until the player correctly guesses the integer
        - for each guess, the program prints whether the player's guess was too low, too high, or correct
        - at the conclusion (when the integer has been guessed):
        - print the total number of guesses for that game
        - print a message regarding how well the player did in that game (e.g the player took way too long to guess the number, the player was awesome, etc.). To do this, you will have to decide on ranges for your messages and give a rationale for your decision in a comment in the program.

3. After all games have been played, print a summary showing the average number of guesses.

42. A CD changer allows you to load more than one CD. Many of these have random buttons, which allow you to play random tracks from a specified CD, or play random tracks from random CDs. You are to simulate a play list from such a CD changer using the **randi** function. The CD changer that we are going to simulate can load 3 different CDs. You are to assume that three CDs have been loaded. To begin with, the program should "decide" how many tracks there are on each of the three CDs, by generating random integers in the range from MIN to MAX. You decide on the values of MIN and MAX (look at some CDs; how many tracks do they have? What's a reasonable range?). The program will print the number of tracks on each CD. Next, the program will ask the user for his or her favorite track; the user must specify which track and which CD it's on. Next, the program will generate a "playlist" of the N random tracks that it will play, where N is an integer. For each of the N songs, the program will first randomly pick one of the 3 CDs, and then randomly pick one of the tracks from that CD. Finally, the program will print whether the user's favorite track was played or not. The output from the program will look something like this depending on the random integers generated and the user's input:

```
There are 15 tracks on CD 1.
There are 22 tracks on CD 2.
There are 13 tracks on CD 3.

What's your favorite track?
Please enter the number of the CD: 4
Sorry, that's not a valid CD.
Please enter the number of the CD: 1
Please enter the track number: 17
Sorry, that's not a valid track on CD 1.
Please enter the track number: -5
Sorry, that's not a valid track on CD 1.
Please enter the track number: 11

Play List:
CD 2 Track 20
CD 3 Track 11
CD 3 Track 8
CD 2 Track 1
CD 1 Track 7
CD 3 Track 8
CD 1 Track 3
CD 1 Track 15
CD 3 Track 12
CD 1 Track 6
Sorry, your favorite track was not played.
```

43. Write your own code to perform matrix multiplication. Recall that to multiply two matrices, the inner dimensions must be the same.

$$[A]_{m\,x\,n}[B]_{n\,x\,p} = [C]_{m\,x\,p}$$

Every element in the resulting matrix C is obtained by:

$$c_{ij} = \sum_{k=1}^{n} a_{ik} b_{kj}.$$

So, three nested loops are required.

CHAPTER 6

MATLAB Programs

KEY TERMS

functions that return more than one value
functions that do not return any values
side effects
call-by-value
modular programs
main program
primary function
subfunction
menu-driven program
variable scope
base workspace
local variable
main function
global variable
persistent variable
declaring variables
bug
debugging
syntax errors
run-time errors
logical errors
tracing
breakpoints
breakpoint alley
function stubs
live script
code cells

CONTENTS

Chapter 3 introduced scripts and user-defined functions. In that chapter, we saw how to write scripts, which are sequences of statements that are stored in MATLAB code files and then executed. We also saw how to write user-defined functions, also stored in MATLAB code files that calculate and return a single value. In this chapter, we will expand on these concepts and introduce other kinds of user-defined functions. We will show how MATLAB® programs consist of combinations of scripts and user-defined functions. The mechanisms for interactions of variables in code files and the Command Window will be explored. Techniques for finding and fixing mistakes in programs will be reviewed. Finally, the use of live scripts created by the Live Editor (new as of R2016a), and using code cells in scripts will be introduced.

6.1 MORE TYPES OF USER-DEFINED FUNCTIONS

We have already seen how to write a user-defined function, stored in a code file, that calculates and returns one value. This is just one type of function. It is also possible for a function to return multiple values and it is possible for a function to return nothing. We will categorize functions as follows:

MATLAB®. http://dx.doi.org/10.1016/B978-0-12-804525-1.00006-4

- Functions that calculate and return one value
- Functions that calculate and return more than one value
- Functions that just accomplish a task, such as printing, without returning any values

Thus, although many functions calculate and return values, some do not. Instead, some functions just accomplish a task. Categorizing the functions as above is somewhat arbitrary, but there are differences between these three types of functions, including the format of the function headers and also the way in which the functions are called. Regardless of what kind of function it is, all functions must be defined and all function definitions consist of the ***header*** and the ***body***. Also, the function must be called for it to be utilized. All functions are stored in code files that have an extension of .m.

In general, any function in MATLAB consists of the following:

- The function header (the first line); this has:
  - the reserved word **function**
  - if the function ***returns*** values, the name(s) of the output argument(s), followed by the assignment operator (=)
  - the name of the function (important: this should be the same as the name of the file in which this function is stored to avoid confusion)
  - the input arguments in parentheses, if there are any (separated by commas if there is more than one).
- A comment that describes what the function does (this is printed if **help** is used).
- The body of the function, which includes all statements, including putting values in all output arguments if there are any.
- **end** at the end of the function.

6.1.1 Functions That Return More Than One Value

Functions that return one value have one output argument, as we saw previously. Functions that return more than one value must, instead, have more than one output argument in the function header in square brackets. That means that in the body of the function, values must be put in all output arguments listed in the function header. The general form of a function definition for a function that calculates and ***returns more than one value*** looks like this:

functionname.m

```
function [output arguments] = functionname(input arguments)
% Comment describing the function
% Format of function call

Statements here; these must include putting values in all of the output
arguments listed in the header

end
```

In the vector of output arguments, the output argument names are by convention separated by commas.

Choosing New, then Function brings up a template in the Editor that can then be filled in:

```
function [ output_args ] = untitled( input_args )
%UNTITLED Summary of this function goes here
%    Detailed explanation goes here


end
```

If this is not desired, it may be easier to start with New Script.

For example, here is a function that calculates two values, both the area and the circumference of a circle; this is stored in a file called *areacirc.m*:

areacirc.m

```
function [area, circum] = areacirc(rad)
% areacirc returns the area and
% the circumference of a circle
% Format: areacirc(radius)

area = pi * rad .* rad;
circum = 2 * pi * rad;
end
```

As this function is calculating two values, there are two output arguments in the function header (*area* and *circum*), which are placed in square brackets [ ]. Therefore, somewhere in the body of the function, values have to be stored in both.

As the function is returning two values, it is important to capture and store these values in separate variables when the function is called. In this case, the first value returned, the area of the circle, is stored in a variable *a* and the second value returned is stored in a variable *c*:

```
>> [a, c] = areacirc(4)
a =
   50.2655
c =
   25.1327
```

If this is not done, only the first value returned is retained—in this case, the area:

```
>> disp(areacirc(4))
    50.2655
```

Note that in capturing the values, the order matters. In this example, the function first returns the area and then the circumference of the circle. The order in which values are assigned to the output arguments within the function, however, does not matter.

QUICK QUESTION!

What would happen if a vector of radii was passed to the function?

Answer: As the .* operator is used in the function to multiply *rad* by itself, a vector can be passed to the input argument *rad*. Therefore, the results will also be vectors, so the variables on the left side of the assignment operator would become vectors of areas and circumferences.

```
>> [a, c] = areacirc(1:4)
a =
   3.1416   12.5664   28.2743   50.2655
c =
   6.2832   12.5664   18.8496   25.1327
```

QUICK QUESTION!

What if you want only the second value that is returned?

Answer: Function outputs can be ignored using the tilde:

```
>> [~, c] = areacirc(1:4)
c =
  6.2832  12.5664  18.8496  25.1327
```

The **help** function shows the comment listed under the function header:

```
>> help areacirc
  This function calculates the area and
  the circumference of a circle
  Format: areacirc(radius)
```

The *areacirc* function could be called from the Command Window as shown here, or from a script. Here is a script that will prompt the user for the radius of just one circle, call the *areacirc* function to calculate and return the area and circumference of the circle, and print the results:

calcareacirc.m

```
% This script prompts the user for the radius of a circle,
%  calls a function to calculate and return both the area
%  and the circumference, and prints the results
% It ignores units and error-checking for simplicity

radius = input('Please enter the radius of the circle: ');
[area, circ] = areacirc(radius);
fprintf('For a circle with a radius of %.1f,\n', radius)
fprintf('the area is %.1f and the circumference is %.1f\n',...
    area, circ)
```

```
>> calcareacirc
Please enter the radius of the circle: 5.2
For a circle with a radius of 5.2,
the area is 84.9 and the circumference is 32.7
```

PRACTICE 6.1

Write a function *perimarea* that calculates and returns the perimeter and area of a rectangle. Pass the length and width of the rectangle as input arguments. For example, this function might be called from the following script:

calcareaperim.m

```
% Prompt the user for the length and width of a rectangle,
%  call a function to calculate and return the perimeter
%  and area, and print the result
% For simplicity it ignores units and error-checking
length = input('Please enter the length of the rectangle: ');
width = input('Please enter the width of the rectangle: ');
[perim, area] = perimarea(length, width);
fprintf('For a rectangle with a length of %.1f and a', length)
fprintf(' width of %.1f,\nthe perimeter is %.1f,', width, perim)
fprintf(' and the area is %.1f\n', area)
```

As another example, consider a function that calculates and returns three output arguments. The function will receive one input argument representing a total number of seconds and returns the number of hours, minutes, and remaining seconds that it represents. For example, 7515 total seconds is 2 h, 5 min, and 15 s because $7515 = 3600 \times 2 + 60 \times 5 + 15$.

The algorithm is as follows.

- Divide the total seconds by 3600, which is the number of seconds in an hour. For example, 7515/3600 is 2.0875. The integer part is the number of hours (e.g., 2).
- The remainder of the total seconds divided by 3600 is the remaining number of seconds; it is useful to store this in a local variable.
- The number of minutes is the remaining number of seconds divided by 60 (again, the integer part).
- The number of seconds is the remainder of the previous division.

breaktime.m

```
function [hours, minutes, secs] = breaktime(totseconds)
% breaktime breaks a total number of seconds into
% hours, minutes, and remaining seconds
% Format: breaktime(totalSeconds)

hours = floor(totseconds/3600);
remsecs = rem(totseconds, 3600);
minutes = floor(remsecs/60);
secs = rem(remsecs,60);
end
```

An example of calling this function is:

```
>> [h, m, s] = breaktime(7515)
h =
     2
m =
     5
s =
    15
```

As before, it is important to store all values that the function returns by using three separate variables.

6.1.2 Functions That Accomplish a Task Without Returning Values

Many functions do not calculate values but rather accomplish a task, such as printing formatted output. As these functions do not return any values, there are no output arguments in the function header.

The general form of a function definition for a ***function that does not return any values*** looks like this:

```
functionname.m
```

```
function functionname(input arguments)
% Comment describing the function

Statements here
end
```

Note what is missing in the function header: there are no output arguments and no assignment operator.

For example, the following function just prints the two arguments, numbers, passed to it in a sentence format:

```
printem.m
```

```
function printem(a,b)
% printem prints two numbers in a sentence format
% Format: printem(num1, num2)

fprintf('The first number is %.1f and the second is %.1f\n',a,b)
end
```

As this function performs no calculations, there are no output arguments in the function header and no assignment operator (=). An example of a call to the *printem* function is:

```
>> printem(3.3, 2)
The first number is 3.3 and the second is 2.0
```

Note that as the function does not return a value, it cannot be called from an assignment statement. Any attempt to do this would result in an error, such as the following:

```
>> x = printem(3, 5) % Error!!
Error using printem
Too many output arguments.
```

We can therefore think of the call to a function that does not return values as a statement by itself, in that the function call cannot be imbedded in another statement such as an assignment statement or an output statement.

The tasks that are accomplished by functions that do not return any values (e.g., output from an **fprintf** statement or a **plot**) are sometimes referred to as ***side effects***. Some standards for commenting functions include putting the side effects in the block comment.

PRACTICE 6.2

Write a function that receives a vector as an input argument and prints the individual elements from the vector in a sentence format.

```
>> printvecelems([5.9   33   11])
Element 1 is 5.9
Element 2 is 33.0
Element 3 is 11.0
```

6.1.3 Functions That Return Values Versus Printing

A function that calculates and ***returns*** values (through the output arguments) does not normally also print them; that is left to the calling script or function. It is a good programming practice to separate these tasks.

If a function just prints a value, rather than returning it, the value cannot be used later in other calculations. For example, here is a function that just prints the circumference of a circle:

calccircum1.m

```
function calccircum1(radius)
% calccircum1 displays the circumference of a circle
%   but does not return the value
% Format: calccircum1(radius)

disp(2 * pi * radius)
end
```

Calling this function prints the circumference, but there is no way to store the value so that it can be used in subsequent calculations:

```
>> calccircum1(3.3)
   20.7345
```

Since no value is returned by the function, attempting to store the value in a variable would be an error:

```
>> c = calccircum1(3.3)
Error using calccircum1
Too many output arguments.
```

By contrast, the following function calculates and returns the circumference, so that it can be stored and used in other calculations. For example, if the circle is the base of a cylinder and we wish to calculate the surface area of the cylinder, we would need to multiply the result from the *calccircum2* function by the height of the cylinder.

calccircum2.m

```
function circle_circum = calccircum2(radius)
% calccircum2 calculates and returns the
%    circumference of a circle
% Format: calccircum2(radius)

circle_circum = 2 * pi * radius;
end
```

```
>> circumference = calccircum2(3.3)
circumference =
   20.7345

>> height = 4;
>> surf_area = circumference * height
surf_area =
   82.9380
```

One possible exception to this rule of not printing when returning is to have a function return a value if possible but throw an error if not.

6.1.4 Passing Arguments to Functions

In all function examples presented thus far, at least one argument was passed in the function call to be the value(s) of the corresponding input argument(s) in the function header. The ***call-by-value*** method is the term for this method of passing the values of the arguments to the input arguments in the functions.

In some cases, however, it is not necessary to pass any arguments to the function. Consider, for example, a function that simply prints a random real number with two decimal places:

printrand.m

```
function printrand()
% printrand prints one random number
% Format: printrand or printrand()

fprintf('The random # is %.2f\n',rand)
end
```

Here is an example of calling this function:

```
>> printrand()
The random # is 0.94
```

As nothing is passed to the function, there are no arguments in the parentheses in the function call and none in the function header, either. The parentheses are not even needed in either the function or the function call, either. The following works as well:

printrandnp.m

```
function printrandnp
% printrandnp prints one random number
% Format: printrandnp or printrandnp()

fprintf('The random # is %.2f\n',rand)
end
```

```
>> printrandnp
The random # is 0.52
```

In fact, the function can be called with or without empty parentheses, whether or not there are empty parentheses in the function header.

This was an example of a function that did not receive any input arguments nor did it return any output arguments; it simply accomplished a task.

The following is another example of a function that does not receive any input arguments, but in this case, it does return a value. The function prompts the user for a string and returns the value entered.

stringprompt.m

```
function outstr = stringprompt
% stringprompt prompts for a string and returns it
% Format stringprompt or stringprompt()

disp('When prompted, enter a string of any length.')
outstr = input('Enter the string here: ', 's');
end
```

```
>> mystring = stringprompt
When prompted, enter a string of any length.
Enter the string here: Hi there

mystring =
Hi there
```

PRACTICE 6.3

Write a function that will prompt the user for a string of at least one character, loop to error-check to make sure that the string has at least one character and return the string.

QUICK QUESTION!

It is important that the number of arguments passed in the call to a function must be the same as the number of input arguments in the function header, even if that number is zero. Also, if a function returns more than one value, it is important to "capture" all values by having an equivalent number of variables in a vector on the left side of an assignment statement. Although it is not an error if there aren't enough variables, some of the values returned will be lost. The following question is posed to highlight this.

Given the following function header (note that this is just the function header, not the entire function definition):

```
function [outa, outb] = qq1(x, y, z)
```

Which of the following proposed calls to this function would be valid?

```
(a) [var1, var2] = qq1(a, b, c);
(b) answer = qq1(3, y, q);
(c) [a, b] = myfun(x, y, z);
(d) [outa, outb] = qq1(x, z);
```

Answer: The first proposed function call, (a), is valid. There are three arguments that are passed to the three input arguments in the function header, the name of the function is *qq1*, and there are two variables in the assignment statement to store the two values returned from the function. Function call (b) is valid, although only the first value returned from the function would be stored in *answer*; the second value would be lost. Function call (c) is invalid because the name of the function is given incorrectly. Function call (d) is invalid because only two arguments are passed to the function, but there are three input arguments in the function header.

6.2 MATLAB PROGRAM ORGANIZATION

Typically, a MATLAB program consists of a script that calls functions to do the actual work.

6.2.1 Modular Programs

A ***modular program*** is a program in which the solution is broken down into modules, and each is implemented as a function. The script that calls these functions is typically called the ***main program***.

To demonstrate the concept, we will use the very simple example of calculating the area of a circle. In Section 6.3 a much longer example will be given. For this example, there are three steps in the algorithm to calculate the area of a circle:

- Get the input (the radius)
- Calculate the area
- Display the results

In a modular program, there would be one main script (or, possibly a function instead) that calls three separate functions to accomplish these tasks:

- A function to prompt the user and read in the radius
- A function to calculate and return the area of the circle
- A function to display the results

As scripts and functions are all stored in code files that have an extension of .m, there would therefore be four separate code files altogether for this program; one script file and three function code files, as follows:

calcandprintarea.m

```
% This is the main script to calculate the
%   area of a circle
% It calls 3 functions to accomplish this
radius = readradius;
area = calcarea(radius);
printarea(radius,area)
```

readradius.m

```
function radius = readradius
% readradius prompts the user and reads the radius
% Ignores error-checking for now for simplicity
% Format: readradius or readradius()

disp('When prompted, please enter the radius in inches.')
radius = input('Enter the radius: ');
end
```

calcarea.m

```
function area = calcarea(rad)
% calcarea returns the area of a circle
% Format: calcarea(radius)

area = pi * rad .* rad;
end
```

printarea.m

```
function printarea(rad,area)
% printarea prints the radius and area
% Format: printarea(radius, area)

fprintf('For a circle with a radius of %.2f inches,\n',rad)
fprintf('the area is %.2f inches squared.\n',area)
end
```

When the program is executed, the following steps will take place:

- the script *calcandprintarea* begins executing
- *calcandprintarea* calls the *readradius* function
  - *readradius* executes and returns the radius

- *calcandprintarea* resumes executing and calls the *calcarea* function, passing the radius to it
  - *calcarea* executes and returns the area
- *calcandprintarea* resumes executing and calls the *printarea* function, passing both the radius and the area to it
  - *printarea* executes and prints
- the script finishes executing

Running the program would be accomplished by typing the name of the script; this would call the other functions:

```
>> calcandprintarea
When prompted, please enter the radius in inches.
Enter the radius: 5.3
For a circle with a radius of 5.30 inches,
the area is 88.25 inches squared.
```

Note how the function calls and the function headers match up. For example:

readradius function:

```
function call: radius = readradius;
function header: function radius = readradius
```

In the *readradius* function call, no arguments are passed so there are no input arguments in the function header. The function returns one output argument so that is stored in one variable.

calcarea function:

```
function call: area = calcarea(radius);
function header: function area = calcarea(rad)
```

In the *calcarea* function call, one argument is passed in parentheses so there is one input argument in the function header. The function returns one output argument so that is stored in one variable.

printarea function:

```
function call: printarea(radius,area)
function header: function printarea(rad,area)
```

In the *printarea* function call, there are two arguments passed, so there are two input arguments in the function header. The function does not return anything, so the call to the function is a statement by itself; it is not in an assignment or output statement.

PRACTICE 6.4

Modify the *readradius* function to error-check the user's input to make sure that the radius is valid. The function should ensure that the radius is a positive number by looping to print an error message until the user enters a valid radius.

6.2.2 Subfunctions

Thus far, every function has been stored in a separate code file. However, it is possible to have more than one function in a given file. For example, if one function calls another, the first (calling) function would be the ***primary function*** and the function that is called is a ***subfunction***. These functions would both be stored in the same code file, first the primary function and then the subfunction. The name of the code file would be the same as the name of the primary function, to avoid confusion.

To demonstrate this, a program that is similar to the previous one, but calculates and prints the area of a rectangle, is shown here. The script, or main program, first calls a function that reads the length and width of the rectangle, and then calls a function to print the results. This function calls a subfunction to calculate the area.

rectarea.m

```
% This program calculates & prints the area of a rectangle

% Call a fn to prompt the user & read the length and width
[length, width] = readlenwid;
% Call a fn to calculate and print the area
printrectarea(length, width)
```

readlenwid.m

```
function [l,w] = readlenwid
% readlenwid reads & returns the length and width
% Format: readlenwid or readlenwid()

l = input('Please enter the length: ');
w = input('Please enter the width: ');
end
```

printrectarea.m

```
function printrectarea(len, wid)
% printrectarea prints the rectangle area
% Format: printrectarea(length, width)

% Call a subfunction to calculate the area
area = calcrectarea(len,wid);
fprintf('For a rectangle with a length of %.2f\n',len)
fprintf('and a width of %.2f, the area is %.2f\n', ...
     wid, area);
end

function area = calcrectarea(len, wid)
% calcrectarea returns the rectangle area
% Format: calcrectarea(length, width)
area = len * wid;
end
```

An example of running this program follows:

```
>> rectarea
Please enter the length: 6
Please enter the width: 3
For a rectangle with a length of 6.00
and a width of 3.00, the area is 18.00
```

Note how the function calls and function headers match up. For example:

readlenwid function:

```
function call: [length, width] = readlenwid;
function header: function [l,w] = readlenwid
```

In the *readlenwid* function call, no arguments are passed so there are no input arguments in the function header. The function returns two output arguments so there is a vector with two variables on the left side of the assignment statement in which the function is called.

printrectarea function:

```
function call: printrectarea(length, width)
function header: function printrectarea(len, wid)
```

In the *printrectarea* function call, there are two arguments passed, so there are two input arguments in the function header. The function does not return anything, so the call to the function is a statement by itself; it is not in an assignment or output statement.

calcrectarea subfunction:

```
function call: area = calcrectarea(len,wid);
function header: function area = calcrectarea(len, wid)
```

In the *calcrectarea* function call, two arguments are passed in parentheses so there are two input arguments in the function header. The function returns one output argument so that is stored in one variable.

The **help** command can be used with the script *rectarea*, the function *readlenwid*, and with the primary function, *printrectarea*. To view the first comment in the subfunction, as it is contained within the *printrectarea.m* file, the operator > is used to specify both the primary and subfunctions:

```
>> help rectarea
This program calculates & prints the area of a rectangle

>> help printrectarea
printrectarea prints the rectangle area
Format: printrectarea(length, width)

>> help printrectarea>calcrectarea
calcrectarea returns the rectangle area
Format: calcrectarea(length, width)
```

PRACTICE 6.5

For a right triangle with sides *a*, *b*, and *c*, where *c* is the hypotenuse and θ is the angle between sides *a* and *c*, the lengths of sides *a* and *b* are given by:

```
a = c * cos(θ)
b = c * sin(θ)
```

Write a script *righttri* that calls a function to prompt the user and read in values for the hypotenuse and the angle (in radians), and then calls a function to calculate and return the lengths of sides *a* and *b* and a function to print out all values in a sentence format. For simplicity, ignore units. Here is an example of running the script; the output format should be exactly as shown here:

```
>> righttri
Enter the hypotenuse: 5
Enter the angle: .7854
For a right triangle with hypotenuse 5.0
 and an angle 0.79 between side a & the hypotenuse,
 side a is 3.54 and side b is 3.54
```

For extra practice, do this using two different program organizations:

- One script that calls three separate functions
- One script that calls two functions; the function that calculates the lengths of the sides will be a subfunction to the function that prints

6.3 APPLICATION: MENU-DRIVEN MODULAR PROGRAM

Many longer, more involved programs that have interaction with the user are ***menu-driven***, which means that the program prints a menu of choices and then continues to loop to print the menu of choices until the user chooses to end the program. A modular menu-driven program would typically have a function that presents the menu and gets the user's choice, as well as functions to implement the action for each choice. These functions may have subfunctions. Also, the functions would error-check all user input.

As an example of such a menu-driven program, we will write a program to explore the constant *e*.

The constant *e*, called the natural exponential base, is used extensively in mathematics and engineering. There are many diverse applications of this constant. The value of the constant e is approximately 2.7183... Raising e to the power of

x, or e^x, is so common that this is called the exponential function. In MATLAB, as we have seen, there is a function for this, **exp**.

One way to determine the value of e is by finding a limit.

$$e = \lim_{n \to \infty} \left(1 + \frac{1}{n}\right)^n$$

As the value of n increases toward infinity, the result of this expression approaches the value of e.

An approximation for the exponential function can be found using what is called a Maclaurin series:

$$e^x \approx 1 + \frac{x^1}{1!} + \frac{x^2}{2!} + \frac{x^3}{3!} + \ldots$$

We will write a program to investigate the value of e and the exponential function. It will be menu-driven. The menu options will be:

- Print an explanation of e.
- Prompt the user for a value of n and then find an approximate value for e using the expression $(1+1/n)^n$.
- Prompt the user for a value for x. Print the value of **exp(x)** using the built-in function. Find an approximate value for e^x using the Maclaurin series just given.
- Exit the program.

The algorithm for the script main program follows:

- Call a function *eoption* to display the menu and return the user's choice.
- Loop until the user chooses to exit the program. If the user has not chosen to exit, the action of the loop is to:
  - Depending on the user's choice, do one of the following:
    - Call a function *explaine* to print an explanation of *e*.
    - Call a function *limite* that will prompt the user for n and calculate an approximate value for e
    - Prompt the user for x and call a function *expfn* that will print both an approximate value for e^x and the value of the built-in **exp(x)**. Note that because any value for x is acceptable, the program does not need to error-check this value.
  - Call the function *eoption* to display the menu and return the user's choice again.

The algorithm for the *eoption* function follows:

- Display the four choices.
- Error-check by looping to display the menu until the user chooses one of the four options.
- Return the integer value corresponding to the choice.

The algorithm for the *explaine* function is:

- Print an explanation of e, the **exp** function, and how to find approximate values.

The algorithm for the *limite* function is:

- Call a subfunction *askforn* to prompt the user for an integer *n*.
- Calculate and print the approximate value of e using *n*.

The algorithm for the subfunction *askforn* is:

- Prompt the user for a positive integer for *n*.
- Loop to print an error message and reprompt until the user enters a positive integer.
- Return the positive integer *n*.

The algorithm for the *expfn* function is:

- Receive the value of *x* as an input argument.
- Print the value of **exp(x)**.
- Assign an arbitrary value for the number of terms *n* (an alternative method would be to prompt the user for this).
- Call a subfunction *appex* to find an approximate value of **exp(x)** using a series with *n* terms.
- Print this approximate value.

The algorithm for the *appex* subfunction is:

- Receive *x* and *n* as input arguments.
- Initialize a variable for the running sum of the terms in the series (to 1 for the first term) and for a running product that will be the factorials in the denominators.
- Loop to add the *n* terms to the running sum.
- Return the resulting sum.

The entire program consists of the following script file and four function code files:

eapplication.m

```
% This script explores e and the exponential function

% Call a function to display a menu and get a choice
choice = eoption;

% Choice 4 is to exit the program
while choice ~= 4
   switch choice
       case 1
           % Explain e
           explaine;
       case 2
           % Approximate e using a limit
           limite;
       case 3
           % Approximate exp(x) and compare to exp
           x = input('Please enter a value for x: ');
           expfn(x);
   end
   % Display menu again and get user's choice
   choice = eoption;
end
```

eoption.m

```
function choice = eoption
% eoption prints a menu of options and error-checks
%   until the user chooses one of the options
% Format: eoption or eoption()

printchoices
choice = input('');
while ~any(choice == 1:4)
      disp('Error - please choose one of the options.')
      printchoices
      choice = input('');
end
end

function printchoices
fprintf('Please choose an option:\n\n');
fprintf('1) Explanation\n')
fprintf('2) Limit\n')
fprintf('3) Exponential function\n')
fprintf('4) Exit program\n\n')
end
```

explaine.m

```
function explaine
% explaine explains a little bit about e
% Format: explaine or explaine()

fprintf('The constant e is called the natural')
fprintf(' exponential base.\n')
fprintf('It is used extensively in mathematics and')
fprintf(' engineering.\n')
fprintf('The value of the constant e is ~ 2.7183\n')
fprintf('Raising e to the power of x is so common that')
fprintf(' this is called the exponential function.\n')
fprintf('An approximation for e is found using a limit.\n')
fprintf('An approximation for the exponential function')
fprintf(' can be found using a series.\n')
end
```

limite.m

```
function limite
% limite returns an approximate of e using a limit
% Format: limite or limite()

% Call a subfunction to prompt user for n
n = askforn;
fprintf('An approximation of e with n = %d is %.2f\n', ...
    n, (1 + 1/n) ^n)
end

function outn = askforn
% askforn prompts the user for n
% Format askforn or askforn()
% It error-checks to make sure n is a positive integer

inputnum = input('Enter a positive integer for n: ');
num2 = int32(inputnum);
while num2 ~= inputnum || num2 < 0
     inputnum = input('Invalid! Enter a positive integer: ');
     num2 = int32(inputnum);
end
outn = inputnum;
end
```

expfn.m

```
function expfn(x)
% expfn compares the built-in function exp(x)
%  and a series approximation and prints
% Format expfn(x)

fprintf('Value of built-in exp(x) is %.2f\n',exp(x))

% n is arbitrary number of terms
n = 10;
fprintf('Approximate exp(x) is %.2f\n', appex(x,n))
end

function outval = appex(x,n)
% appex approximates e to the x power using terms up to
%  x to the nth power
% Format appex(x,n)

% Initialize the running sum in the output argument
% outval to 1 (for the first term)
outval = 1;

for i = 1:n
   outval = outval + (x^i)/factorial(i);
end
end
```

Running the script will bring up the menu of options.

```
>> eapplication
Please choose an option:

(1) Explanation
(2) Limit
(3) Exponential function
(4) Exit program
```

Then, what happens will depend on which option(s) the user chooses. Every time the user chooses, the appropriate function will be called and then this menu will appear again. This will continue until the user chooses 4 for 'Exit Program.' Examples will be given of running the script, with different sequences of choices.

In the following example, the user

- Chose 1 for 'Explanation'
- Chose 4 for 'Exit Program'

```
>> eapplication
Please choose an option:
```

```
(1) Explanation
(2) Limit
(3) Exponential function
(4) Exit program

1
The constant e is called the natural exponential base.
It is used extensively in mathematics and engineering.
The value of the constant e is ~ 2.7183
Raising e to the power of x is so common that this is called the exponential
function.
An approximation for e is found using a limit.
An approximation for the exponential function can be found using a series.
Please choose an option:

(1) Explanation
(2) Limit
(3) Exponential function
(4) Exit program

4
```

In the following example, the user

- Chose 2 for 'Limit'
  - When prompted for n, entered two invalid values before finally entering a valid positive integer
- Chose 4 for 'Exit Program'

```
>> eapplication
Please choose an option:

(1) Explanation
(2) Limit
(3) Exponential function
(4) Exit program

2
Enter a positive integer for n: -4
Invalid! Enter a positive integer: 5.5
Invalid! Enter a positive integer: 10
An approximation of e with n = 10 is 2.59
Please choose an option:

(1) Explanation
(2) Limit
(3) Exponential function
(4) Exit program

4
```

To see the difference in the approximate value for e as *n* increases, the user kept choosing 2 for 'Limit,' and entering larger and larger values each time in the following example (the menu is not shown for simplicity):

```
>> eapplication
Enter a positive integer for n: 4
An approximation of e with n = 4 is 2.44
Enter a positive integer for n: 10
An approximation of e with n = 10 is 2.59
Enter a positive integer for n: 30
An approximation of e with n = 30 is 2.67
Enter a positive integer for n: 100
An approximation of e with n = 100 is 2.70
```

In the following example, the user

- Chose 3 for 'Exponential function'
  - When prompted, entered 4.6 for x
- Chose 3 for 'Exponential function' again
  - When prompted, entered −2.3 for x
- Chose 4 for 'Exit Program'

Again, for simplicity, the menu options and choices are not shown.

```
>> eapplication
Please enter a value for x: 4.6
Value of built-in exp(x) is 99.48
Approximate exp(x) is 98.71
Please enter a value for x: -2.3
Value of built-in exp(x) is 0.10
Approximate exp(x) is 0.10
```

6.4 VARIABLE SCOPE

The ***scope*** of any variable is the workspace in which it is valid. The workspace created in the Command Window is called the ***base workspace***.

As we have seen before, if a variable is defined in any function, it is a ***local variable*** to that function, which means that it is only known and used within that function. Local variables only exist while the function is executing; they cease to exist when the function stops executing. For example, in the following function that calculates the sum of the elements in a vector, there is a local loop variable *i*.

mysum.m

```
function runsum = mysum(vec)
% mysum returns the sum of a vector
% Format: mysum(vector)

runsum = 0;
for i=1:length(vec)
    runsum = runsum + vec(i);
end
end
```

Running this function does not add any variables to the base workspace, as demonstrated in the following:

```
>> clear
>> who
>> disp(mysum([5  9  1]))
    15
>> who
>>
```

In addition, variables that are defined in the Command Window cannot be used in a function (unless passed as arguments to the function).

However, scripts (as opposed to functions) *do* interact with the variables that are defined in the Command Window. For example, the previous function is changed to be a script *mysumscript*.

mysumscript.m

```
% This script sums a vector
vec = 1:5;
runsum = 0;
for i=1:length(vec)
    runsum = runsum + vec(i);
end
disp(runsum)
```

The variables defined in the script do become part of the base workspace:

```
>> clear
>> who
>> mysumscript
    15
>> who
Your variables are:
i     runsum     vec
```

Variables that are defined in the Command Window can be used in a script, but cannot be used in a function. For example, the vector *vec* could be defined in the Command Window (instead of in the script), but then used in the script:

mysumscriptii.m

```
% This script sums a vector from the Command Window

runsum = 0;
for i=1:length(vec)
    runsum = runsum + vec(i);
end
disp(runsum)
```

```
>> clear
>> vec = 1:7;
>> who
Your variables are:
vec

>> mysumscriptii
    28
>> who
Your variables are:
i    runsum    vec
```

Note

This however, is very poor programming style. It is much better to pass the vector *vec* to a function.

Because variables created in scripts and in the Command Window both use the base workspace, many programmers begin scripts with a **clear** command to eliminate variables that may have already been created elsewhere (either in the Command Window or in another script).

Instead of a program consisting of a script that calls other functions to do the work, in some cases programmers will write a ***main function*** to call the other functions. So, the program consists of all functions rather than one script and the rest functions. The reason for this is again because both scripts and the Command Window use the base workspace. By using only functions in a program, no variables are added to the base workspace.

It is possible in MATLAB as well in other languages, to have ***global variables*** that can be shared by functions without passing them. Although there are some cases in which using **global** variables is efficient, it is generally regarded as poor programming style and therefore will not be explained further here.

6.4.1 Persistent Variables

Normally, when a function stops executing, the local variables from that function are cleared. That means that every time a function is called, memory is allocated and used while the function is executing, but released when it ends. With variables that are declared as ***persistent variables***, however, the value is not cleared so the next time the function is called, the variable still exists and retains its former value.

The following program demonstrates this. The script calls a function *func1* which initializes a variable *count* to 0, increments it, and then prints the value. Every time this function is called, the variable is created, initialized to 0, changed to 1, and then cleared when the function exits. The script then calls a function *func2* which first declares a **persistent** variable *count*. If the variable has not yet been initialized, which will be the case the first time the function is called, it is initialized to 0. Then, like the first function, the variable is incremented and the value is printed. With the second function however, the variable remains with its value when the function exits, so the next time the function is called, the variable is incremented again.

persistex.m

```
% This script demonstrates persistent variables

% The first function has a variable "count"
fprintf('This is what happens with a "normal" variable:\n')
func1
func1

% The second function has a persistent variable "count"
fprintf('\nThis is what happens with a persistent variable:\n')
func2
func2
```

func1.m

```
function func1
% func1 increments a normal variable "count"
% Format func1 or func1()

count = 0;
count = count + 1;
fprintf('The value of count is %d\n',count)
end
```

func2.m

```
function func2
% func2 increments a persistent variable "count"
% Format func2 or func2()

persistent count % Declare the variable

if isempty(count)
    count = 0;
end
count = count + 1;
fprintf('The value of count is %d\n',count)
end
```

The line

```
persistent count
```

declares the variable *count* which allocates space for it but does not initialize it. The **if** statement then initializes it (the first time the function is called). In many languages, variables always have to be declared before they can be used; in MATLAB, this is true only for **persistent** variables.

The functions can be called from the script or from the Command Window, as shown. For example, the functions are called first from the script. With the **persistent** variable, the value of *count* is incremented. Then, *func1* is called from the Command Window and *func2* is also called from the Command Window. As the value of the **persistent** variable had the value 2, this time it is incremented to 3.

```
>> persistex
This is what happens with a "normal" variable:
The value of count is 1
The value of count is 1

This is what happens with a persistent variable:
The value of count is 1
The value of count is 2

>> func1
The value of count is 1

>> func2
The value of count is 3
```

As can be seen from this, every time the function *func1* is called, whether from *persistex* or from the Command Window, the value of 1 is printed. However, with *func2* the variable *count* is incremented every time it is called. It is first called in this example from *persistex* twice, so *count* is 1 and then 2. Then, when called from the Command Window, it is incremented to 3 (so it is counting how many times the function is called).

The way to restart a **persistent** variable is to use the **clear** function. The command

```
>> clear functions
```

will reinitialize all **persistent** variables (see **help clear** for more options). However, as of Version 2015a, a better method has been established which is to clear an individual function rather than all of them, e.g.,

```
>> clear func2
```

PRACTICE 6.6

The following function *posnum* prompts the user to enter a positive number and loops to error-check. It returns the positive number entered by the user. It calls a subfunction in the loop to print an error message. The subfunction has a **persistent** variable to count the number of times an error has occurred. Here is an example of calling the function:

```
>> enteredvalue = posnum
Enter a positive number: -5
Error # 1 ...    Follow instructions!
Does -5.00 look like a positive number to you?
Enter a positive number: -33
Error # 2 ...    Follow instructions!
Does -33.00 look like a positive number to you?
Enter a positive number: 6
enteredvalue =
     6
```

Fill in the subfunction below to accomplish this.

posnum.m

```
function num = posnum
% Prompt user and error-check until the
%  user enters a positive number
% Format posnum or posnum()

num = input('Enter a positive number: ');
while num < 0
        errorsubfn(num)
        num = input('Enter a positive number: ');
end
end

function errorsubfn(num)
% Fill this in


end
```

Of course, the numbering of the error messages will continue if the function is executed again without clearing it first.

6.5 DEBUGGING TECHNIQUES

Any error in a computer program is called a ***bug***. This term is thought to date back to the 1940s, when a problem with an early computer was found to have been caused by a moth in the computer's circuitry! The process of finding errors in a program and correcting them, is still called ***debugging***.

As we have seen, the **checkcode** function can be used to help find mistakes or potential problems in script and function files.

6.5.1 Types of Errors

There are several different kinds of errors that can occur in a program, which fall into the categories of ***syntax errors***, ***runtime errors***, and ***logical errors***.

Syntax errors are mistakes in using the language. Examples of syntax errors are missing a comma or a quotation mark, or misspelling a word. MATLAB itself will flag syntax errors and give an error message. For example, the following string is missing the end quote:

```
>> mystr = 'how are you;
mystr = 'how are you;
          |
Error: A MATLAB string constant is not terminated properly.
```

If this type of error is typed in a script or function using the Editor, the Editor will flag it.

Another common mistake is to spell a variable name incorrectly; MATLAB will also catch this error. Newer versions of MATLAB will typically be able to correct this for you, as in the following:

```
>> value = 5;
>> newvalue = valu + 3;
Undefined function or variable 'valu'.

Did you mean:
>> newvalue = value + 3;
```

Runtime or execution-time errors are found when a script or function is executing. With most languages, an example of a runtime error would be attempting to divide by zero. However, in MATLAB, this will return the constant Inf. Another example would be attempting to refer to an element in an array that does not exist.

runtimeEx.m

```
% This script shows an execution-time error

vec = 3:5;

for i = 1:4
    disp(vec(i))
end
```

The previous script initializes a vector with three elements, but then attempts to refer to a fourth. Running it prints the three elements in the vector, and then an error message is generated when it attempts to refer to the fourth element. Note that MATLAB gives an explanation of the error and gives the line number in the script in which the error occurred.

```
>> runtimeEx
     3
     4
     5
Attempted to access vec(4); index out of bounds because numel(vec)=3.

 Error in runtimeEx (line 6)
     disp(vec(i))
```

Logical errors are more difficult to locate because they do not result in any error message. A logical error is a mistake in reasoning by the programmer, but it is not a mistake in the programming language. An example of a logical error would be dividing by 2.54 instead of multiplying to convert inches to centimeters. The results printed or returned would be incorrect, but this might not be obvious.

All programs should be robust and should wherever possible anticipate potential errors, and guard against them. For example, whenever there is an input into a program, the program should error-check and make sure that the input is in the correct range of values. Also, before dividing, any denominator should be checked to make sure that it is not zero.

Despite the best precautions, there are bound to be errors in programs.

6.5.2 Tracing

Many times, when a program has loops and/or selection statements and is not running properly, it is useful in the debugging process to know exactly which statements have been executed. For example, the following is a function that attempts to display "In middle of range" if the argument passed to it is in the range from 3 to 6, and "Out of range" otherwise.

testifelse.m

```
function testifelse(x)
% testifelse will test the debugger
% Format: testifelse(Number)

if 3 < x < 6
    disp('In middle of range')
else
    disp('Out of range')
end
end
```

However, it seems to print "In middle of range" for all values of *x*:

```
>> testifelse(4)
In middle of range

>> testifelse(7)
In middle of range

>> testifelse(-2)
In middle of range
```

One way of following the flow of the function or ***tracing*** it, is to use the **echo** function. The **echo** function, which is a toggle, will display every statement as it is executed as well as results from the code. For scripts, just **echo** can be typed, but for functions, the name of the function must be specified. For example, the general form is

```
echo functionname on/off
```

For the *testifelse* function, it can be called as:

```
>> echo testifelse on
>> testifelse(-2)
% This function will test the debugger
if 3 < x < 6
    disp('In middle of range')
In middle of range
end
```

We can see from this result that the action of the if clause was executed.

6.5.3 Editor/Debugger

MATLAB has many useful functions for debugging, and debugging can also be done through its Editor, which is more properly called the Editor/Debugger.

Typing **help debug** at the prompt in the Command Window will show some of the debugging functions. Also, in the Help Documentation, typing "debugging" in the Search Documentation will display basic information about the debugging processes.

It can be seen in the previous example that the action of the if clause was executed and it printed "In middle of range," but just from that it cannot be determined why this happened. There are several ways to set ***breakpoints*** in a file (script or function) so that the variables or expressions can be examined. These can be done from the Editor/Debugger or commands can be typed from the Command Window. For example, the following **dbstop** command will set a breakpoint in the sixth line of this function (which is the action of the if clause), which allows the values of variables and/or expressions to be examined at that point in the execution. The function **dbcont** can be used to continue the execution and **dbquit** can be used to quit the debug mode. Note that the prompt becomes K>> in debug mode.

```
>> dbstop testifelse 6
>> testifelse(-2)
5         disp('In middle of range')
K>> x
x =
    -2

K>> 3 < x
ans =
     0

K>> 3 < x < 6
ans =
     1

K>> dbcont
In middle of range
end
>>
```

By typing the expressions $3 < x$ and then $3 < x < 6$, we can determine that the expression $3 < x$ will return either 0 or 1. Both 0 and 1 are less than 6, so the expression will always be **true** regardless of the value of x! Once in the debug mode, instead of using **dbcont** to continue the execution, **dbstep** can be used to step through the rest of the code one line at a time.

Breakpoints can also be set and cleared through the Editor. When a file is open in the Editor, in between the line numbers on the left and the lines of code is a thin gray strip which is the ***breakpoint alley***. In this, there are underscore marks next to the executable lines of code (as opposed to comments, for example). Clicking the mouse in the alley next to a line will create a breakpoint at that line (and then clicking on the red dot that indicates a breakpoint will clear it).

PRACTICE 6.7

The following script is bad code in several ways. Use **checkcode** first to check it for potential problems and then use the techniques described in this section to set breakpoints and check values of variables.

debugthis.m

```
for i = 1:5
    i = 3;
    disp(i)
end

for j = 2:4
    vec(j) = j
end
```

6.5.4 Function Stubs

Another common debugging technique that is used when there is a script main program that calls many functions is to use ***function stubs***. A function stub is a place holder, used so that the script will work even though that particular function hasn't been written yet. For example, a programmer might start with a script main program which consists of calls to three functions that accomplish all of the tasks.

mainscript.m

```
% This program gets values for x and y, and
%     calculates and prints z

[x, y] = getvals;
z = calcz(x,y);
printall(x,y,z)
```

The three functions have not yet been written however, so function stubs are put in place so that the script can be executed and tested. The function stubs consist of the proper function headers, followed by a simulation of what the function will eventually do. For example, the first two functions put arbitrary values in for the output arguments and the last function prints.

getvals.m

```
function [x, y] = getvals
x = 33;
y = 11;
end
```

calcz.m

```
function z = calcz(x,y)
z = x + y;
end
```

printall.m

```
function printall(x,y,z)
disp(x)
disp(y)
disp(z)
end
```

Then, the functions can be written and debugged one at a time. It is much easier to write a working program using this method than to attempt to write everything at once—then, when errors occur, it is not always easy to determine where the problem is!

6.6 LIVE SCRIPTS, CODE CELLS, AND PUBLISHING CODE

An entirely new type of script has been introduced in MATLAB as of Version 2016a. The script is called a ***live script*** and is created using the ***Live Editor***. A live script is much more dynamic than a simple script; it can embed equations, images, and hyperlinks in addition to formatted text. Instead of having graphs in separate windows, the graphics are shown next to the code that created them. It is also possible to put the graphs inline, under the code.

The scripts that we have seen so far have been simple scripts, stored in files that have an extension of .m. Live scripts are instead stored using the .mlx file format.

An example of a live script is shown in Fig. 6.1. In this live script named "sintest.mlx," there is text, followed by an equation, then code, more text, another equation, and more code. All of these are in separate sections. The sections are created by clicking on "code," "text," "equation" and so forth from the "Insert" section, with "section break" in between each type of section. All output, including error messages if there are any, are shown to the right of the code. In this case, the graphs produced by the code sections are shown to the right of the code. Clicking on the icon above the graphs will move them inline.

There are several ways to create a live script. The simplest is to click on New, then Live Script. A simple script can also be transformed into a live script by choosing Save As and then Live Script. Right clicking on a set of commands from the Command History window also pops up an option to save as a Live Script.

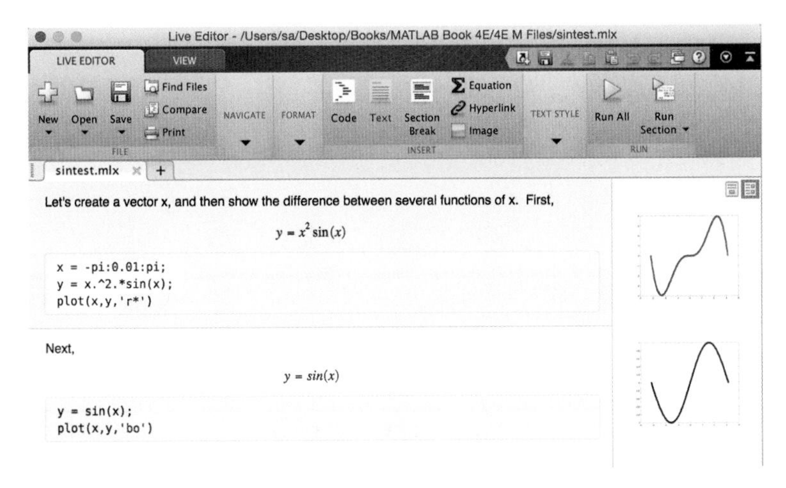

FIGURE 6.1
The Live Editor.

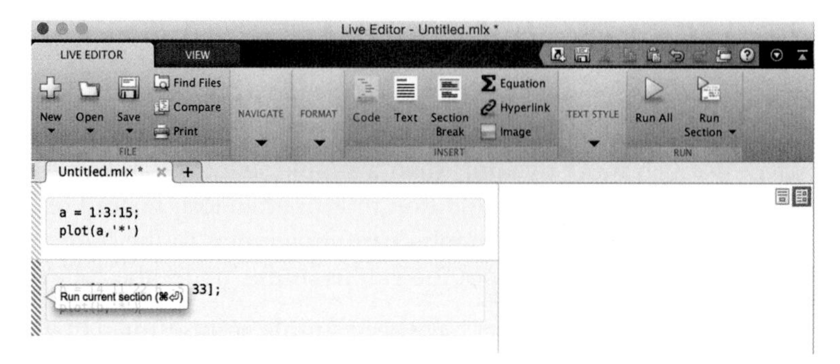

FIGURE 6.2
"Run current section" from Live Editor.

All of the code from the live script can be executed by choosing the Run All button. Alternatively, individual sections can be executed by clicking on the bar to the left of the section as seen in Fig. 6.2.

Once a live script has been completed, it can be shared with others as an.mlx file, or it can be converted to a PDF or HTML format. To do this, click on the down arrow under "Save," and then either "Export to PDF" or "Export to HTML."

Live scripts can also be converted to code files with the .m extension by choosing Save As and then choosing MATLAB Code file from the drop down menu for the Format.

Using the **type** command on a live script will show just the code sections. The entire contents of the .mlx file can be seen from the Live Editor.

```
>> type sintest.mlx

x = -pi:0.01:pi;
y = x.^2.*sin(x);
plot(x,y,'r*')
y = sin(x);
plot(x,y,'bo')
```

PRACTICE 6.8

If you have R2016a or later, try creating a live script with text, equations, and code to produce at least one plot.

6.6.1 Code Cells

With simple code file scripts, one can break the code into sections called ***code cells***. With code cells, you can run one cell at a time and you can also publish the code in an HTML format with plots embedded and with formatted equations.

To break code into cells, create comment lines that begin with two % symbols; these become the cell titles. For example, a script from Chapter 3 that plots sin and cos has been modified to have two cells: one that creates vectors for sin(x) and cos(x) and plots them; and a second that adds a legend, title, and axis labels to the plot.

sinncosCells.m

```
% This script plots sin(x) and cos(x) in the same Figure
% Window for values of x ranging from 0 to 2pi

%% Create vectors and plot
clf
x = 0: 2*pi/40: 2*pi;
y = sin(x);
plot(x,y,'ro')
hold on
y = cos(x);
plot(x,y,'b+')

%% Add legends, axis labels, and title
legend('sin', 'cos')
xlabel('x')
ylabel('sin(x) or cos(x)')
title('sin and cos on one graph')
```

When viewing this script in the Editor, the individual cells can be chosen by clicking the mouse anywhere within the cell. This will highlight the cell with a background color. Then from the Editor tab, you can choose "Run Section" to run just that one code cell and remain within that cell, or you can choose "Run and Advance" to run that code cell and then advance to the next.

By choosing the "Publish" tab and then "Publish," the code is published by default in HTML document. For the *plot1ptCells* script, this creates a document in which there is a table of contents (consisting of the two cell titles), the first code block which plots, followed by the actual plot, and then the second code block that annotates the Figure Window, followed by the modified plot.

■ Explore Other Interesting Features

From the Command Window, type **help debug** to find out more about the debugging and **help dbstop** in particular to find out more options for stopping code. Breakpoints can be set at specified locations in a file, only when certain condition(s) apply, and when errors occur.

Investigate the **dbstatus** function.

Explore the use of the functions **mlock** and **munlock** to block functions from being cleared using **clear**.

It is also possible to create code cells in functions. Investigate this. ■

SUMMARY

COMMON PITFALLS

- Not matching up arguments in a function call with the input arguments in a function header.
- Not having enough variables in an assignment statement to store all of the values returned by a function through the output arguments.
- Attempting to call a function that does not return a value from an assignment statement, or from an output statement.
- Not using the same name for the function and the file in which it is stored.
- Not thoroughly testing functions for all possible inputs and outputs.
- Forgetting that **persistent** variables are updated every time the function in which they are declared is called—whether from a script or from the Command Window.

PROGRAMMING STYLE GUIDELINES

- If a function is calculating one or more values, return these value(s) from the function by assigning them to output variable(s).
- Give the function and the file in which it is stored the same name.
- Function headers and function calls must correspond. The number of arguments passed to a function must be the same as the number of input arguments in the function header. If the function returns values, the number of variables in the left side of an assignment statement should match the number of output arguments returned by the function.
- If arguments are passed to a function in the function call, do not replace these values by using **input** in the function itself.
- Functions that calculate and return value(s) will not normally also print them.
- Functions should not normally be longer than one page in length.
- Do not declare variables in the Command Window and then use them in a script, or vice versa.
- Pass all values to be used in functions to input arguments in the functions.
- When writing large programs with many functions, start with the main program script and use function stubs, filling in one function at a time while debugging.

MATLAB Reserved Words	
global	persistent

MATLAB Functions and Commands	
echo	dbquit
dbstop	dbstep
dbcont	

MATLAB Operator	
> path for subfunction	%% code cell title

Exercises

1. Given the following function header:

```
function [x, y] = calcem(a, b, c)
```

   Which of the following function calls would be valid—and why?

```
[num, val] = calcem(4, 11, 2)
result = calcem(3, 5, 7)
```

2. Write a function that will receive as an input argument a number of kilometers (K). The function will convert the kilometers to miles and to US nautical miles, and return both results. The conversions are: 1 K=0.621 miles and 1 US nautical mile=1.852 K.
3. Write a function "splitem" that will receive one vector of numbers as an input argument, and will return two vectors: one with the positive (>=0) numbers from the original vector, and the second, the negative numbers from the original vector. Use vectorized code (no loops) in your function.
4. Write a function to calculate the volume and surface area of a hollow cylinder. It receives as input arguments the radius of the cylinder base and the height of the cylinder. The volume is given by $\pi r^2 h$, and the surface area is $2\pi rh$.

   Satellite navigation systems have become ubiquitous. Navigation systems based in space such as the Global Positioning System (GPS) can send data to handheld personal devices. The coordinate systems that are used to represent locations present this data in several formats.

5. The geographic coordinate system is used to represent any location on Earth as a combination of latitude and longitude values. These values are angles that can be written in the decimal degrees (DD) form or the degrees, minutes,

and seconds (DMS) form just like time. For example, 24.5° is equivalent to 24° 30′0″. Write a script that will prompt the user for an angle in DD form and will print in sentence format the same angle in DMS form. The script should error-check for invalid user input. The angle conversion is to be done by calling a separate function from the script.

6. Given the following function header:

```
function doit(a, b)
```

Which of the following function calls would be valid—and why?

```
fprintf('The result is %.1f\n', doit(4,11))
```

```
doit(5, 2, 11.11)
```

```
x = 11;
y = 3.3;
doit(x,y)
```

7. Write a function that prints the area and circumference of a circle for a given radius. Only the radius is passed to the function. The function does not return any values. The area is given by πr^2 and the circumference is $2\pi r$.
8. Write a function that will receive an integer *n* and a character as input arguments and will print the character *n* times.
9. Write a function that receives a matrix as an input argument and prints a random row from the matrix.
10. Write a function that receives a count as an input argument and prints the value of the count in a sentence that would read "It happened 1 time." if the value of the count is 1, or "It happened *xx* times." if the value of count (*xx*) is greater than 1.
11. Write a function that receives an *x* vector, a minimum value, and a maximum value, and plots **sin(x)** from the specified minimum to the specified maximum.
12. Write a function that prompts the user for a value of an integer *n* and returns the value of *n*. No input arguments are passed to this function. Error-check to make sure that an integer is entered.
13. Write a script that will:
    - Call a function to prompt the user for an angle in degrees
    - Call a function to calculate and return the angle in radians. (Note: π radians = 180°)
    - Call a function to print the result

    Write all of the functions, also. Note that the solution to this problem involves four code files: one which acts as a main program (the script), and three for the functions.
14. Modify the program in Exercise 13 so that the function to calculate the angle is a subfunction to the function that prints.

15. In 3D space, the Cartesian coordinates (x,y,z) can be converted to spherical coordinates (radius r, inclination θ, azimuth ϕ) by the following equations:

$$r = \sqrt{x^2 + y^2 + z^2}, \quad \theta = \cos^{-1}\left(\frac{z}{r}\right), \quad \phi = \tan^{-1}\left(\frac{y}{x}\right)$$

A program is being written to read in Cartesian coordinates, convert to spherical, and print the results. So far, a script *pracscript* has been written that calls a function *getcartesian* to read in the Cartesian coordinates and a function *printspherical* that prints the spherical coordinates. Assume that the *getcartesian* function exists and reads the Cartesian coordinates from a file. The function *printspherical* calls a subfunction *convert2spher* that converts from Cartesian to spherical coordinates. You are to write the *printspherical* function. Here is an example:

```
>> pracscript
The radius is 5.46
The inclination angle is 1.16
The azimuth angle is 1.07
```

pracscript.m

```
[x,y,z] = getcartesian();
printspherical(x,y,z)
```

getcartesian.m

```
function [x,y,z] = getcartesian()
% Assume this gets x,y,z from a file
end
```

16. The lump sum S to be paid when interest on a loan is compounded annually is given by $S = P(1 + i)^n$ where P is the principal invested, i is the interest rate, and n is the number of years. Write a program that will plot the amount S as it increases through the years from 1 to n. The main script will call a function to prompt the user for the number of years (and error-check to make sure that the user enters a positive integer). The script will then call a function that will plot S for years 1 through n. It will use 0.05 for the interest rate and $10,000 for P.
17. Write a program to write a length conversion chart to a file. It will print lengths in feet from 1 to an integer specified by the user, in one column and the corresponding length in meters (1 foot = 0.3048 m) in a second column. The main script will call one function that prompts the user for the maximum length in feet; this function must error-check to make sure that the user enters a valid positive integer. The script then calls a function to write the lengths to a file.
18. The script *circscript* loops n times to prompt the user for the circumference of a circle (where n is a random integer). Error-checking is ignored to focus on functions in this program. For each, it calls one function to calculate the radius and area of that circle, and then calls another function to print these values. The formulas are $r = c/(2\pi)$ and $a = \pi r^2$ where r is the radius, c is the circumference, and a is the area. Write the two functions.

circscript.m

```
n = randi(4);
for i = 1:n
   circ = input('Enter the circumference of the circle: ');
   [rad, area] = radarea(circ);
   dispra(rad,area)
end
```

19. The distance between any two points (x_1,y_1) and (x_2,y_2) is given by:

$$\text{distance} = \sqrt{(x_1 - x_2)^2 + (y_1 - y_2)^2}$$

The area of a triangle is:

$$\text{area} = \sqrt{s \times (s - a) \times (s - b) \times (s - c)}$$

where *a*, *b*, and *c* are the lengths of the sides of the triangle, and *s* is equal to half the sum of the lengths of the three sides of the triangle. Write a script that will prompt the user to enter the coordinates of three points that determine a triangle (e.g., the *x* and *y* coordinates of each point). The script will then calculate and print the area of the triangle. It will call one function to calculate the area of the triangle. This function will call a subfunction that calculates the length of the side formed by any two points (the distance between them).

20. Write a program to write a temperature conversion chart to a file. The main script will:
   - call a function that explains what the program will do
   - call a function to prompt the user for the minimum and maximum temperatures in degrees Fahrenheit and return both values. This function checks to make sure that the minimum is less than the maximum and calls a subfunction to swap the values if not.
   - call a function to write temperatures to a file: the temperature in degrees *F* from the minimum to the maximum in one column and the corresponding temperature in degrees *C* in another column. The conversion is $C = (F - 32) \times 5/9$.

21. Modify the function *func2* from Section 6.4.1 that has a **persistent** variable *count*. Instead of having the function print the value of *count*, the value should be returned.

22. Write a function *per2* that receives one number as an input argument. The function has a **persistent** variable that sums the values passed to it. Here are the first two times the function is called:

```
>> per2(4)
ans =
     4
```

```
>> per2(6)
ans =
    10
```

23. What would be the output from the following program? Think about it, write down your answer and then type it in to verify.

testscope.m

```
answer = 5;
fprintf('Answer is %d\n',answer)
pracfn
pracfn
fprintf('Answer is %d\n',answer)
printstuff
fprintf('Answer is %d\n',answer)
```

pracfn.m

```
function pracfn
persistent count
if isempty(count)
    count = 0;
end
count = count + 1;
fprintf('This function has been called %d times.\n',count)
end
```

printstuff.m

```
function printstuff
answer = 33;
fprintf('Answer is %d\n',answer)
pracfn
fprintf('Answer is %d\n',answer)
end
```

24. Assume a matrix variable *mat*, as in the following example:

```
mat =
     4     2     4     3     2
     1     3     1     0     5
     2     4     4     0     2
```

The following **for** loop

```
[r, c] = size(mat);
for i = 1:r
    sumprint(mat(i,:))
end
```

prints this result:

```
The sum is now 15
The sum is now 25
The sum is now 37
```

Write the function *sumprint*.

25. The following script *land* calls functions to:
    - prompt the user for a land area in acres
    - calculate and return the area in hectares and in square miles
    - print the results

    One acre is 0.4047 hectares. One square mile is 640 acres. Assume that the last function, that prints, exists—you do not have to do anything for that function. You are to write the entire function that calculates and returns the area in hectares and in square miles, and write just a function stub for the function that prompts the user and reads. Do not write the actual contents of this function; just write a stub!

    land.m

```
inacres = askacres;
[sqmil, hectares] = convacres(inacres);
dispareas(inacres, sqmil, hectares) % Assume this exists
```

26. The braking distance of a car depends on its speed as the brakes are applied and on the car's braking efficiency. A formula for the braking distance is

$$b_d = \frac{s^2}{2Rg}$$

    where b_d is the braking distance, s is the car's speed, R is the braking efficiency, and g is the acceleration due to gravity (9.81). A script has been written that calls a function to prompt the user for s and R, calls another function to calculate the braking distance and calls a third function to print the braking distance in a sentence format with one decimal place. You are to write a function stub for the function that prompts for s and R and the actual function definitions for the other two functions.

```
[s, R] = promptSandR;
brakDist = calcbd(s, R);
printbd(brakDist)
```

27. Write a menu-driven program to convert a time in seconds to other units (minutes, hours, and so on). The main script will loop to continue until the user chooses to exit. Each time in the loop, the script will generate a random time in seconds, call a function to present a menu of options, and print the converted time. The conversions must be made by individual functions (e.g., one to convert from seconds to minutes). All user-entries must be error-checked.
28. Write a menu-driven program to investigate the constant π. Model it after the program that explores the constant e. Pi (π) is the ratio of a circle's circumference to its diameter. Many mathematicians have found ways to approximate π. For example, Machin's formula is:

$$\frac{\pi}{4} = 4\arctan\left(\frac{1}{5}\right) - \arctan\left(\frac{1}{239}\right)$$

Leibniz found that π can be approximated by:

$$\pi = \frac{4}{1} - \frac{4}{3} + \frac{4}{5} - \frac{4}{7} + \frac{4}{9} - \frac{4}{11} + \cdots$$

This is called a sum of a series. There are six terms shown in this series. The first term is 4, the second term is −4/3, the third term is 4/5, and so forth. For example, the menu-driven program might have the following options:

- Print the result from Machin's formula.
- Print the approximation using Leibniz' formula, allowing the user to specify how many terms to use.
- Print the approximation using Leibniz' formula, looping until a "good" approximation is found.
- Exit the program.

29. Write a program to calculate the position of a projectile at a given time t. For an initial velocity v_0 and angle of departure θ_0, the position is given by x and y coordinates as follows (note: the gravity constant g is 9.81 m/s^2):

$$x = v_0 \cos(\theta_0) t$$
$$y = v_0 \sin(\theta_0) t - \frac{1}{2} g t^2$$

The program should initialize the variables for the initial velocity, time, and angle of departure. It should then call a function to find the x and y coordinates and then another function to print the results. If you have version R2016a or later, make the script into a live script.

CHAPTER 7

String Manipulation

KEY TERMS

string
substring
control characters
white space characters
leading blanks
trailing blanks
vectors of characters
empty string
string concatenation
delimiter
token

CONTENTS

A ***string*** in the MATLAB® software consists of any number of characters and is contained in single quotes. Actually, strings are vectors in which every element is a single character, which means that many of the vector operations and functions that we have already seen work with strings.

MATLAB also has many built-in functions that are written specifically to manipulate strings. In some cases strings contain numbers, and it is useful to convert from strings to numbers and vice versa; MATLAB has functions to do this as well.

There are many applications for using strings, even in fields that are predominantly numerical. For example, when data files consist of combinations of numbers and characters, it is often necessary to read each line from the file as a string, break the string into pieces, and convert the parts that contain numbers to number variables that can be used in computations. In this chapter the string manipulation techniques necessary for this will be introduced, and applications in file input/output will be demonstrated in Chapter 9.

7.1 CREATING STRING VARIABLES

A string consists of any number of characters (including, possibly, none). The following are examples of strings:

```
''
' '
'x'
'cat'
'Hello there'
'123'
```


MATLAB®. http://dx.doi.org/10.1016/B978-0-12-804525-1.00007-6

A ***substring*** is a subset or part of a string. For example, 'there' is a substring within the string 'Hello there'.

Characters include letters of the alphabet, digits, punctuation marks, white space, and control characters. ***Control characters*** are characters that cannot be printed, but accomplish a task (e.g., a backspace or tab). ***White space characters*** include the space, tab, newline (which moves the cursor down to the next line), and carriage return (which moves the cursor to the beginning of the current line). ***Leading blanks*** are blank spaces at the beginning of a string, for example, ' hello', and ***trailing blanks*** are blank spaces at the end of a string.

There are several ways by which string variables can be created. One is by using an assignment statement:

```
>> word = 'cat';
```

Another method is to read into a string variable. Recall that to read into a string variable using the **input** function, the second argument 's' must be included:

```
>> strvar = input('Enter a string: ', 's')
Enter a string: xyzabc
strvar =
xyzabc
```

If leading or trailing blanks are typed by the user, these will be stored in the string. For example, in the following, the user entered four blanks and then 'xyz':

```
>> s = input('Enter a string: ','s')
Enter a string:     xyz
s =
        xyz
```

7.1.1 Strings as Vectors

Strings are treated as ***vectors of characters*** or, in other words, a vector in which every element is a single character, so many vector operations can be performed. For example, the number of characters in a string can be found using the **length** function:

```
>> length('cat')
ans =
     3

>> length(' ')  % one space
ans =
     1

>> length('')  % empty string
ans =
     0
```

Note

There is a difference between an ***empty string***, which has a length of 0, and a string consisting of a blank space, which has a length of 1.

Expressions can refer to an individual element (a character within the string), or a subset of a string, or a transpose of a string:

```
>> mystr = 'Hi';
>> mystr(1)
ans =
H

>> mystr'
ans =
H
i

>> sent = 'Hello there';
>> length(sent)
ans =
      11
>> sent(4:8)
ans =
lo th
```

Note

A blank space in a string is a valid character within the string.

A character matrix can be created that consists of strings in every row. The following is created as a column vector of strings, but the end result is that it is a matrix in which every element is a character:

```
>> wordmat = ['Hello';'Howdy']
wordmat =
Hello
Howdy
>> size(wordmat)
ans =
    2     5
```

This created a *2 x 5* matrix of characters.

With a character matrix we can refer to an individual element (a single character) or an individual row (one of the strings):

```
>> wordmat(2,4)
ans =
d

>> wordmat(1,:)
ans =
Hello
```

As rows within a matrix must always be of the same length, the shorter strings must be padded with blanks so that all strings have the same length; otherwise, an error will occur.

```
>> greetmat = ['Hello'; 'Goodbye']      % Incorrect
Error using vertcat
Dimensions of matrices being concatenated are not consistent.
```

```
>> greetmat = ['Hello '; 'Goodbye']
greetmat =
Hello
Goodbye
>> size(greetmat)
ans =
     2     7
```

PRACTICE 7.1

Prompt the user for a string. Print the length of the string and also the first and last characters in the string. Make sure that this works regardless of what the user enters.

7.2 OPERATIONS ON STRINGS

MATLAB has many built-in functions that work with strings. Some of the string manipulation functions that perform the most common operations will be described here.

7.2.1 Concatenation

String concatenation means to join strings together. Of course, as strings are just vectors of characters, the method of concatenating vectors also works for strings. For example, to create one long string from two strings, it is possible to join them by putting them in square brackets:

```
>> first = 'Bird';
>> last = 'house';
>> [first last]
ans =
Birdhouse
```

Note that the variable names (or strings) must be separated by a blank space in the brackets, but there is no space in between the strings when they are concatenated.

The function **strcat** concatenates horizontally, meaning that it creates one longer string from the inputs.

```
>> first = 'Bird';
>> last = 'house';
>> strcat(first,last)
ans =
Birdhouse
```

If there are leading or trailing blanks in the strings, there is a difference between these two methods of concatenating. The method of using the square brackets will concatenate all of the characters in the strings, including all leading and trailing blanks.

```
>> str1 = 'xxx    ';
>> str2 = '    yyy';
>> [str1 str2]
ans =
xxx        yyy

>> length(ans)
ans =
    12
```

The **strcat** function, however, will remove trailing blanks (but not leading blanks) from strings before concatenating. Note that in these examples, the trailing blanks from *str1* are removed, but the leading blanks from *str2* are not:

```
>> strcat(str1,str2)
ans =
xxx    yyy

>> length(ans)
ans =
     9

>> strcat(str2,str1)
ans =
    yyyxxx

>> length(ans)
ans =
     9
```

We have seen already that the **char** function can be used to convert from an ASCII code to a character. The **char** function can also be used to concatenate vertically, meaning that it will create a column vector of strings (or, in other words, create a matrix of characters). When using the **char** function to create a matrix, it will automatically pad the strings in the rows with trailing blanks as necessary so that they are all of the same length.

```
>> clear greetmat
>> greetmat = char('Hello','Goodbye')
greetmat =
Hello
Goodbye
>> size(greetmat)
ans =
     2     7
```

PRACTICE 7.2

Create a variable that stores the word 'northeast'. From this, create two separate variables *v1* and *v2* that store 'north' and 'east'. Then, create a matrix consisting of the values of *v1* and *v2* in separate rows.

7.2.2 Creating Customized Strings

There are several built-in functions that create customized strings, including **blanks** and **sprintf**.

The **blanks** function will create a string consisting of *n* blank characters (which are kind of hard to see!). However, in MATLAB, if the mouse is moved to highlight the result in *ans*, the blanks can be seen.

```
>> b = blanks(4)
b =

>> length(b)
ans =
     4
```

It is usually most useful to use this function when concatenating strings, and a number of blank spaces is desired in between. For example, this will insert five blank spaces in between the words:

```
>> [first blanks(5) last]
ans =
Bird     house
```

Displaying the transpose of the string resulting from the **blanks** function can also be used to move the cursor down. In the Command Window it would look like this:

```
>> disp(blanks(4)')




>>
```

This is useful in a script or function to create space in output, and is equivalent to printing the newline character four times.

The **sprintf** function works exactly like the **fprintf** function, *but instead of printing it creates a string*. Here are several examples in which the output is not suppressed so the value of the string variable is shown:

```
>> sent1 = sprintf('The value of pi is %.2f', pi)
sent1 =
The value of pi is 3.14

>> sent2 = sprintf('Some numbers: %5d, %2d', 33, 6)
sent2 =
Some numbers:    33,  6
```

```
>> length(sent2)
ans =
    23
```

In the following example, however, the output of the assignment is suppressed so the string is created including a random integer and stored in the string variable. Then, some exclamation points are concatenated to that string.

```
>> phrase = sprintf('A random integer is %d', ...
                   randi([5,10]));
>> strcat(phrase, '!!!')
ans =
A random integer is 7!!!
```

All of the formatting options that can be used in the **fprintf** function can also be used in the **sprintf** function.

7.2.2.1 Applications of Customized Strings: Prompts, Labels, and Arguments to Functions

One very useful application of the **sprintf** function is to create customized strings, including formatting and/or numbers that are not known ahead of time (e.g., entered by the user or calculated). These customized strings can then be passed to other functions, for example, for plot titles or axis labels. For example, assume that a file "expnoanddata.dat" stores an experiment number, followed by the experiment data. In this case the experiment number is "123," and then the rest of the file consists of the actual data.

```
123   4.4   5.6   2.5   7.2   4.6   5.3
```

The following script would load these data and plot them with a title that includes the experiment number.

plotexpno.m

```
% This script loads a file that stores an experiment number
% followed by the actual data. It plots the data and puts
% the experiment # in the plot title

load expnoanddata.dat
experNo = expnoanddata(1);
data = expnoanddata(2:end);
plot(data,'ko')
xlabel('Sample #')
ylabel('Weight')
title(sprintf('Data from experiment %d', experNo))
```

The script loads all numbers from the file into a row vector. It then separates the vector; it stores the first element, which is the experiment number, in a variable *experNo*, and the rest of the vector in a variable *data* (the rest being from the second element to the end). It then plots the data, using **sprintf** to create the title, which includes the experiment number as seen in Fig. 7.1.

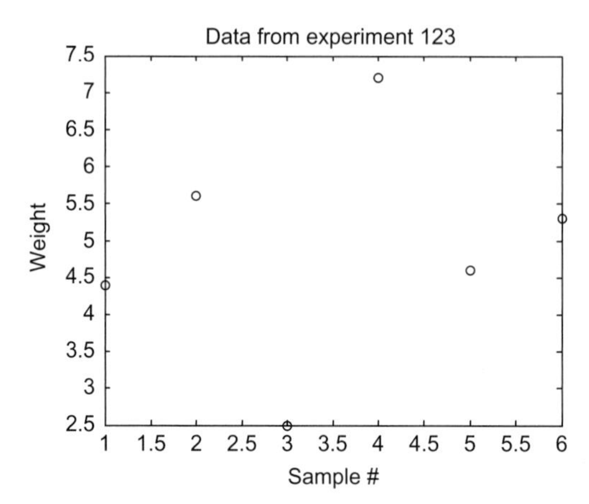

FIGURE 7.1
Customized title in plot using **sprintf**.

PRACTICE 7.3

In a loop, create and print strings with file names "file1.dat," "file2.dat," and so on for file numbers 1 through 5.

QUICK QUESTION!

How could we use the **sprintf** function to customize prompts for the **input** function?

Answer: For example, if it is desired to have the contents of a string variable printed in a prompt, **sprintf** can be used:

```
>> username = input('Please enter your name: ', 's');
Please enter your name: Bart

>> prompt = sprintf('%s, Enter your id #: ',username);
>> id_no = input(prompt)
Bart, Enter your id #: 177
id_no =
   177
```

Another way of accomplishing this (in a script or function) would be:

```
fprintf('%s, Enter your id #: ',username);
id_no = input('');
```

Note that the calls to the **sprintf** and **fprintf** functions are identical except that the **fprintf** prints (so there is no need for a prompt in the **input** function), whereas the **sprintf** creates a string that can then be displayed by the **input** function. In this case using **sprintf** seems cleaner than using **fprintf** and then having an empty string for the prompt in **input**.

As another example, the following program prompts the user for endpoints (x_1, y_1) and (x_2, y_2) of a line segment, and calculates the midpoint of the line segment, which is the point (x_m, y_m). The coordinates of the midpoint are found by:

$$x_m = \frac{1}{2}(x_1 + x_2) \qquad y_m = \frac{1}{2}(y_1 + y_2)$$

The script *midpoint* calls a function *entercoords* to separately prompt the user for the x and y coordinates of the two endpoints, calls a function *findmid* twice to calculate separately the x and y coordinates of the midpoint, and then prints this midpoint. When the program is executed, the output looks like this:

```
>> midpoint
Enter the x coord of the first endpoint: 2
Enter the y coord of the first endpoint: 4
Enter the x coord of the second endpoint: 3
Enter the y coord of the second endpoint: 8
The midpoint is (2.5, 6.0)
```

In this example, the word 'first' or 'second' is passed to the *entercoords* function so that it can use whichever word is passed in the prompt. The prompt is customized using **sprintf**.

midpoint.m

```
% This program finds the midpoint of a line segment

[x1, y1] = entercoords('first');
[x2, y2] = entercoords('second');

midx = findmid(x1,x2);
midy = findmid(y1,y2);

fprintf('The midpoint is (%.1f, %.1f)\n',midx,midy)
```

entercoords.m

```
function [xpt, ypt] = entercoords(word)
% entercoords reads in & returns the coordinates of
%   the specified endpoint of a line segment
% Format: entercoords(word) where word is 'first'
%         or 'second'

prompt = sprintf('Enter the x coord of the %s endpoint: ', ...
     word);
xpt = input(prompt);

prompt = sprintf('Enter the y coord of the %s endpoint: ', ...
     word);
ypt = input(prompt);
end
```

findmid.m

```
function mid = findmid(pt1,pt2)
% findmid calculates a coordinate (x or y) of the
%   midpoint of a line segment
% Format: findmid(coord1, coord2)

mid = 0.5 * (pt1 + pt2);
end
```

7.2.3 Removing White Space Characters

MATLAB has functions that will remove trailing blanks from the end of a string and/or leading blanks from the beginning of a string.

The **deblank** function will remove blank spaces from the end of a string. For example, if some strings are padded in a character matrix so that all are of the same length, it is frequently desired to then remove those extra blank spaces to use the string in its original form.

```
>> names = char('Sue', 'Cathy','Xavier')
names =
Sue
Cathy
Xavier

>> name1 = names(1,:)
name1 =
Sue
>> length(name1)
ans =
     6
>> name1 = deblank(name1);
>> length(name1)
ans =
     3
```

Note
The **deblank** function only removes trailing blanks from a string, not leading blanks.

The **strtrim** function will remove both leading and trailing blanks from a string, but not blanks in the middle of the string. In the following example, the three blanks in the beginning and four blanks in the end are removed, but not the two blanks in the middle. Highlighting the result in the Command Window with the mouse would show the blank spaces.

```
>> strvar = [blanks(3) 'xx' blanks(2) 'yy' blanks(4)]
strvar =
   xx  yy
```

```
>> length(strvar)
ans =
    13
>> strtrim(strvar)
ans =
xx  yy
>> length(ans)
ans =
     6
```

7.2.4 Changing Case

MATLAB has two functions that convert strings to all uppercase letters, or lowercase, called **upper** and **lower**.

```
>> mystring = 'AbCDEfgh';
>> lower(mystring)
ans =
abcdefgh
>> upper(ans)
ans =
ABCDEFGH
```

PRACTICE 7.4

Assume that these expressions are typed sequentially in the Command Window. Think about it, write down what you think the results will be, and then verify your answers by actually typing them.

```
lnstr = '1234567890';
mystr = '  abc   xy';
newstr = strtrim(mystr)
length(newstr)
upper(newstr(1:3))
sprintf('Number is %4.1f', 3.3)
```

7.2.5 Comparing Strings

There are several functions that compare strings and return **logical true** if they are equivalent, or **logical false** if not. The function **strcmp** compares strings, character by character. It returns **logical true** if the strings are completely identical (which infers that they must also be of the same length), or **logical false** if the strings are not of the same length or any corresponding characters are not identical. Note that for strings, these functions are used to determine whether strings are equal to each other or not, not the equality operator ==. Here are some examples of these comparisons:

```
>> word1 = 'cat';
>> word2 = 'car';
>> word3 = 'cathedral';
>> word4 = 'CAR';
>> strcmp(word1,word3)
ans =
     0
>> strcmp(word1,word1)
ans =
     1
>> strcmp(word2,word4)
ans =
     0
```

The function **strncmp** compares only the first *n* characters in strings and ignores the rest. The first two arguments are the strings to compare and the third argument is the number of characters to compare (the value of *n*).

```
>> strncmp(word1,word3,3)
ans =
     1
```

QUICK QUESTION!

How can we compare strings, ignoring whether the characters in the string are uppercase or lowercase?

Answer: See the following Programming Concept and Efficient Method.

THE PROGRAMMING CONCEPT

Ignoring the case when comparing strings can be done by changing all characters in the strings to either upper- or lowercase, for example, in MATLAB using the **upper** or **lower** function:

```
>> strcmp(upper(word2),upper(word4))
ans =
     1
```

THE EFFICIENT METHOD

The function **strcmpi** compares the strings but ignores the case of the characters.

```
>> strcmpi(word2,word4)
ans =
     1
```

There is also a function **strncmpi,** which compares *n* characters, ignoring the case.

7.2.6 Finding, Replacing, and Separating Strings

There are functions that find and replace strings, or parts of strings, within other strings and functions that separate strings into substrings.

The function **strfind** receives two strings as input arguments. The general form is **strfind(string, substring)**; it finds all occurrences of the substring within the string, and returns the subscripts of the beginning of the occurrences. The substring can consist of one character, or any number of characters. If there is more than one occurrence of the substring within the string, **strfind** returns a vector with all indices. Note that what is returned is the index of the beginning of the substring.

```
>> strfind('abcde', 'd')
ans =
     4
>> strfind('abcde', 'bc')
ans =
     2
>> strfind('abcdeabcdedd', 'd')
ans =
     4     9    11    12
```

If there are no occurrences, the empty vector is returned.

```
>> strfind('abcdeabcde','ef')
ans =
     []
```

QUICK QUESTION!

How can you find how many blanks there are in a string (e.g., 'how are you')?

Answer: The **strfind** function will return an index for every occurrence of a substring within a string, so the result is a vector of indices. The **length** of this vector of indices would be the number of occurrences. For example, the following finds the number of blank spaces in the variable *phrase*.

```
>> phrase = 'Hello, and how are you doing?';
>> length(strfind(phrase,' '))
ans =
     5
```

If it is desired to get rid of any leading and trailing blanks first (in case there are any), the **strtrim** function would be used first.

```
>> phrase = ' Well, hello there! ';
>> length(strfind(strtrim(phrase),' '))
ans =
     2
```

Let's expand this and write a script that creates a vector of strings that are phrases. The output is not suppressed so that the strings can be seen when the script is executed. It loops through this vector and passes each string to a function *countblanks*. This function counts the number of blank spaces in the string, not including any leading or trailing blanks.

phraseblanks.m

```
% This script creates a column vector of phrases
% It loops to call a function to count the number
%  of blanks in each one and prints that

phrasemat = char('Hello and how are you?', ...
  'Hi there everyone!', 'How is it going?', 'Whazzup?')
[r c]  = size(phrasemat);

for i = 1:r
    % Pass each row (each string) to countblanks function
    howmany = countblanks(phrasemat(i,:));
    fprintf('Phrase %d had %d blanks\n',i,howmany)
end
```

countblanks.m

```
function num = countblanks(phrase)
% countblanks returns the # of blanks in a trimmed string
% Format: countblanks(string)

num = length(strfind(strtrim(phrase), ' '));
end
```

For example, running this script would result in:

```
>> phraseblanks
phrasemat =
Hello and how are you?
Hi there everyone!
How is it going?
Whazzup?

Phrase 1 had 4 blanks
Phrase 2 had 2 blanks
Phrase 3 had 3 blanks
Phrase 4 had 0 blanks
```

The function **strrep** finds all occurrences of a substring within a string, and replaces them with a new substring. The order of the arguments matters. The format is

```
strrep(string, oldsubstring, newsubstring)
```

The following replaces all occurrences of the substring 'e' with the substring 'x':

```
>> strrep('abcdeabcde','e','x')
ans =
abcdxabcdx
```

All strings can be of any length, and the lengths of the old and new substrings do not have to be the same. If the old substring is not found, nothing is changed in the original string.

In addition to the string functions that find and replace, there is a function that separates a string into two substrings. The **strtok** function breaks a string into two pieces; it can be called several ways. The function receives one string as an input argument. It looks for the first ***delimiter***, which is a character or set of characters that act as a separator within the string.

By default, the delimiter is any white space character. The function returns a ***token*** that is the beginning of the string, up to (but not including) the first delimiter. It also returns the rest of the string, which includes the delimiter. Assigning the returned values to a vector of two variables will capture both of these. The format is

```
[token, rest] = strtok(string)
```

where *token* and *rest* are variable names. For example,

```
>> sentence1 = 'Hello there';
>> [word, rest] = strtok(sentence1)
word =
Hello
rest =
 there

>> length(word)
ans =
     5
>> length(rest)
ans =
     6
```

Note that the rest of the string includes the blank space delimiter.

Alternate delimiters can be defined. The format

```
[token, rest] = strtok(string, delimeters)
```

returns a token that is the beginning of the string, up to the first character contained within the delimiters string, and also the rest of the string. In the following example, the delimiter is the character 'l'.

```
>> [word, rest] = strtok(sentence1,'l')
word =
He
rest =
llo there
```

Leading delimiter characters are ignored, whether it is the default white space or a specified delimiter. For example, the leading blanks are ignored here:

```
>> [firstpart, lastpart] = strtok(' materials science')
firstpart =
materials

lastpart =
 science
```

QUICK QUESTION!

What do you think **strtok** returns if the delimiter is not in the string?

Answer: The first result returned will be the entire string, and the second will be the empty string.

```
>> [first, rest] = strtok('ABCDE')
first =
ABCDE

rest =
   Empty string: 1-by-0
```

PRACTICE 7.5

Think about what would be returned by the following sequence of expressions and statements, and then type them into MATLAB to verify your results.

```
dept = 'Electrical';
strfind(dept,'e')

strfind(lower(dept),'e')

phone_no = '703-987-1234';
[area_code, rest] = strtok(phone_no,'-')

rest = rest(2:end)

strcmpi('Hi','HI')
```

QUICK QUESTION!

The function **date** returns the current date as a string (e.g., '10-Dec-2015'). How could we write a function to return the day, month, and year as separate output arguments?

Answer: We could use **strrep** to replace the '-' characters with blanks and then use **strtok** with the blank as the default delimiter to break up the string (twice) or, more simply, we could just use **strtok** and specify the '-' character as the delimiter.

separatedate.m

```
function [todayday, todaymo, todayyr] = separatedate
% separatedate separates the current date into day,
% month, and year
% Format: separatedate or separatedate()

[todayday, rest] = strtok(date,'-');
[todaymo, todayyr] = strtok(rest,'-');
todayyr = todayyr(2:end);
end
```

As we need to separate the string into three parts, we need to use the **strtok** function twice. The first time the string is separated into '10' and '-Dec-2015' using **strtok**. Then, the second string is separated into 'Dec' and '-2015' using **strtok**. (As leading delimiters are ignored the second '-' is found as the delimiter in '-Dec-2015'.) Finally, we need to remove the '-' from the string '-2015'; this can be done by just indexing from the second character to the end of the string.

An example of calling this function follows:

```
>> [d, m, y] = separatedate
d =
10
m =
Dec
y =
2015
```

Note that no input arguments are passed to the *separatedate* function; instead, the **date** function returns the current date as a string. Also, note that all three output arguments are strings.

7.2.7 Evaluating a String

The function **eval** is used to evaluate a string. If the string contains a call to a function, then that function call is executed. For example, in the following, the string 'plot(x)' is evaluated to be a call to the **plot** function, and it produces the plot shown in Fig. 7.2.

```
>> x = [2 6 8 3];
>> eval('plot(x)')
```

The **eval** function is frequently used when input is used to create a customized string. In the following example, the user chooses the type of plot to use for some quiz grades. The string that the user enters (in this case, 'bar') is concatenated with the string '(x)' to create the string 'bar(x)'; this is then evaluated as a call to the **bar** function as seen in Fig. 7.3. The name of the plot type is also used in the title.

```
>> x = [9 7 10 9];
>> whatplot = input('What type of plot?: ', 's');
What type of plot?: bar
>> eval([whatplot '(x)'])
>> title(whatplot)
>> xlabel('Student #')
>> ylabel('Quiz Grade')
```

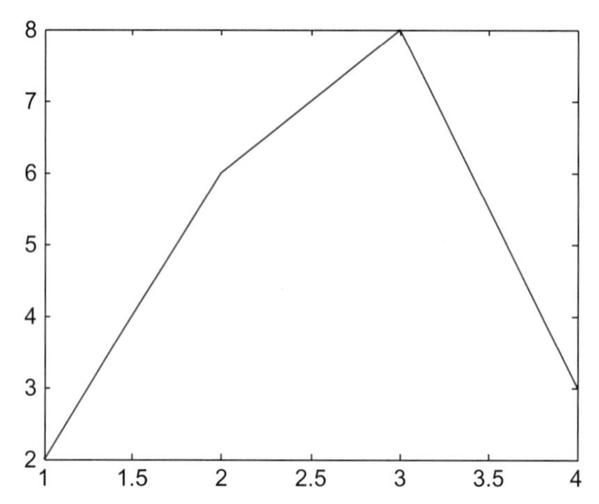

FIGURE 7.2
Plot type passed to the **eval** function.

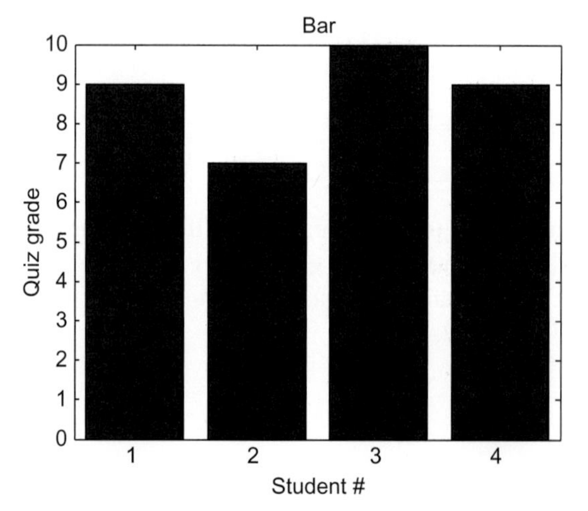

FIGURE 7.3
Plot type entered by the user.

PRACTICE 7.6

Create an *x* vector. Prompt the user for 'sin', 'cos', or 'tan' and create a string with that function of *x* (e.g., 'sin(x)' or 'cos(x)'). Use **eval** to create a *y* vector using the specified function.

The **eval** function is very powerful, but it is usually more efficient to avoid using it.

7.3 THE "IS" FUNCTIONS FOR STRINGS

There are several "is" functions for strings, which return **logical true** or **false**. The function **isletter** returns **logical true** for every character in a string if the character is a letter of the alphabet or **false** if not. The function **isspace** returns **logical true** for every character that is a white space character.

```
>> isletter('EK127')
ans =
     1    1    0    0    0

>> isspace('a b')
ans =
     0    1    0
```

The **isstrprop** function determines whether the characters in a string are in a category specified by a second argument. For example, the following tests to see whether or not the characters are alphanumeric; all are except for the dot '.'.

```
>> isstrprop('AB123.4','alphanum')
ans =
     1    1    1    1    1    0    1
```

The **ischar** function will return **logical true** if the vector argument is a character vector (in other words, a string), or **logical false** if not.

```
>> vec = 'EK127';
>> ischar(vec)
ans =
     1

>> vec = 3:5;
>> ischar(vec)
ans =
     0
```

7.4 CONVERTING BETWEEN STRING AND NUMBER TYPES

MATLAB has several functions that convert numbers to strings in which each character element is a separate digit and vice versa.

Note
These are different from the functions such as **char** and **double** that convert characters to ASCII equivalents and vice versa.

To convert numbers to strings, MATLAB has the functions **int2str** for integers and **num2str** for real numbers (which also works with integers). The function **int2str** would convert, for example, the integer 38 to the string '38'.

```
>> num = 38;
num =
    38
>> s1 = int2str(num)
s1 =
38

>> length(num)
ans =
     1
>> length(s1)
ans =
     2
```

The variable *num* is a scalar that stores one number, whereas *s1* is a string that stores two characters, '3' and '8'.

Even though the result of the first two assignments is "38," note that the indentation in the Command Window is different for the number and the string.

The **num2str** function, which converts real numbers, can be called in several ways. If only the real number is passed to the **num2str** function, it will create a string that has four decimal places, which is the default in MATLAB for displaying real numbers. The precision can also be specified (which is the number of digits), and format strings can also be passed, as shown in the following:

```
>> str2 = num2str(3.456789)
str2 =
3.4568
>> length(str2)
ans =
     6

>> str3 = num2str(3.456789,3)
str3 =
3.46
```

```
>> str = num2str(3.456789,'%6.2f')
str =
3.46
```

Note that in the last example, MATLAB removed the leading blanks from the string.

The functions **str2double** and **str2num** do the reverse; they takes a string in which number(s) are stored and converts them to the type **double**:

```
>> num = str2double('123.456')
num =
   123.4560
```

If there is a string in which there are numbers separated by blanks, the **str2num** function will convert this to a vector of numbers (of the default type **double**). For example,

```
>> mystr = '66 2 111';
>> numvec = str2num(mystr)
numvec =
     66     2   111

>> sum(numvec)
ans =
   179
```

The **str2double** function is a better function to use in general than **str2num**, but it can only be used when a scalar is passed; it would not work, for example, for the variable *mystr* above.

PRACTICE 7.7

Think about what would be returned by the following sequence of expressions and statements, and then type them into MATLAB to verify your results.

```
vec = 'yes or no';
isspace(vec)

all(isletter(vec) ~= isspace(vec))

ischar(vec)

nums = [33 1.5];
num2str(nums)

nv = num2str(nums)

sum(nums)
```

QUICK QUESTION!

Let's say that we have a string that consists of an angle followed by either 'd' for degrees or 'r' for radians. For example, it may be a string entered by the user:

```
degrad = input('Enter angle and d/r: ', 's');
Enter angle and d/r: 54r
```

How could we separate the string into the angle and the character, and then get the sine of that angle using either **sin** or **sind**, as appropriate (**sin** for radians or **sind** for degrees)?

Answer: First, we could separate this string into its two parts:

```
>> angle = degrad(1:end-1)
angle =
54
>> dorr = degrad(end)
dorr =
r
```

Then, using an **if-else** statement, we would decide whether to use the **sin** or **sind** function, based on the value of the variable *dorr*. Let's assume that the value is 'r' so we want to use **sin**. The variable *angle* is a string so the following would not work:

```
>> sin(angle)
Undefined function 'sin' for input arguments
of type 'char'.
```

Instead, we could either use **str2double** to convert the string to a number. A complete script to accomplish this is shown here.

angleDorR.m

```
% Prompt the user for angle and 'd' for degrees
% or 'r' for radians; print the sine of the angle

% Read in the response as a string and then
% separate the angle and character
degrad = input('Enter angle and d/r: ', 's');
angle = degrad(1:end-1);
dorr = degrad(end);

% Error-check to make sure user enters 'd' or 'r'
while dorr ~= 'd' && dorr ~= 'r'
   disp('Error! Enter d or r with the angle.')
   degrad = input('Enter angle and d/r: ', 's');
   angle = degrad(1:end-1);
   dorr = degrad(end);
end
% Convert angle to number
anglenum = str2double(angle);
fprintf('The sine of %.1f ', anglenum)
% Choose sin or sind function
if dorr == 'd'
   fprintf('degrees is %.3f.\n', sind(anglenum))
else
   fprintf('radians is %.3f.\n', sin(anglenum))
end
```

```
>> angleDorR
Enter angle and d/r: 3.1r
The sine of 3.1 radians is 0.042.
```

```
>> angleDorR
Enter angle and d/r: 53t
Error! Enter d or r with the angle.
Enter angle and d/r: 53d
The sine of 53.0 degrees is 0.799.
```

■ Explore Other Interesting Features

In many of the search and replace functions, search patterns can be specified which use ***regular expressions***. Use **help** to find out about these patterns.

Explore the **sscanf** function, which reads data from a string.

Explore the **strjust** function, which justifies a string.

Explore the **mat2str** function, to convert from a matrix to a string.

Explore the properties that can be examined with the **isstrprop** function.

Investigate why the string compare functions are used to compare strings, rather than the equality operator. ■

SUMMARY

COMMON PITFALLS

- Putting arguments to **strfind** in incorrect order.
- Trying to use == to compare strings for equality, instead of the **strcmp** function (or its variations).
- Confusing **sprintf** and **fprintf**. The syntax is the same, but **sprintf** creates a string whereas **fprintf** prints.
- Trying to create a vector of strings with varying lengths (the easiest way is to use **char** which will pad with extra blanks automatically).
- Forgetting that when using **strtok**, the second argument returned (the "rest" of the string) contains the delimiter.
- When breaking a string into pieces, forgetting to convert the numbers in the strings to actual numbers that can then be used in calculations.

PROGRAMMING STYLE GUIDELINES

- Trim trailing blanks from strings that have been stored in matrices before using.
- Make sure that the correct string comparison function is used, for example, **strcmpi** if ignoring case is desired.

MATLAB Functions and Commands			
strcat	lower	strrep	isstrprop
blanks	strcmp	strtok	ischar
sprintf	strncmp	date	int2str
deblank	strcmpi	eval	num2str
strtrim	strncmpi	isletter	str2double
upper	strfind	isspace	str2num

Exercises

1. A file name is supposed to be in the form *filename.ext*. Write a function that will determine whether a string is in the form of a name followed by a dot followed by a three-character extension, or not. The function should return 1 for **logical true** if it is in that form, or 0 for **false** if not.
2. The following script calls a function *getstr* that prompts the user for a string, error-checking until the user enters something (the error would occur if the user just hits the Enter key without any other characters first). The script then prints the length of the string. Write the *getstr* function.

```
thestring = getstr();
fprintf('Thank you, your string is %d characters long\n', ...
    length(thestring))
```

3. Write a script that will, in a loop, prompt the user for four course numbers. Each will be a string of length 5 of the form 'CS101'. These strings are to be stored in a character matrix.
4. Write a function that will generate two random integers, each in the inclusive range from 10 to 30. It will then return a string consisting of the two integers joined together, for example, if the random integers are 11 and 29, the string that is returned will be '1129'.
5. Write a script that will create *x* and *y* vectors. Then, it will ask the user for a color ('red,' 'blue,' or 'green') and for a plot style (circle or star). It will then create a string *pstr* that contains the color and plot style, so that the call to the **plot** function would be: **plot(x,y,pstr)**. For example, if the user enters 'blue' and '*', the variable *pstr* would contain 'b*'.
6. Assume that you have the following function and that it has not yet been called.

strfunc.m

```
function strfunc(instr)
persistent mystr
if isempty(mystr)
    mystr = '';
end
mystr = strcat(instr,mystr);
fprintf('The string is %s\n',mystr)
end
```

What would be the result of the following sequential expressions?

```
strfunc('hi')
strfunc('hello')
```

7. Explain in words what the following function accomplishes (not step-by-step, but what the end result is).

dostr.m

```
function out = dostr(inp)
persistent str
[w, r] = strtok(inp);
str = strcat(str,w);
out = str;
end
```

8. Write a function that will receive a name and department as separate strings and will create and return a code consisting of the first two letters of the name and the last two letters of the department. The code should be upper-case letters. For example,

```
>> namedept('Robert','Mechanical')
ans =
ROAL
```

9. Write a function "createUniqueName" that will create a series of unique names. When the function is called, a string is passed as an input argument. The functions adds an integer to the end of the string, and returns the resulting string. Every time the function is called, the integer that it adds is incremented. Here are some examples of calling the function:

```
>> createUniqueName('hello')
ans =
hello1
>> varname = createUniqueName('variable')
varname =
variable2
```

10. What does the **blanks** function return when a 0 is passed to it? A negative number? Write a function *myblanks* that does exactly the same thing as the **blanks** function, using the programming method. Here are some examples of calling it:

```
>> fprintf('Here is the result:%s!\n', myblanks(0))
Here is the result:!

>> fprintf('Here is the result:%s!\n', myblanks(7))
Here is the result:       !
```

11. Write a function that will prompt the user separately for a filename and extension and will create and return a string with the form 'filename.ext'.
12. Write a function that will receive one input argument, which is an integer *n*. The function will prompt the user for a number in the range from 1 to *n* (the actual value of *n* should be printed in the prompt) and return the user's input. The function should error-check to make sure that the user's input is in the correct range.
13. Write a script that will generate a random integer, ask the user for a field width, and print the random integer with the specified field width. The script will use **sprintf** to create a string such as 'The # is %4d\n' (for e.g., if, the user entered 4

for the field width) which is then passed to the **fprintf** function. To print (or create a string using **sprintf**) either the % or \ character, there must be two of them in a row.

14. Write a function called *plotsin* that will graphically demonstrate the difference in plotting the sin function with a different number of points in the range from 0 to 2 Π. The function will receive two arguments, which are the number of points to use in two different plots of the sin function. For example, the following call to the function:

```
>> plotsin(5,30)
```

will result in Fig. 7.4 in which the first plot has five points altogether in the range from 0 to 2 Π, inclusive, and the second has 30:

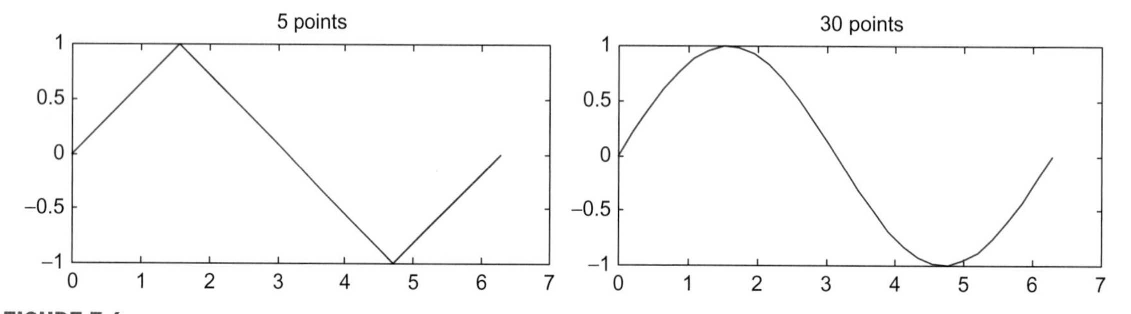

FIGURE 7.4
Subplot with sin.

15. If the strings passed to **strfind** are of the same length, what are the only two possible results that could be returned?
16. Vectorize this:

```
while mystrn(end) == ' '  % Note one space in quotes
     mystrn = mystrn(1:end-1);
end
```

17. Vectorize this:

```
loc = findstr(sentence, ' '); % one blank space
where = loc(1);
first = sentence(1:where-1);
last = sentence(where:end);
```

18. Vectorize this:

```
vec = [];
for i = 1:8
   vec = [ vec ' ']; % one blank space
end
vec   % just for display
```

19. Vectorize this:

```
if length(str1) ~= length(str2)
    outlog = false;
else
    outlog = true;
    for i=1:length(str1)
        if str1(i)  ~= str2(i)
            outlog = false;
        end
    end
end
outlog % Just to display the value
```

20. Write a function *nchars* that will create a string of *n* characters, without using any loops or selection statements.

```
>> nchars('*', 6)
ans =
******
```

21. Write a function that will receive two input arguments: a character matrix that is a column vector of strings, and a string. It will loop to look for the string within the character matrix. The function will return the row number in which the string is found if it is in the character matrix, or the empty vector if not. Use the programming method.
22. Write a function *rid_multiple_blanks* that will receive a string as an input argument. The string contains a sentence that has multiple blank spaces in between some of the words. The function will return the string with only one blank in between words. For example,

```
>> mystr = 'Hello  and how  are  you?';
>> rid_multiple_blanks(mystr)
ans =
Hello and how are you?
```

23. Words in a string variable are separated by right slashes (/) instead of blank spaces. Write a function *slashtoblank* that will receive a string in this form and will return a string in which the words are separated by blank spaces. This should be general and work regardless of the value of the argument. No loops allowed in this function; the built-in string function(s) must be used.
24. Two variables store strings that consist of a letter of the alphabet, a blank space, and a number (in the form 'R 14.3'). Write a script that would initialize two such variables. Then, use string manipulating functions to extract the numbers from the strings and add them together.

Cryptography, or encryption, is the process of converting plaintext (e.g., a sentence or paragraph), into something that should be unintelligible, called the ciphertext. The reverse process is code-breaking, or cryptanalysis, which relies on searching the encrypted message for weaknesses and deciphering it from that point. Modern security systems are heavily reliant on these processes.

25. In cryptography, the intended message sometimes consists of the first letter of every word in a string. Write a function *crypt* that will receive a string with the encrypted message and return the message.

```
>> estring = 'The early songbird tweets';
>> m = crypt(estring)
m =
Test
```

26. Using the functions **char** and **double**, one can shift words. For example, one can convert from lower case to upper case by subtracting 32 from the character codes:

```
>> orig = 'ape';
>> new = char(double(orig)-32)
new =
APE

>> char(double(new)+32)
ans =
ape
```

We've "encrypted" a string by altering the character codes. Figure out the original string. Try adding and subtracting different values (do this in a loop) until you decipher it:

```
Jmkyvih$mx$syx$}ixC
```

27. Load files named *file1.dat*, *file2.dat*, and so on in a loop. To test this, create just two files with these names in your Current Folder first.
28. Either in a script or in the Command Window, create a string variable that stores a string in which numbers are separated by the character 'x', for example, '12x3x45x2'. Create a vector of the numbers, and then get the sum (Note: For the example given it would be 62 but the solution should be general).
29. Create the following two variables:

```
>> var1 = 123;
>> var2 = '123';
```

Then, add 1 to each of the variables. What is the difference?

30. The built-in **clock** function returns a vector with six elements representing the year, month, day, hours, minutes, and seconds. The first five elements are integers whereas the last is a **double** value, but calling it with **fix** will convert all to integers. The built-in **date** function returns the day, month, and year as a string. For example,

```
>> fix(clock)
ans =
     2013    4    25    14    25    49
>> date
ans =
25-Apr-2013
```

Write a script that will call both of these built-in functions, and then compare results to make sure that the year is the same. The script will have to convert one from a string to a number, or the other from a number to a string in order to compare.

31. Use **help isstrprop** to find out what properties can be tested; try some of them on a string variable.
32. Find out how to pass a vector of integers to **int2str** or real numbers to **num2str**.
33. Write a script that will first initialize a string variable that will store *x* and *y* coordinates of a point in the form 'x 3.1 y 6.4'. Then, use string manipulating functions to extract the coordinates and plot them.
34. Write a function *wordscramble* that will receive a word in a string as an input argument. It will then randomly scramble the letters and return the result. Here is an example of calling the function:

```
>> wordscramble('fantastic')
ans =
safntcait
```

Massive amounts of temperature data have been accumulated and stored in files. To be able to comb through this data and gain insights into global temperature variations, it is often useful to visualize the information.

35. A file called *avehighs.dat* stores for three locations, the average high temperatures for each month for a year (rounded to integers). There are three lines in the file; each stores the location number followed by the 12 temperatures (this format may be assumed). For example, the file might store:

```
432   33 37 42 45 53 72 82 79 66 55 46 41
777   29 33 41 46 52 66 77 88 68 55 48 39
567   55 62 68 72 75 79 83 89 85 80 77 65
```

Write a script that will read these data in and plot the temperatures for the three locations separately in one Figure Window. A **for** loop must be used to accomplish this. For example, if the data are as shown in the previous data block, the Figure Window would appear as in Fig. 7.5. The axis labels and titles should be as shown.

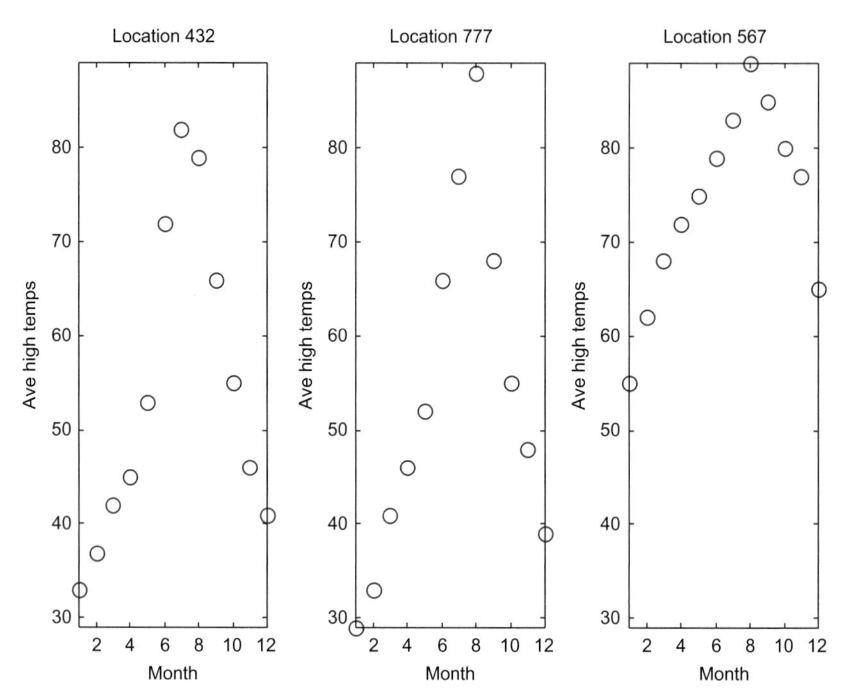

FIGURE 7.5
Subplot to display data from file using a **for** loop.

If you have version R2016a or later, write the script as a live script.

CHAPTER 8

Data Structures

KEY TERMS

data structures	tables	nested structure
cell array	sorting	ordinal categorical arrays
structures	content indexing	descending order
fields	cell indexing	ascending order
database	comma-separated list	selection sort
record	dot operator	dynamic field names
categorical arrays	vector of structures	index vector

CONTENTS

Data structures are variables that store more than one value. For it to make sense to store more than one value in a variable, the values should somehow be logically related. There are many different kinds of data structures. We have already been working with one kind, arrays (e.g., vectors and matrices). An array is a data structure in which all of the values are logically related in that they are of the same type and represent, in some sense, "the same thing." So far, that has been true for the vectors and matrices that we have used. We use vectors and matrices when we want to be able to loop through them (or, essentially, have this done for us using vectorized code).

A ***cell array*** is a kind of data structure that stores values of different types. Cell arrays can be vectors or matrices; the different values are referred to as the elements of the array. One very common use of a cell array is to store strings of different lengths. Cell arrays actually store pointers to the stored data.

Structures are data structures that group together values that are logically related, but are not the same thing and not necessarily the same type. The different values are stored in separate ***fields*** of the structure.

One use of structures is to set up a ***database*** of information. For example, a professor might want to store for every student in a class: the student's name, university identifier number, grades on all assignments and quizzes, and so

MATLAB®. http://dx.doi.org/10.1016/B978-0-12-804525-1.00008-8

forth. In many programming languages and database programs, the terminology is that within a database file there would be one ***record*** of information for each student; each separate piece of information (name, quiz 1 score, and so on) would be called a ***field*** of the record. In the MATLAB® software these records are called **structs**.

Both cell arrays and structures can be used to store values that are of different types in a single variable. The main difference between them is that cell arrays are indexed, and can therefore be used with loops or vectorized code. Structures, however, are not indexed; the values are referenced using the names of the fields, which can be more mnemonic than indexing.

Other, more advanced, data structures are also covered in this chapter. These include ***categorical arrays*** and ***tables***. Categorical arrays are a type of array that allows one to store a finite, countable number of different possible values. A table is a data structure that stores information in a table format with rows and columns, each of which can be mnemonically labeled.

Finally, ***sorting*** the various types of data structures will be covered, both programmatically and using built-in sort functions. With a database, it is frequently useful to have it sorted on multiple fields, but this can be time-consuming. The use of index vectors is also presented as an alternative to physically sorting a database.

8.1 CELL ARRAYS

One type of data structure that MATLAB has, but is not found in many programming languages, is a ***cell array***. A cell array in MATLAB is an array, but, unlike the vectors and matrices we have used so far, elements in cell arrays are cells that can store different types of values.

8.1.1 Creating Cell Arrays

There are several ways to create cell arrays. For example, we will create a cell array in which one element will store an integer, one element will store a character, one element will store a vector, and one element will store a string. Just like with the arrays we have seen so far, this could be a 1×4 row vector, a 4×1 column vector, or a 2×2 matrix. Some of the syntax for creating vectors and matrices is the same as before in that values within rows are separated by spaces or commas, and rows are separated by semicolons. However, for cell arrays, curly braces are used rather than square brackets. For example, the following creates a row vector cell array with four different types of values:

```
>> cellrowvec = {23, 'a', 1:2:9, 'hello'}
cellrowvec =
    [23]    'a'     [1x5 double]     'hello'
```

To create a column vector cell array, the values are instead separated by semicolons:

```
>> cellcolvec = {23; 'a'; 1:2:9; 'hello'}
cellcolvec =
     [        23]
     'a'
     [1x5 double]
     'hello'
```

This method creates a 2×2 cell array matrix:

```
>> cellmat = {23 'a'; 1:2:9 'hello'}
cellmat =
     [        23]    'a'
     [1x5 double]    'hello'
```

The type of cell arrays is **cell**.

```
>> class(cellmat)
ans =
cell
```

Another method of creating a cell array is to simply assign values to specific array elements and build it up element-by-element. However, as explained before, extending an array element-by-element is a very inefficient and time-consuming method.

It is much more efficient, if the size is known ahead of time, to preallocate the array. For cell arrays, this is done with the **cell** function. For example, to preallocate a variable *mycellmat* to be a 2×2 cell array, the **cell** function would be called as follows:

```
>> mycellmat = cell(2,2)
mycellmat =
       []      []
       []      []
```

Note that this is a function call, so the arguments to the function are in parentheses; a matrix is created in which all of the elements are empty vectors. Then, each element can be replaced by the desired value.

How to refer to each element to accomplish this will be explained next.

8.1.2 Referring to and Displaying Cell Array Elements and Attributes

Just like with the other vectors we have seen so far, we can refer to individual elements of cell arrays. However, with cell arrays, there are two different ways to do this. The elements in cell arrays are cells. These cells can contain different types of values. With cell arrays, you can refer to the cells, or to the contents of the cells.

Using curly braces for the subscripts will reference the contents of a cell; this is called ***content indexing***. For example, this refers to the contents of the second element of the cell array *cellrowvec; ans* will have the type **char**:

```
> cellrowvec{2}
ans =
a
```

Row and column subscripts are used to refer to the contents of an element in a matrix (again using curly braces):

```
>> cellmat{1,1}
ans =
    23
```

Values can be assigned to cell array elements. For example, after preallocating the variable *mycellmat* in the previous section, the elements can be initialized:

```
>> mycellmat{1,1} = 23
mycellmat =
    [23]      []
      []      []
```

Using parentheses for the subscripts references the cells; this is called ***cell indexing***. For example, this refers to the second cell in the cell array *cellrowvec; ans* will be a *1* × *1* **cell** array:

```
>> onec = cellcolvec(2)
onec =
    'a'
>> class(onec)
ans =
cell
```

When an element of a cell array is itself a data structure, only the type of the element is displayed when the cells are shown. For example, in the previous cell arrays, the vector is shown just as "*1* × *5* double" (this is a high-level view of the cell array). This is what would be displayed with cell indexing; content indexing would display its contents:

```
>> cellmat(2,1)
ans =
    [1x5 double]

>> cellmat{2,1}
ans =
     1     3     5     7     9
```

Since this results in a vector, parentheses can be used to refer to its elements. For example, the fourth element of the vector is:

```
>> cellmat{2,1}(4)
ans =
     7
```

Note that the index into the cell array is given in curly braces; parentheses are then used to refer to an element of the vector.

One can also refer to subsets of cell arrays, such as in the following:

```
>> cellcolvec{2:3}
ans =
a

ans =
     1     3     5     7     9
```

Note, however, that MATLAB stored *cellcolvec*{2} in the default variable *ans*, and then replaced that with the value of *cellcolvec*{3}. Using content indexing returns them as a ***comma-separated list***. However, they could be stored in two separate variables by having a vector of variables on the left side of an assignment:

```
>> [c1, c2] = cellcolvec{2:3}
c1 =
a

c2 =
     1     3     5     7     9
```

Using cell indexing, the two cells would be put in a new cell array (in this case, in *ans*):

```
>> cellcolvec(2:3)
ans =
    'a'
    [1x5 double]
```

There are several methods for displaying cell arrays. The **celldisp** function displays the contents of all elements of the cell array:

```
>> celldisp(cellrowvec)
cellrowvec{1} =
    23

cellrowvec{2} =
a

cellrowvec{3} =
     1     3     5     7     9

cellrowvec{4} =
hello
```

The function **cellplot** puts a graphical display of the cell array into a Figure Window; however, it is a high-level view and basically just displays the same information as typing the name of the variable (so, for instance, it would not show the contents of the vector in the previous example). In other words, it shows the cells, not their contents.

Many of the functions and operations on arrays that we have already seen also work with cell arrays. For example, here are some related to dimensioning:

```
>> length(cellrowvec)
ans =
     4

>> size(cellcolvec)
ans =
     4     1

>> cellrowvec{end}
ans =
hello
```

To delete an element from a vector cell array, use cell indexing:

```
>> cellrowvec
mycell =

    [23]      'a'      [1x5 double]      'hello'

>> cellrowvec(2) = []
cellrowvec =

    [23]      [1x5 double]      'hello'
```

For a matrix, an entire row or column can be deleted using cell indexing:

```
>> cellmat
mycellmat =
    [         23]   'a'
    [1x5 double]   'hello'

>> cellmat(1,:) = []
mycellmat =
    [1x5 double]   'hello'
```

8.1.3 Storing Strings in Cell Arrays

One useful application of a cell array is to store strings of different lengths. As cell arrays can store different types of values, strings of different lengths can be stored in the elements.

```
>> names = {'Sue', 'Cathy', 'Xavier'}
names =

    'Sue'    'Cathy'    'Xavier'
```

This is extremely useful because, unlike vectors of strings created using **char**, these strings do not have extra trailing blanks. The length of each string can be displayed using a for loop to loop through the elements of the cell array:

```
>> for i = 1:length(names)
      disp(length(names{i}))
   end
    3

    5

    6
```

It is possible to convert from a cell array of strings to a character array, and vice versa. MATLAB has several functions that facilitate this. For example, the function **cellstr** converts from a character array padded with blanks to a cell array in which the trailing blanks have been removed.

```
>> greetmat = char('Hello','Goodbye');

>> cellgreets = cellstr(greetmat)
cellgreets =
     'Hello'
     'Goodbye'
```

The **char** function can convert from a cell array to a character matrix:

```
>> names = {'Sue', 'Cathy', 'Xavier'};
>> cnames = char(names)
cnames =

Sue
Cathy
Xavier

>> size(cnames)
ans =
     3     6
```

The functions **strjoin** and **strsplit** were introduced in R2013a. The function **strjoin** will concatenate, or join, all strings from a cell array into one string separated by one space each by default (but other delimiters can be specified).

```
>> longer = strjoin(names)
longer =
Sue Cathy Xavier
>> thisday = strjoin({'August','14','2016'},'-')
thisday =
August-14-2016
```

The function **strsplit** will essentially do the opposite; it splits a string into elements in a cell array with either a specified delimiter or a blank space by default.

```
>> ca = strsplit(thisday, '-')
ca =
    'August'    '14'    '2016'
>> strsplit(longer)
 ans =
    'Sue'    'Cathy'    'Xavier'
```

The function **iscellstr** will return **logical true** if a cell array is a cell array of all strings or **logical false** if not.

```
>> iscellstr(names)
ans =
     1

>> iscellstr(cellcolvec)
ans =
     0
```

We will see several examples that utilize cell arrays containing strings of varying lengths in later chapters, including advanced file input functions and customizing plots.

PRACTICE 8.1

Write an expression that would display a random element from a cell array (without assuming that the number of elements in the cell array is known). Create two different cell arrays and try the expression on them to make sure that it is correct.

For more practice, write a function that will receive one cell array as an input argument and will display a random element from it.

8.2 STRUCTURES

Structures are data structures that group together values that are logically related in what are called ***fields*** of the structure. An advantage of structures is that the fields are named, which helps to make it clear what values are stored in the structure. However, structure variables are not arrays. They do not have elements that are indexed, so it is not possible to loop through the values in a structure or to use vectorized code.

8.2.1 Creating and Modifying Structure Variables

Creating structure variables can be accomplished by simply storing values in fields using assignment statements or by using the **struct** function.

In our first example, assume that the local computer super mart wants to store information on the software packages that it sells. For each one, they will store the following:

- item number
- cost to the store
- price to the customer
- character code indicating the type of software

An individual structure variable for a given software package might look like this:

package			
item_no	cost	price	code
123	19.99	39.95	g

The name of the structure variable is *package*; it has four fields: *item_no, cost, price,* and *code.*

One way to initialize a structure variable is to use the **struct** function. The names of the fields are passed as strings; each one is followed by the value for that field (so, pairs of field names and values are passed to **struct**).

Note

Some programmers use names that begin with an uppercase letter for structure variables (e.g., *Package*) to make them easily distinguishable.

```
>> package = struct('item_no',123,'cost',19.99,...
      'price',39.95,'code','g')

package =
     item_no: 123
        cost: 19.9900
       price: 39.9500
        code: 'g'
```

Note that in the Workspace Window, the variable *package* is listed as a 1×1 struct; the type of the variable is **struct**.

```
>> class(package)
ans =
struct
```

MATLAB, as it is written to work with arrays, assumes the array format. Just like a single number is treated as a 1×1 double, a single structure is treated as a 1×1 struct. Later in this chapter we will see how to work more generally with vectors of structs.

An alternative method of creating this structure, which is not as efficient, involves using the ***dot operator*** to refer to fields within the structure. The name of the structure variable is followed by a dot, or period, and then the name of the field within that structure. Assignment statements can be used to assign values to the fields.

```
>> package.item_no = 123;
>> package.cost = 19.99;
>> package.price = 39.95;
>> package.code = 'g';
```

By using the dot operator in the first assignment statement, a structure variable is created with the field *item_no*. The next three assignment statements add more fields to the structure variable. Again, extending the structure in this manner is not as efficient as using **struct**.

Adding a field to a structure later is accomplished as shown here, by using an assignment statement.

An entire structure variable can be assigned to another. This would make sense, for example, if the two structures had some values in common. Here, for example, the values from one structure are copied into another and then two fields are selectively changed, referring to them using the dot operator.

```
>> newpack = package;
>> newpack.item_no = 111;
>> newpack.price = 34.95
newpack =
    item_no: 111
       cost: 19.9900
      price: 34.9500
       code: 'g'
```

To print from a structure, the **disp** function will display either the entire structure or an individual field.

```
>> disp(package)
    item_no: 123
       cost: 19.9900
      price: 39.9500
       code: 'g'

>> disp(package.cost)
   19.9900
```

However, using **fprintf,** only individual fields can be printed; the entire structure cannot be printed without referring to all fields individually.

```
>> fprintf('%d %c\n', package.item_no, package.code)
123 g
```

The function **rmfield** removes a field from a structure. It returns a new structure with the field removed, but does not modify the original structure (unless the returned structure is assigned to that variable). For example, the following would remove the *code* field from the *newpack* structure, but store the resulting structure in the default variable *ans*. The value of *newpack* remains unchanged.

```
>> rmfield(newpack, 'code')
ans =
   item_no: 111
      cost: 19.9900
     price: 34.9500
```

```
>> newpack
newpack =
     item_no: 111
        cost: 19.9000
       price: 34.9500
        code: 'g'
```

To change the value of *newpack*, the structure that results from calling **rmfield** must be assigned to *newpack*.

```
>> newpack = rmfield(newpack, 'code')
newpack =
     item_no: 111
        cost: 19.9000
       price: 34.9500
```

PRACTICE 8.2

A silicon wafer manufacturer stores, for every part in its inventory, a part number, quantity in the factory, and the cost for each.

onepart

part_no	quantity	costper
123	4	33.95

Create this structure variable using **struct**. Print the cost in the form $xx.xx.

8.2.2 Passing Structures to Functions

An entire structure can be passed to a function or individual fields can be passed. For example, here are two different versions of a function that calculates the profit on a software package. The profit is defined as the price minus the cost.

In the first version, the entire structure variable is passed to the function, so the function must use the dot operator to refer to the *price* and *cost* fields of the input argument.

calcprof.m

```
function profit = calcprof(packstruct)
% calcprofit calculates the profit for a
% software package
% Format: calcprof(structure w/ price & cost fields)

profit = packstruct.price - packstruct.cost;
end
```

```
>> calcprof(package)
ans =
   19.9600
```

In the second version, just the *price* and *cost* fields are passed to the function using the dot operator in the function call. These are passed to two scalar input arguments in the function header, so there is no reference to a structure variable in the function itself, and the dot operator is not needed in the function.

calcprof2.m

```
function profit = calcprof2(oneprice, onecost)
% Calculates the profit for a software package
% Format: calcprof2(price, cost)

profit = oneprice - onecost;
end
```

```
>> calcprof2(package.price, package.cost)
ans =
    19.9600
```

It is important, as always with functions, to make sure that the arguments in the function call correspond one-to-one with the input arguments in the function header. In the case of *calcprof*, a structure variable is passed to an input argument, which is a structure. For the second function *calcprof2*, two individual fields, which are **double** values, are passed to two **double** input arguments.

8.2.3 Related Structure Functions

There are several functions that can be used with structures in MATLAB. The function **isstruct** will return **logical** 1 for **true** if the variable argument is a structure variable or 0 if not. The **isfield** function returns **logical true** if a fieldname (as a string) is a field in the structure argument or **logical false** if not.

```
>> isstruct(package)
ans =
      1

>> isfield(package,'cost')
ans =
      1
```

The **fieldnames** function will return the names of the fields that are contained in a structure variable.

```
>> pack_fields = fieldnames(package)
pack_fields =
      'item_no'
      'cost'
      'price'
      'code'
```

As the names of the fields are of varying lengths, the **fieldnames** function returns a cell array with the names of the fields.

Curly braces are used to refer to the elements, as *pack_fields* is a cell array. For example, we can refer to the length of one of the field names:

```
>> length(pack_fields{2})
ans =
     4
```

QUICK QUESTION!

How can we ask the user for a field in a structure and either print its value or an error if it is not actually a field?

Answer: The **isfield** function can be used to determine whether or not it is a field of the structure. Then, by concatenating that field name to the structure variable and a dot, and then passing the entire string to **eval**, the expression would be evaluated as the actual field in the structure. The following is the code for the variable *package*:

```
inputfield = input('Which field would you like to see: ','s');
if isfield(package, inputfield)
    fprintf('The value of the %s field is: ', ...inputfield)
    disp(eval(['package.' inputfield]))
else
    fprintf('Error: %s is not a valid field\n', inputfield)
end
```

The code would produce this output (assuming the *package* variable was initialized as shown previously):

```
Which field would you like to see: cost
The value of the cost field is: 19.99
```

PRACTICE 8.3

Modify the code from the preceding Quick Question! to use **sprintf**.

8.2.4 Vectors of Structures

In many applications, including database applications, information would normally be stored in a ***vector of structures***, rather than in individual structure variables. For example, if the computer super mart is storing information on all of the software packages that it sells, it would likely be in a vector of structures such as the following:

packages

	item_no	cost	price	code
1	123	19.99	39.95	g
2	456	5.99	49.99	l
3	587	11.11	33.33	w

In this example, *packages* is a vector that has three elements. It is shown as a column vector. Each element is a structure consisting of four fields: *item_no*, *cost*, *price*, and *code*. It may look like a matrix, which has rows and columns, but it is, instead, a vector of structures.

This vector of structures can be created in several ways. One method is to create a structure variable, as shown earlier, to store information on one software package.

This can then be expanded to be a vector of structures.

```
>> packages = struct('item_no',123,'cost',19.99,...
     'price',39.95,'code','g');
>> packages(2) = struct('item_no',456,'cost', 5.99,...
     'price',49.99,'code','l');
>> packages(3) = struct('item_no',587,'cost',11.11,...
     'price',33.33,'code','w');
```

The first assignment statement shown here creates the first structure in the structure vector, the next one creates the second structure, and so on. This actually creates a 1×3 row vector.

Alternatively, the first structure could be treated as a vector to begin with, for example

```
>> packages(1) = struct('item_no',123,'cost',19.99,...
     'price',39.95,'code','g');
>> packages(2) = struct('item_no',456,'cost', 5.99,...
     'price',49.99,'code','l');
>> packages(3) = struct('item_no',587,'cost',11.11,...
     'price',33.33,'code','w');
```

Both of these methods, however, involve extending the vector. As we have already seen, preallocating any vector in MATLAB is more efficient than extending it. There are several methods of preallocating the vector. By starting with the last element, MATLAB would create a vector with that many elements. Then, the elements from 1 through end-1 could be initialized. For example, for a vector of structures that has three elements, start with the third element.

```
>> packages(3) = struct('item_no',587,'cost',11.11,...
     'price',33.33,'code','w');
>> packages(1) = struct('item_no',123,'cost',19.99,...
     'price',39.95,'code','g');
>> packages(2) = struct('item_no',456,'cost', 5.99,...
     'price',49.99,'code','l');
```

Another method is to create one element with the values from one structure, and use **repmat** to replicate it to the desired size. The remaining elements can then be modified. The following creates one structure and then replicates this into a 1×3 matrix.

```
>> packages = repmat(struct('item_no',123,'cost',19.99,...
     'price',39.95,'code','g'),1,3);
>> packages(2) = struct('item_no',456,'cost', 5.99,...
     'price',49.99,'code','l');
>> packages(3) = struct('item_no',587,'cost',11.11,...
     'price',33.33,'code','w');
```

Also, the vector of structures can be preallocated without assigning any values.

```
>> packages(3) = ...
   struct('item_no',[],'cost',[],'price',[],'code',[])
packages =
1x3 struct array with fields:
    item_no
    cost
    price
    code
```

Then, the values in the individual structures could be replaced in any order as above.

Typing the name of the variable will display only the size of the structure vector and the names of the fields:

```
>> packages
packages =
1x3 struct array with fields:
    item_no
    cost
    price
    code
```

The variable *packages* is now a vector of structures, so each element in the vector is a structure. To display one element in the vector (one structure), an index into the vector would be specified. For example, to refer to the second element:

```
>> packages(2)
ans =
    item_no: 456
       cost: 5.9900
      price: 49.9900
       code: 'l'
```

To refer to a field, it is necessary to refer to the particular structure, and then the field within it. This means using an index into the vector to refer to the structure, and then the dot operator to refer to a field. For example:

```
>> packages(1).code
ans =
g
```

Thus, there are essentially three levels to this data structure. The variable *packages* is the highest level, which is a vector of structures. Each of its elements is an individual structure. The fields within these individual structures are the lowest level. The following loop displays each element in the *packages* vector.

```
>> for i = 1:length(packages)
      disp(packages(i))
   end

   item_no: 123
      cost: 19.9900
     price: 39.9500
      code: 'g'

   item_no: 456
      cost: 5.9900
     price: 49.9900
      code: 'l'

   item_no: 587
      cost: 11.1100
     price: 33.3300
      code: 'w'
```

To refer to a particular field for all structures, in most programming languages it would be necessary to loop through all elements in the vector and use the dot operator to refer to the field for each element. However, this is not the case in MATLAB.

THE PROGRAMMING CONCEPT

For example, to print all of the costs, a **for** loop could be used:

```
>> for i=1:3
      fprintf('%f\n',packages(i).cost)
   end
19.990000
5.990000
11.110000
```

THE EFFICIENT METHOD

However, **fprintf** would do this automatically in MATLAB:

```
>> fprintf('%f\n',packages.cost)
19.990000
5.990000
11.110000
```

Using the dot operator in this manner to refer to all values of a field would result in the values being stored successively in the default variable *ans* as this method results in a comma-separated list:

```
>> packages.cost
ans =
   19.9900

ans =
    5.9900

ans =
   11.1100
```

However, the values can all be stored in a vector:

```
>> pc = [packages.cost]
pc =
   19.9900    5.9900   11.1100
```

Using this method, MATLAB allows the use of functions on all of the same fields within a vector of structures. For example, to sum all three cost fields, the vector of cost fields is passed to the **sum** function:

```
>> sum([packages.cost])
ans =
   37.0900
```

For vectors of structures, the entire vector (e.g., *packages*) could be passed to a function, or just one element (e.g., *packages(1)*) which would be a structure, or a field within one of the structures (e.g., *packages(2).price*).

The following is an example of a function that receives the entire vector of structures as an input argument, and prints all of it in a nice table format.

printpackages.m

```
function printpackages(packstruct)
% printpackages prints a table showing all
% values from a vector of 'packages' structures
% Format: printpackages(package structure)

fprintf('\nItem # Cost Price Code\n\n')
no_packs = length(packstruct);
for i = 1:no_packs
    fprintf('%6d %6.2f %6.2f %3c\n', ...
        packstruct(i).item_no, ...
        packstruct(i).cost, ...
        packstruct(i).price, ...
        packstruct(i).code)
end
end
```

The function loops through all of the elements of the vector, each of which is a structure, and uses the dot operator to refer to and print each field. An example of calling the function follows:

```
>> printpackages(packages)

Item #   Cost   Price   Code

   123  19.99   39.95    g
   456   5.99   49.99    l
   587  11.11   33.33    w
```

PRACTICE 8.4

A silicon wafer manufacturer stores, for every part in their inventory, a part number, how many are in the factory, and the cost for each. First, create a vector of structs called *parts* so that when displayed it has the following values:

```
>> parts
parts =
1x3 struct array with fields:
    partno
    quantity
    costper

>> parts(1)
ans =
      partno: 123
    quantity: 4
     costper: 33

>> parts(2)
ans =
      partno: 142
    quantity: 1
     costper: 150

>> parts(3)
ans =
      partno: 106
    quantity: 20
     costper: 7.5000
```

Next, write general code that will, for any values and any number of structures in the variable *parts*, print the part number and the total cost (quantity of the parts multiplied by the cost of each) in a column format.

For example, if the variable *parts* stores the previous values, the result would be:

```
123 132.00
142 150.00
106 150.00
```

The previous example involved a vector of structs. In the next example, a somewhat more complicated data structure will be introduced: a vector of structs in which some fields are vectors themselves. The example is a database of information that a professor might store for the class. This will be implemented as a vector of structures. The vector will store all of the class information.

Every element in the vector will be a structure, representing all information about one particular student. For every student, the professor wants to store (for now, this would be expanded later):

- name (a string)
- university identifier (ID) number
- quiz scores (a vector of four quiz scores)

The vector variable, called *student*, might look like the following:

	student					
	name	id_no	quiz			
			1	2	3	4
1	C, Joe	999	10.0	9.5	0.0	10.0
2	Hernandez, Pete	784	10.0	10.0	9.0	10.0
3	Brownnose, Violet	332	7.5	6.0	8.5	7.5

Each element in the vector is a struct with three fields (*name, id_no, quiz*). The *quiz* field is a vector of quiz grades. The *name* field is a string.

This data structure could be defined as follows.

```
>> student(3) = struct('name','Brownnose, Violet',...
   'id_no',332,'quiz', [7.5  6  8.5  7.5]);
>> student(1) = struct('name','C, Joe',...
   'id_no',999,'quiz', [10  9.5  0  10]);
>> student(2) = struct('name','Hernandez, Pete',...
   'id_no',784,'quiz', [10  10  9  10]);
```

Once the data structure has been initialized, in MATLAB we could refer to different parts of it. The variable *student* is the entire array; MATLAB just shows the names of the fields.

```
>> student
student =
1x3 struct array with fields:
    name
    id_no
    quiz
```

To see the actual values, one would have to refer to individual structures and/or fields.

```
>> student(1)
ans =
    name: 'C, Joe'
   id_no: 999
    quiz: [10 9.5000 0 10]

>> student(1).quiz
ans =
  10.0000    9.5000    0   10.0000

>> student(1).quiz(2)
ans =
   9.5000

>> student(3).name(1)
ans =
B
```

With a more complicated data structure like this, it is important to be able to understand different parts of the variable. The following are examples of expressions that refer to different parts of this data structure:

- `student` is the entire data structure, which is a vector of structs
- `student(1)` is an element from the vector, which is an individual struct
- `student(1).quiz` is the *quiz* field from the structure, which is a vector of **double** values
- `student(1).quiz(2)` is an individual **double** quiz grade
- `student(3).name(1)` is the first letter of the third student's name

One example of using this data structure would be to calculate and print the quiz average for each student. The following function accomplishes this. The *student* structure, as defined before, is passed to this function. The algorithm for the function is:

- Print column headings
- Loop through the individual students; for each:
  - Sum the quiz grades
  - Calculate the average
  - Print the student's name and quiz average

With the programming method, a second (nested) loop would be required to find the running sum of the quiz grades. However, as we have seen, the **sum** function can be used to sum the vector of all quiz grades for each student. The function is defined as follows:

printAves.m

```
function printAves(student)
% This function prints the average quiz grade
% for each student in the vector of structs
% Format: printAves(student array)

fprintf('%-20s %-10s\n', 'Name', 'Average')
for i = 1:length(student)
    qsum = sum([student(i).quiz]);
    no_quizzes = length(student(i).quiz);
    ave = qsum / no_quizzes;
    fprintf('%-20s %.1f\n', student(i).name, ave);
end
```

Here is an example of calling the function:

```
>> printAves(student)
Name                Average
C, Joe              7.4
Hernandez, Pete     9.8
Brownnose, Violet   7.4
```

8.2.5 Nested Structures

A ***nested structure*** is a structure in which at least one member is itself a structure. For example, a structure for a line segment might consist of fields representing the two points at the ends of the line segment. Each of these points would be represented as a structure consisting of the x and y coordinates.

lineseg			
endpoint1		endpoint2	
x	y	x	y
2	4	1	6

This shows a structure variable called *lineseg* that has two fields for the endpoints of the line segment, *endpoint1* and *endpoint2*. Each of these is a structure consisting of two fields for the x and y coordinates of the individual points, *x* and *y*.

One method of defining this is to nest calls to the **struct** function:

```
>> lineseg = struct('endpoint1',struct('x',2,'y',4), ...
                    'endpoint2',struct('x',1,'y',6))
```

This method is the most efficient.

Another method would be to create structure variables first for the points, and then use these for the fields in the **struct** function (instead of using another **struct** function).

```
>> pointone = struct('x', 5, 'y', 11);
>> pointtwo = struct('x', 7, 'y', 9);
>> lineseg = struct('endpoint1', pointone,...
                    'endpoint2', pointtwo);
```

A third method, the least efficient, would be to build the nested structure one field at a time. As this is a nested structure with one structure inside of another, the dot operator must be used twice here to get to the actual *x* and *y* coordinates.

```
>> lineseg.endpoint1.x = 2;
>> lineseg.endpoint1.y = 4;
>> lineseg.endpoint2.x = 1;
>> lineseg.endpoint2.y = 6;
```

Once the nested structure has been created, we can refer to different parts of the variable *lineseg*. Just typing the name of the variable shows only that it is a structure consisting of two fields, *endpoint1* and *endpoint2*, each of which is a structure.

```
>> lineseg
lineseg =
     endpoint1: [1x1 struct]
     endpoint2: [1x1 struct]
```

Typing the name of one of the nested structures will display the field names and values within that structure:

```
>> lineseg.endpoint1
ans =
    x: 2
    y: 4
```

Using the dot operator twice will refer to an individual coordinate, such as in the following example:

```
>> lineseg.endpoint1.x
ans =
     2
```

QUICK QUESTION!

How could we write a function *strpoint* that returns a string "(x,y)" containing the x and y coordinates? For example, it might be called separately to create strings for the two endpoints and then printed as shown here:

```
>> fprintf('The line segment consists of %s and %s\n',...
    strpoint(lineseg.endpoint1), ...
    strpoint(lineseg.endpoint2))
The line segment consists of (2, 4) and (1, 6)
```

QUICK QUESTION!—CONT'D

Answer: As an *endpoint* structure is passed to an input argument in the function, the dot operator is used within the function to refer to the x and y coordinates. The **sprintf** function is used to create the string that is returned.

strpoint.m

```
function ptstr = strpoint(ptstruct)
% strpoint receives a struct containing x and y
% coordinates and returns a string '(x,y)'
% Format: strpoint(structure with x and y fields)

ptstr = sprintf('(%d, %d)', ptstruct.x, ptstruct.y);
end
```

8.2.6 Vectors of Nested Structures

Combining vectors and nested structures, it is possible to have a vector of structures in which some fields are structures themselves. Here is an example in which a company manufactures cylinders from different materials for industrial use. Information on them is stored in a data structure in a program. The variable *cyls* is a vector of structures, each of which has fields *code, dimensions,* and *weight*. The *dimensions* field is a structure itself consisting of fields *rad* and *height* for the radius and height of each cylinder.

cyls

	code	dimensions		weight
		rad	height	
1	x	3	6	7
2	a	4	2	5
3	c	3	6	9

The following is an example of initializing the data structure by preallocating:

```
>> cyls(3) = struct('code', 'c', 'dimensions',...
   struct('rad', 3, 'height', 6), 'weight', 9);
>> cyls(1) = struct('code', 'x', 'dimensions',...
   struct('rad', 3, 'height', 6), 'weight', 7);
>> cyls(2) = struct('code', 'a', 'dimensions',...
   struct('rad', 4, 'height', 2), 'weight', 5);
```

There are several layers in this variable. For example:

- `cyls` is the entire data structure, which is a vector of structs
- `cyls(1)` is an individual element from the vector, which is a struct
- `cyls(2).code` is the *code* field from the struct *cyls(2)*; it is a character

- `cyls(3).dimensions` is the *dimensions* field from the struct *cyls(3)*; it is a struct itself
- `cyls(1).dimensions.rad` is the *rad* field from the struct *cyls(1).dimensions*; it is a **double** number

For these cylinders, one desired calculation may be the volume of each cylinder, which is defined as $\pi * r^2 * h$, where r is the radius and h is the height. The following function *printcylvols* prints the volume of each cylinder, along with its code for identification purposes. It calls a subfunction to calculate each volume.

printcylvols.m

```
function printcylvols(cyls)
% printcylvols prints the volumes of each cylinder
% in a specialized structure
% Format: printcylvols(cylinder structure)

% It calls a subfunction to calculate each volume

for i = 1:length(cyls)
    vol = cylvol(cyls(i).dimensions);
    fprintf('Cylinder %c has a volume of %.1f in^3\n', ...
        cyls(i).code, vol);
end
end

function cvol = cylvol(dims)
% cylvol calculates the volume of a cylinder
% Format: cylvol(dimensions struct w/ fields 'rad', 'height')

cvol = pi * dims.rad ^2 * dims.height;
end
```

The following is an example of calling this function.

```
>> printcylvols(cyls)
Cylinder x has a volume of 169.6 in^3
Cylinder a has a volume of 100.5 in^3
Cylinder c has a volume of 169.6 in^3
```

Note that the entire data structure, *cyls*, is passed to the function. The function loops through every element, each of which is a structure. It prints the *code* field for each, which is given by *cyls(i).code*. To calculate the volume of each cylinder, only the radius and height are needed, so rather than passing the entire structure to the subfunction *cylvol* (which would be *cyls(i)*), only the *dimensions* field is passed (*cyls(i).dimensions*). The function then receives the *dimensions* structure as an input argument, and uses the dot operator to refer to the *rad* and *height* fields within it.

PRACTICE 8.5

Modify the function *cylvol* to calculate the surface area of the cylinder in addition to the volume (2 pi r^2+2 pi r h).

8.3 ADVANCED DATA STRUCTURES

MATLAB has several types of data structures in addition to the arrays, cell arrays, and structures that we have already seen. These can be found in the documentation under data types.

8.3.1 Categorical Arrays

Categorical arrays are a type of array that allows one to store a finite, countable number of different possible values. Categorical arrays are fairly new in MATLAB, introduced in R2013b. Categorical arrays are defined using the **categorical** function.

For example, a group is polled on their favorite ice cream flavors; the results are stored in a categorical array:

```
>> icecreamfaves = categorical({'Vanilla', 'Chocolate', ...
'Chocolate', 'Rum Raisin', 'Vanilla', 'Strawberry', ...
'Chocolate', 'Rocky Road', 'Chocolate', 'Rocky Road', ...
'Vanilla', 'Chocolate', 'Strawberry', 'Chocolate'});
```

Another way to create this would be to store the strings in a cell array, and then convert using the **categorical** function:

```
>> cellicecreamfaves = {'Vanilla', 'Chocolate', ...
'Chocolate', 'Rum Raisin', 'Vanilla', 'Strawberry', ... 'Chocolate',
'Rocky Road', 'Chocolate', 'Rocky Road', ...
'Vanilla', 'Chocolate', 'Strawberry', 'Chocolate'}
>> icecreamfaves = categorical(cellicecreamfaves);
```

There are several functions that can be used with categorical arrays. The function **categories** will return the list of possible categories as a cell column vector, sorted in alphabetical order.

```
>> cats = categories(icecreamfaves)
cats =
    'Chocolate'
    'Rocky Road'
    'Rum Raisin'
    'Strawberry'
    'Vanilla'
```

The functions **countcats** and **summary** will show the number of occurrences of each of the categories.

```
>> countcats(icecreamfaves)
ans =
     6     2     1     2     3
>> summary(icecreamfaves)
     Chocolate  Rocky Road  Rum Raisin  Strawberry  Vanilla
             6           2           1           2        3
```

In the case of the favorite ice cream flavors, there is no natural order for them, so they are listed in alphabetical order. It is also possible to have ***ordinal categorical arrays***, however, in which an order is given to the categories.

For example, a person has a wearable fitness tracker that tracks the days on which a personal goal for the number of steps taken is reached; these are stored in a file. To simulate this, a variable *stepgoalsmet* stores these data for a few weeks. Another cell array stores the possible days of the week.

```
>> stepgoalsmet = {'Tue', 'Thu', 'Sat', 'Sun', 'Tue', ...
'Sun', 'Thu', 'Sat', 'Wed', 'Sat', 'Sun'};
>> daynames = {'Mon','Tue','Wed','Thu','Fri','Sat','Sun'};
```

Then, an ordinal categorical array, *ordgoalsmet*, is created. This allows days to be compared using relational operators.

```
>> ordgoalsmet= categorical(stepgoalsmet,daynames,'Ordinal',true);
>> summary(ordgoalsmet)
     Mon      Tue      Wed      Thu      Fri      Sat      Sun
       0        2        1        2        0        3        3
>> ordgoalsmet(1) < ordgoalsmet(3)
ans =
     1
>> ordgoalsmet(4) < ordgoalsmet(3)
ans =
     0
```

8.3.2 Tables

A ***table*** is a data structure that stores information in a table format with rows and columns, each of which can be mnemonically labeled. For example, the following uses the **table** function to store some simple information for a doctor's patients.

```
>> names = {'Harry','Sally','Jose'};
>> weights = [185; 133; 210]; % Note column vectors
>> heights = [74; 65.4; 72.2];
>> patients = table(weights, heights, 'RowNames', names)
patients =
             weights    heights
             _______    _______
     Harry   185          74
     Sally   133        65.4
     Jose    210        72.2
```

This created a 3×2 table, with two variables named *weights* and *heights*.

There are many ways to index into tables, to either create a new table that is a subset of the original, or to extract information from the table into other types of data structures. Using parentheses to index, the result is another table; the indexing can be done using integers (as with arrays we have seen so far) or by using row or variable names.

```
>> patients(1:2, 1)
ans =
                weights
                _______
        Harry   185
        Sally   133
>> patients({'Harry' 'Jose'}, :)
ans =
                weights     heights
                _______     _______
        Harry   185           74
        Jose    210           72.2
```

Using curly braces to index, the data can be extracted; in the following example, into a **double** matrix.

```
>> mat = patients{{'Harry' 'Jose'}, :}
mat =
   185.0000   74.0000
   210.0000   72.2000
```

The **summary** function can be used for tables; it shows the variables and some statistical data for each.

```
>> summary(patients)
Variables:
    weights: 3x1 double
        Values:
            min        133
            median     185
            max        210
    heights: 3x1 double
        Values:
            min        65.4
            median     72.2
            max          74
```

8.4 SORTING

Sorting is the process of putting a list in order—either ***descending*** (highest to lowest) or ***ascending*** (lowest to highest) order. For example, here is a list of n integers, visualized as a column vector.

1	85
2	70
3	100
4	95
5	80
6	91

What is desired is to sort this in ascending order in place—by rearranging this vector, not creating another. The following is one basic algorithm.

- Look through the vector to find the smallest number and then put it in the first element in the vector. How? By exchanging it with the number currently in the first element.
- Then, scan the rest of the vector (from the second element down) looking for the next smallest (or, the smallest in the rest of the vector). When found, put it in the first element of the rest of the vector (again, by exchanging).
- Continue doing this for the rest of the vector. Once the next-to-last number has been placed in the correct location in the vector, the last number, by default, has been as well.

What is important in each pass through the vector is *where* the smallest value is so the elements to be exchanged are known (not what the actual smallest number is).

This table shows the progression. The left column shows the original vector. The second column (from the left) shows that the smallest number, the 70, is now in the first element in the vector. It was put there by exchanging with what had been in the first element, 85. This continues element-by-element, until the vector has been sorted.

85	70	70	70	70	70
70	85	80	80	80	80
100	100	100	85	85	85
95	95	95	95	91	91
80	80	85	100	100	95
91	91	91	91	95	100

This is called the ***selection sort***; it is one of many different sorting algorithms.

THE PROGRAMMING CONCEPT

The following function implements the selection sort to sort a vector:

mysort.m

```
function outv = mysort(vec)
% mysort sorts a vector using the selection sort
% Format: mysort(vector)

% Loop through the elements in the vector to end-1
for i = 1:length(vec)-1
    indlow = i; % stores the index of the smallest
    % Find where the smallest number is
    %    in the rest of the vector
    for j=i+1:length(vec)
        if vec(j) < vec(indlow)
            indlow = j;
        end
    end
    % Exchange elements
    temp = vec(i);
    vec(i) = vec(indlow);
    vec(indlow) = temp;
end
outv = vec;
end
```

```
>> vec = [85 70 100 95 80 91];
>> vec = mysort(vec)
vec =
    70    80    85    91     95    100
```

THE EFFICIENT METHOD

MATLAB has a built-in function, **sort**, that will sort a vector in ascending order:

```
>> vec = [85 70 100 95 80 91];
>> vec = sort(vec)
vec =
    70    80    85    91    95    100
```

Descending order can also be specified. For example,

```
>> sort(vec,'descend')
ans =
   100    95    91    85    80    70
```

Sorting a row vector results in another row vector. Sorting a column vector results in another column vector. Note that if we did not have the 'descend' option, **fliplr** (for a row vector) or **flipud** (for a column vector) could be used after sorting.

For matrices, the **sort** function will by default sort each column. To sort by rows, the dimension 2 is specified. For example,

```
>> mat
mat =
       4     6     2
       8     3     7
       9     7     1
>> sort(mat) % sorts by column
ans =
       4     3     1
       8     6     2
       9     7     7

>> sort(mat,2) % sorts by row
ans =
       2     4     6
       3     7     8
       1     7     9
```

8.4.1 Sorting Vectors of Structures

When working with a vector of structures, it is common to sort based on a particular field of the structures. For example, recall the vector of structures used to store information on different software packages that was created in Section 8.2.4.

packages

	item_no	cost	price	code
1	123	19.99	39.95	g
2	456	5.99	49.99	l
3	587	11.11	33.33	w

Here is a function that sorts this vector of structures in ascending order based on the *price* field.

mystructsort.m

```
function outv = mystructsort(structarr)
% mystructsort sorts a vector of structs on the price field
% Format: mystructsort(structure vector)

for i = 1:length(structarr)-1
    indlow = i;
    for j=i+1:length(structarr)
        if structarr(j).price < structarr(indlow).price
```

```
            indlow = j;
        end
    end
    % Exchange elements
    temp = structarr(i);
    structarr(i) = structarr(indlow);
    structarr(indlow) = temp;
end
outv = structarr;
end
```

Note that only the *price* field is compared in the sort algorithm, but the entire structure is exchanged. Consequently, each element in the vector, which is a structure of information about a particular software package, remains intact.

Recall that we created a function *printpackages* that prints the information in a nice table format. Calling the *mystructsort* function and also the function to print will demonstrate this:

```
>> printpackages(packages)

Item #   Cost   Price   Code

   123  19.99   39.95    g
   456   5.99   49.99    l
   587  11.11   33.33    w

>> packByPrice = mystructsort(packages);
>> printpackages(packByPrice)

Item #   Cost   Price   Code

   587  11.11   33.33    w
   123  19.99   39.95    g
   456   5.99   49.99    l
```

This function only sorts the structures based on the *price* field. A more general function is shown in the following, which receives a string that is the name of the field. The function checks first to make sure that the string that is passed is a valid field name for the structure. If it is, it sorts based on that field, and if not, it returns an empty vector.

The function uses ***dynamic field names*** to refer to the field in a structure. A static field name is

```
struct.fieldname
```

whereas a dynamic field name uses a string

```
struct.('fieldname')
```

which means that it could be a string expression, as the input argument.

generalPackSort.m

```
function outv = generalPackSort(inputarg, fname)
% generalPackSort sorts a vector of structs
% based on the field name passed as an input argument

if isfield(inputarg,fname)
   for i = 1:length(inputarg)-1
       indlow = i;
       for j=i+1:length(inputarg)
           if [inputarg(j).(fname)] < ...
              [inputarg(indlow).(fname)]
               indlow = j;

           end
       end
       % Exchange elements
       temp = inputarg(i);
       inputarg(i) = inputarg(indlow);
       inputarg(indlow) = temp;
   end
   outv = inputarg;
else
   outv = [];
end
end
```

The following are examples of calling the function:

```
>> packByPrice = generalPackSort(packages,'price');
>> printpackages(packByPrice)

Item #  Cost   Price  Code

   587  11.11  33.33   w
   123  19.99  39.95   g
   456   5.99  49.99   l

>> packByCost = generalPackSort(packages,'cost');
>> printpackages(packByCost)

Item #  Cost   Price  Code

   456   5.99  49.99   l
   587  11.11  33.33   w
   123  19.99  39.95   g

>> packByProfit = generalPackSort(packages,'profit')
packByProfit =
      []
```

QUICK QUESTION!

Is this *generalPackSort* function truly general? Would it work for any vector of structures, not just one configured like *packages*?

Answer: It is fairly general. It will work for any vector of structures. However, the comparison will only work for numerical or character fields. Thus, as long as the field is a number or character, this function will work for any vector of structures. If the field is a vector itself (including a string), it will not work.

8.4.2 Sorting Strings

For a matrix of strings, the **sort** function works exactly as shown previously for numbers. For example:

```
>> words = char('Hello', 'Howdy', 'Hi', 'Goodbye', 'Ciao')
words =
Hello
Howdy
Hi
Goodbye
Ciao
```

The following sorts column by column using the ASCII equivalents of the characters. It can be seen from the results that the space character comes before the letters of the alphabet in the character encoding:

```
>> sort(words)
ans =
Ce
Giad
Hildb
Hoolo
Howoyye
```

To sort on the rows instead, the second dimension must be specified.

```
>> sort(words,2)
ans =
  Hello
  Hdowy
     Hi
Gbdeooy
   Caio
```

It can be seen here that the uppercase letters come before the lowercase letters.

How could the strings be sorted alphabetically? MATLAB has a function **sortrows** that will do this. The way it works is that it examines the strings column by column, starting from the left. If it can determine which letter comes first, it picks up the entire string and puts it in the first row. In this example, the first two strings are placed based on the first character, 'C' and 'G.' For the other three strings, they all begin with 'H' so the next column is examined. In this case the strings are placed based on the second character, 'e,' 'i,' 'o.'

```
>> sortrows(words)
ans =
Ciao
Goodbye
Hello
Hi
Howdy
```

The **sortrows** function sorts each row as a block, or group, and it will also work on numbers. In this example the rows beginning with 3 and 4 are placed first. Then, for the rows beginning with 5, the values in the second column (6 and 7) determine the order.

```
>> mat = [5 7 2; 4 6 7; 3 4 1; 5 6 2]
mat =
     5     7     2
     4     6     7
     3     4     1
     5     6     2
>> sortrows(mat)
ans =
     3     4     1
     4     6     7
     5     6     2
     5     7     2
```

In order to sort a cell array of strings, the **sort** function can be used. If the cell array is a row vector, a sorted row vector is returned and if the cell array is a column vector, a sorted column vector is returned. For example, note the transpose operator below which makes this a column vector.

```
>> engcellnames = {'Chemical','Mechanical',...
      'Biomedical','Electrical', 'Industrial'};
>> sort(engcellnames')
ans =
    'Biomedical'
    'Chemical'
    'Electrical'
    'Industrial'
    'Mechanical'
```

Categorical arrays can also be sorted using the **sort** function. A nonordinal categorical array, such as *icecreamfaves*, will be sorted in alphabetical order.

```
>> sort(icecreamfaves)
ans =
Chocolate    Chocolate    Chocolate    Chocolate
Chocolate    Chocolate    Rocky Road   Rocky Road
```

```
Rum Raisin   Strawberry   Strawberry Vanilla
Vanilla      Vanilla
```

An ordinal categorical array, however, will be sorted using the order specified. For example, the ordinal categorical array *ordgoalsmet* was created using *daynames*:

```
>> ordgoalsmet
ordgoalsmet =
      Tue      Thu      Sat      Sun      Tue      Sun
      Thu      Sat      Wed      Sat      Sun

>> daynames = ...
       {'Mon','Tue','Wed','Thu','Fri','Sat','Sun'};
```

Thus, the sorting is done in the order given by *daynames*.

```
>> sort(ordgoalsmet)
ans =
      Tue       Tue       Wed       Thu       Thu       Sat
      Sat       Sat       Sun       Sun       Sun
```

8.5 INDEX VECTORS

Using ***index vectors*** is an alternative to sorting a vector. With indexing, the vector is left in its original order. An index vector is used to "point" to the values in the original vector in the desired order.

For example, here is a vector of exam grades:

grades

1	2	3	4	5	6
85	70	100	95	80	91

In ascending order, the lowest grade is in element 2, the next lowest grade is in element 5, and so on. The index vector *grade_index* gives the order for the vector *grades*.

grade_index

1	2	3	4	5	6
2	5	1	6	4	3

The elements in the index vector are then used as the indices for the original vector. To get the *grades* vector in ascending order, the indices used would be *grades(2)*, *grades(5)*, and so on. Using the index vector to accomplish this, `grades(grade_index(1))` would be the lowest grade, 70, and `grades(grade_index(2))` would be the second-lowest grade. In general, `grades(grade_index(i))` would be the *i*th lowest grade.

To create these in MATLAB:

```
>> grades = [85 70 100 95 80 91];
>> grade_index = [2 5 1 6 4 3];
>> grades(grade_index)
ans =
    70    80    85    91    95   100
```

Note that this is a particular type of index vector in which all of the indices of the original vector appear, in the desired order.

In general, instead of creating the index vector manually as shown here, the procedure to initialize the index vector is to use a sort function. The following is the algorithm:

- initialize the values in the index vector to be the indices 1,2, 3, ... to the length of the original vector
- use any sort algorithm, but compare the elements in the original vector using the index vector to index into it (e.g., using *grades(grade_index(i))* as previously shown)
- when the sort algorithm calls for exchanging values, exchange the elements in the index vector, not in the original vector

Here is a function that implements this algorithm:

createind.m

```
function indvec = createind(vec)
% createind returns an index vector for the
% input vector in ascending order
% Format: createind(inputVector)

% Initialize the index vector
len = length(vec);
indvec = 1:len;

for i = 1:len-1
    indlow = i;
    for j=i+1:len
        % Compare values in the original vector
        if vec(indvec(j)) < vec(indvec(indlow))
            indlow = j;
        end
      end
    % Exchange elements in the index vector
    temp = indvec(i);
    indvec(i) = indvec(indlow);
    indvec(indlow) = temp;
end
end
```

For example, for the grades vector just given:

```
>> clear grade_index
>> grade_index = createind(grades)
grade_index =
     2     5     1     6     4     3
>> grades(grade_index)
ans =
    70    80    85    91    95   100
```

8.5.1 Indexing into Vectors of Structures

Often, when the data structure is a vector of structures, it is necessary to iterate through the vector in order by different fields. For example, for the *packages* vector defined previously, it may be necessary to iterate in order by the *cost* or by the *price* fields.

Rather than sorting the entire vector of structures based on these fields, it may be more efficient to index into the vector based on these fields; so, for example, to have an index vector based on *cost* and another based on *price*.

packages

	item_no	cost	price	code
1	123	19.99	39.95	g
2	456	5.99	49.99	l
3	587	11.11	33.33	w

	cost_ind
1	2
2	3
3	1

	price_ind
1	3
2	1
3	2

These index vectors would be created as before, comparing the fields, but exchanging the entire structures. Once the index vectors have been created, they can be used to iterate through the *packages* vector in the desired order. For example, the function to print the information from *packages* has been modified so that, in addition to the vector of structures, the index vector is also passed and the function iterates using that index vector.

printpackind.m

```
function printpackind(packstruct, indvec)
% printpackind prints a table showing all
% values from a vector of packages structures
% using an index vector for the order
% Format: printpackind(vector of packages, index vector)

fprintf('Item # Cost Price Code\n')
no_packs = length(packstruct);
for i = 1:no_packs
    fprintf('%6d %6.2f %6.2f %3c\n', ...
        packstruct(indvec(i)).item_no, ...
        packstruct(indvec(i)).cost, ...
        packstruct(indvec(i)).price, ...
        packstruct(indvec(i)).code)
end
end
```

```
>> printpackind(packages,cost_ind)
Item #  Cost  Price  Code
   456   5.99  49.99   l
   587  11.11  33.33   w
   123  19.99  39.95   g
```

```
>> printpackind(packages,price_ind)
Item #   Cost   Price   Code
   587  11.11   33.33    w
   123  19.99   39.95    g
   456   5.99   49.99    l
```

PRACTICE 8.6

Modify the function *createind* to create the *cost_ind* index vector.

■ Explore Other Interesting Features

The built-in functions **cell2struct**, which converts a cell array into a vector of structs, and **struct2cell**, which converts a struct to a cell array.

Find the functions that convert from cell arrays to number arrays and vice versa.

Explore the **orderfields** function.

MATLAB has an entire category of data types and built-in functions that operate on dates and times. Find this under Language Fundamentals, then Data Types.

Explore the "is" functions for categorical arrays, such as **iscategorical**, **iscategory**, and **isordinal**.

Explore the table functions **array2table** and **struct2table**.

Explore the functions **deal** (which assigns values to variables) and **orderfields** (which puts structure fields in alphabetical order).

Investigate the **randperm** function. ■

SUMMARY

COMMON PITFALLS

- Confusing the use of parentheses (cell indexing) versus curly braces (content indexing) for a cell array
- Forgetting to index into a vector using parentheses or referring to a field of a structure using the dot operator
- When sorting a vector of structures on a field, forgetting that although only the field in question is compared in the sort algorithm, entire structures must be interchanged.

PROGRAMMING STYLE GUIDELINES

- Use arrays when values are the same type and represent in some sense the same thing.
- Use cell arrays or structures when the values are logically related but not the same type nor the same thing.
- Use cell arrays, rather than character matrices, when storing strings of different lengths.
- Use cell arrays, rather than structures, when it is desired to loop through the values or to vectorize the code.
- Use structures rather than cell arrays when it is desired to use names for the different values rather than indices.
- Use **sortrows** to sort strings stored in a matrix alphabetically; for cell arrays, **sort** can be used.
- When it is necessary to iterate through a vector of structures in order based on several different fields, it may be more efficient to create index vectors based on these fields rather than sorting the vector of structures multiple times.

MATLAB Functions and Commands			
cell	strsplit	isfield	summary
celldisp	iscellstr	fieldnames	table
cellplot	struct	categorical	sort
cellstr	rmfield	categories	sortrows
strjoin	isstruct	countcats	

MATLAB Operators
cell arrays { }
dot operator for structs.
parentheses for dynamic field names ( )

Exercises

1. Create the following cell array:

```
>> ca = {'abc', 11, 3:2:9, zeros(2)}
```

   Use the **reshape** function to make it a 2×2 matrix. Then, write an expression that would refer to just the last column of this cell array.
2. Create a 2×2 cell array using the **cell** function and then put values in the individual elements. Then, insert a row in the middle so that the cell array is now 3×2.
3. Create a row vector cell array to store the string 'xyz,' the number 33.3, the vector 2:6, and the **logical** expression 'a' < 'c.' Use the transpose operator to make this a

column vector, and use **reshape** to make it a 2 × 2 matrix. Use **celldisp** to display all elements.

4. Create a cell array that stores phrases, such as:

```
exclaimcell = {'Bravo', 'Fantastic job'};
```

Pick a random phrase to print.

5. Create three cell array variables that store people's names, verbs, and nouns. For example,

```
names = {'Harry', 'Xavier', 'Sue'};
verbs = {'loves', 'eats'};
nouns = {'baseballs', 'rocks', 'sushi'};
```

Write a script that will initialize these cell arrays, and then print sentences using one random element from each cell array (e.g., 'Xavier eats sushi').

6. Write a script that will prompt the user for strings and read them in, store them in a cell array (in a loop), and then print them out.
7. When would you loop through the elements of a cell array?
8. Write a function *buildstr* that will receive a character and a positive integer n. It will create and return a cell array with strings of increasing lengths, from 1 to the integer *n*. It will build the strings with successive characters in the ASCII encoding.

```
>> buildstr('a',4)
ans =
    'a'     'ab'     'abc'     'abcd'
```

9. Write a function "catit" that will receive one input argument which is a cell array. If the cell array contains only strings, it will return one string which is all of the strings from the cell array concatenated together—otherwise, it will return an empty string. Here is one example of calling the function:

```
>> fishies = {'tuna','shark','salmon','cod'};
>> catit(fishies)
ans =
tunasharksalmoncod
```

Do this using the programming method.

10. Modify the previous function to use the **strjoin** function. Is there a difference?
11. Create a cell array variable that would store for a student his or her name, university id number, and GPA. Print this information.
12. Create a structure variable that would store for a student his or her name, university id number, and GPA. Print this information.
13. Here is an inefficient way of creating a structure variable to store a person's name as first, middle, and last:

```
>> myname.first = 'Homer';
>> myname.middle = 'James';
>> myname.last = 'Fisch';
```

Rewrite this more efficiently using the **struct** function:

14. What would be an advantage of using cell arrays over structures?
15. What would be an advantage of using structures over cell arrays?
16. A complex number is a number of the form a + ib, where a is called the real part, b is called the imaginary part, and $i = \sqrt{-1}$. Write a script that prompts the user separately to enter values for the real and imaginary parts, and stores them in a structure variable. It then prints the complex number in the form a + ib. The script should just print the value of a, then the string '+i,' and then the value of b. For example, if the script is named "compnumstruct," running it would result in:

```
>> compnumstruct
Enter the real part: 2.1
Enter the imaginary part: 3.3
The complex number is 2.1+i3.3
```

17. Create a data structure to store information about the elements in the periodic table of elements. For every element, store the name, atomic number, chemical symbol, class, atomic weight, and a seven-element vector for the number of electrons in each shell. Create a structure variable to store the information, for example for lithium:

```
Lithium 3 Li alkali_metal 6.94 2 1 0 0 0 0 0
```

18. Write a function *separatethem* that will receive one input argument which is a structure containing fields named 'length' and 'width,' and will return the two values separately. Here is an example of calling the function:

```
>> myrectangle = struct('length',33,'width',2);
>> [l w] = separatethem(myrectangle)
l =
    33
w =
     2
```

19. In chemistry, the pH of an aqueous solution is a measure of its acidity. A solution with a pH of 7 is said to be *neutral*, a solution with a pH greater than 7 is *basic*, and a solution with a pH less than 7 is *acidic*. Create a vector of structures with various solutions and their pH values. Write a function that will determine acidity. Add another field to every structure for this.
20. A script stores information on potential subjects for an experiment in a vector of structures called *subjects*. The following shows an example of what the contents might be:

```
>> subjects(1)
ans =
      name: 'Joey'
    sub_id: 111
    height: 6.7000
    weight: 222.2000
```

For this particular experiment, the only subjects who are eligible are those whose height or weight is lower than the average height or weight of all subjects. The script will print the names of those who are eligible. Create a vector with sample data in a script, and then write the code to accomplish this. Don't assume that the length of the vector is known; the code should be general.

21. Quiz data for a class is stored in a file. Each line in the file has the student ID number (which is an integer) followed by the quiz scores for that student. For example, if there are four students and three quizzes for each, the file might look like this:

```
44    7     7.5     8
33    5.5   6       6.5
37    8     8       8
24    6     7       8
```

First create the data file, and then store the data in a script in a vector of structures. Each element in the vector will be a structure that has 2 members: the integer student ID number and a vector of quiz scores. The structure will look like this:

students

	id_no	quiz 1	2	3
1	44	7	7.5	8
2	33	5.5	6	6.5
3	37	8	8	8
4	24	6	7	8

To accomplish this, first use the **load** function to read all information from the file into a matrix. Then, using nested loops, copy the data into a vector of structures as specified. Then, the script will calculate and print the quiz average for each student.

22. Create a nested struct to store a person's name, address, and phone number. The struct should have 3 fields for the name, address, and phone number. The address fields and phone fields will be structs.
23. Design a nested structure to store information on constellations for a rocket design company. Each structure should store the constellation's name and information on the stars in the constellation. The structure for the star information should include the star's name, core temperature, distance from the sun, and whether it is a binary star or not. Create variables and sample data for your data structure.
24. Write a script that creates a vector of line segments (where each is a nested structure as shown in this chapter). Initialize the vector using any method. Print a table showing the values, such as shown in the following:

```
Line  From     To
====  =======  =======
 1    ( 3, 5)  ( 4, 7)
 2    ( 5, 6)  ( 2, 10)
      etc.
```

25. Given a vector of structures defined by the following statements:

```
kit(2).sub.id = 123;
kit(2).sub.wt = 4.4;
kit(2).sub.code = 'a';
kit(2).name = 'xyz';
kit(2).lens = [4 7];
kit(1).name = 'rst';
kit(1).lens = 5:6;
kit(1).sub.id = 33;
kit(1).sub.wt = 11.11;
kit(1).sub.code = 'q';
```

Which of the following expressions are valid? If the expression is valid, give its value. If it is not valid, explain why.

```
>> kit(1).sub

>> kit(2).lens(1)

>> kit(1).code

>> kit(2).sub.id == kit(1).sub.id

>> strfind(kit(1).name, 's')
```

26. Create a vector of structures *experiments* that stores information on subjects used in an experiment. Each struct has four fields: *num*, *name*, *weights*, and *height*. The field *num* is an integer, *name* is a string, *weights* is a vector with two values (both of which are double values), and *height* is a struct with fields *feet* and *inches* (both of which are integers). The following is an example of what the format might look like.

experiments

	num	name	weights		height	
			1	2	feet	inches
1	33	Joe	200.34	202.45	5	6
2	11	Sally	111.45	111.11	7	2

Write a function *printhts* that will receive a vector in this format and will print the name and height of each subject in inches (1 foot=12 in.). This function calls another function *howhigh* that receives a height struct and returns the total height in inches. This function could also be called separately.

27. A team of engineers is designing a bridge to span the Podunk River. As part of the design process, the local flooding data must be analyzed. The following information on each storm that has been recorded in the last 40 years is stored in a file: a code for the location of the source of the data, the amount of rainfall (in inches), and the duration of the storm (in hours), in that order. For example, the file might look like this:

```
321   2.4   1.5
111   3.3   12.1
    etc.
```

Create a data file. Write the first part of the program: design a data structure to store the storm data from the file, and also the intensity of each storm. The intensity is the rainfall amount divided by the duration. Write a function to read the data from the file (use **load**), copy from the matrix into a vector of structs, and then calculate the intensities. Write another function to print all of the information in a neatly organized table. Add a function to the program to calculate the average intensity of the storms. Add a function to the program to print all of the information given on the most intense storm. Use a subfunction for this function which will return the index of the most intense storm.

28. Create an ordinal categorical array to store the four seasons.
29. Create a **table** to store information on students; for each, their name, id number, and major.
30. Write a function *mydsort* that sorts a vector in descending order (using a loop, not the built-in sort function).
31. Write a function *matsort* to sort all of the values in a matrix (decide whether the sorted values are stored by row or by column). It will receive one matrix argument and return a sorted matrix. Do this without loops, using the built-in functions **sort** and **reshape**. For example:

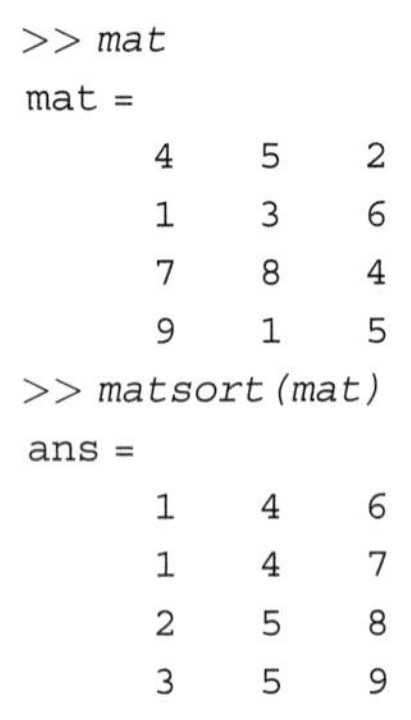

```
>> mat
mat =
     4     5     2
     1     3     6
     7     8     4
     9     1     5
>> matsort(mat)
ans =
     1     4     6
     1     4     7
     2     5     8
     3     5     9
```

32. DNA is a double stranded helical polymer that contains basic genetic information in the form of patterns of nucleotide bases. The patterns of the base molecules A, T, C, and G encode the genetic information. Construct a cell array to store some DNA sequences as strings; such as
    TACGGCAT
    ACCGTAC
    and then sort these alphabetically. Next, construct a matrix to store some DNA sequences of the same length and then sort them alphabetically.
33. Trace this; what will it print?

parts

	code	quantity	weight
1	'x'	11	4.5
2	'z'	33	3.6
3	'a'	25	4.1
4	'y'	31	2.2

	ci	qi	wi
1	3	1	4
2	1	3	2
3	4	4	3
4	2	2	1

```
for i = 1:length(parts)
   fprintf('Part %c weight is %.1f\n',...
        parts(qi(i)).code, parts(qi(i)).weight)
end
```

34. When would you use sorting versus indexing?
35. Write a function that will receive a vector and will return two index vectors: one for ascending order and one for descending order. Check the function by writing a script that will call the function and then use the index vectors to print the original vector in ascending and descending order.

2

PART

Advanced Topics for Problem Solving With MATLAB

CHAPTER 9

Advanced File Input and Output

KEY TERMS

file input and output
lower-level file I/O functions
file types
open the file
close the file
file identifier
permission strings
standard input
standard output
standard error
end of the file

CONTENTS

This chapter extends the input and output concepts (I/O) that were introduced in Chapter 3. In that chapter, we saw how to read values entered by the user using the **input** function, and also the output functions **disp** and **fprintf** that display information in windows on the screen.

For ***file input and output*** (file I/O), we used the **load** and **save** functions that can read from a data file into a matrix, and write from a matrix to a data file. If the data to be written or file to be read are not in a simple matrix format, ***lower-level file I/O functions*** must be used.

The **save** and **load** functions can also be used to store MATLAB® variables, both their name and their contents, into special MATLAB binary files called MAT-files, and to read those variables into the base workspace.

The MATLAB software has functions that can read and write data from different ***file types*** such as spreadsheets. For example, it can read from and write to Excel spreadsheets that have filename extensions such as .xls or .xlsx. There is also an Import Tool, which allows one to import data from a variety of file formats.

In this chapter, we will introduce some functions that work with different file types, as well as the programmatic methods using some of the lower-level file input and output functions.

MATLAB®. http://dx.doi.org/10.1016/B978-0-12-804525-1.00009-X

9.1 USING MAT-FILES FOR VARIABLES

In addition to the functions that manipulate data files, MATLAB has functions that allow reading variables from and saving variables to files. These files are called MAT-files (because the extension on the file name is .mat), and they store the names and contents of variables. Variables can be written to MAT-files, appended to them, and read from them.

Note that MAT-files are very different from the data files that we have worked with so far. Rather than just storing data, MAT-files store the variable names in addition to their values. These files are typically used only within MATLAB; they are not used to share data with other programs.

9.1.1 Writing Variables to a File

The **save** command can be used to write variables to a MAT-file, or to append variables to a MAT-file. By default, the **save** function writes to a MAT-file. It can either save the entire current workspace (all variables that have been created) or a subset of the workspace (including, e.g., just one variable). The **save** function will save the MAT-file in the Current Folder, so it is important to set that correctly first.

To save all workspace variables in a file, the command is:

```
save filename
```

The '.mat' extension is added to the filename automatically. The contents of the file can be displayed using **who** with the '-file' qualifier:

```
who -file filename
```

For example, in the following session in the Command Window, three variables are created; these are then displayed using **who**. Then, the variables are saved to a file named "sess1.mat". The **who** function is then used to display the variables stored in that file.

```
>> mymat = rand(3,5);
>> x = 1:6;
>> y = x.^2;
>> who
Your variables are:
mymat  x    y

>> save sess1
>> who -file sess1
Your variables are:
mymat  x    y
```

To save just one variable to a file, the format is

```
save filename variablename
```

For example, just the matrix variable *mymat* is saved to a file called *sess2*:

```
>> save sess2 mymat
>> who -file sess2
Your variables are:
mymat
```

9.1.2 Appending Variables to a MAT-File

Appending to a file adds to what has already been saved in a file, and is accomplished using the –append option. For example, assuming that the variable *mymat* has already been stored in the file "sess2.mat" as just shown, this would append the variable *x* to the file:

```
>> save -append sess2 x
>> who -file sess2
Your variables are:
mymat  x
```

Without specifying variable(s), just **save –append** would add all variables from the base workspace to the file. When this happens, if the variable is not in the file, it is appended. If there is a variable with the same name in the file, it is replaced by the current value from the base workspace.

9.1.3 Reading From a MAT-file

The **load** function can be used to read from different types of files. As with the **save** function, by default the file will be assumed to be a MAT-file, and **load** can load all variables from the file or only a subset. For example, in a new Command Window session in which no variables have yet been created, the **load** function could load from the files created in the previous section:

```
>> who
>> load sess2
>> who
Your variables are:
mymat x
```

A subset of the variables in a file can be loaded by specifying them in the form:

```
load filename variable list
```

9.2 WRITING AND READING SPREADSHEET FILES

MATLAB has functions **xlswrite** and **xlsread** that will write to and read from Excel spreadsheet files that have extensions such as '.xls'. (Note that this works under Windows environments provided that Excel is loaded. Under other environments, problems may be encountered if Excel cannot be loaded as a COM server.) For example, the following will create a 5×3 matrix of random integers, and then write it to a spreadsheet file "ranexcel.xls" that has five rows and three columns:

```
>> ranmat = randi(100,5,3)
ranmat =
    96    77    62
    24    46    80
    61     2    93
    49    83    74
    90    45    18

>> xlswrite('ranexcel',ranmat)
```

The **xlsread** function will read from a spreadsheet file. For example, use the following to read from the file 'ranexcel.xls':

```
>> ssnums = xlsread('ranexcel')
ssnums =
    96    77    62
    24    46    80
    61     2    93
    49    83    74
    90    45    18
```

In both cases the '.xls' extension on the file name is the default, so it can be omitted.

These are shown in their most basic forms, when the matrix and/or spreadsheet contains just numbers and the entire spreadsheet is read or matrix is written. There are many qualifiers that can be used for these functions, however. For example, the following would read from the spreadsheet file "texttest.xls" that contains:

a	123	Cindy
b	333	Suzanne
c	432	David
d	987	Burt

```
>> [nums, txt] = xlsread('texttest.xls')
nums =
   123
   333
   432
   987
txt =
    'a'    ''    'Cindy'
    'b'    ''    'Suzanne'
    'c'    ''    'David'
    'd'    ''    'Burt'
```

This reads the numbers into a **double** vector variable *nums* and the text into a cell array *txt* (the **xlsread** function always returns the numbers first and then the text). The cell array is 4×3. It has three columns as the file had three columns, but as the middle column had numbers (which were extracted and stored in the vector *nums*), the middle column in the cell array *txt* consists of empty strings.

A loop could then be used to echo print the values from the spreadsheet in the original format:

```
>> for i = 1:length(nums)
       fprintf('%c %d %s\n', txt{i,1}, ...
          nums(i), txt{i,3})
   end
a 123 Cindy
b 333 Suzanne
c 432 David
d 987 Burt
```

These are just examples; MATLAB has many other functions that read from and write to different file formats.

9.3 LOWER-LEVEL FILE I/O FUNCTIONS

When reading from a data file, the **load** function works as long as the data in the file are "regular"—in other words the same kind of data on every line and in the same format on every line—so that they can be read into a matrix. However, data files are not always set up in this manner. When it is not possible to use **load**, MATLAB has what are called lower-level file input functions that can be used. The file must be opened first, which involves finding or creating the file and positioning an indicator at the beginning of the file. This indicator then moves through the file as it is being read from. When the reading has been completed, the file must be closed.

Similarly, the **save** function can write or append matrices to a file, but if the output is not a simple matrix, there are lower-level functions that write to files. Again, the file must be opened first and closed when the writing has been completed.

In general, the steps involved are:

- ***open the file***
- read from the file, write to the file, or append to the file
- ***close the file***

First, the steps involved in opening and closing the file will be described. Several functions that perform the middle step of reading from or writing to the file will be described subsequently.

9.3.1 Opening and Closing a File

Files are opened with the **fopen** function. By default, the **fopen** function opens a file for reading. If another mode is desired, a "permission string" is used to specify which, for example, writing or appending. The **fopen** function returns −1 if it is not successful in opening the file or an integer value that becomes the ***file identifier*** if it is successful. This file identifier is then used to refer to the file when calling other file I/O functions. The general form is

```
fid = fopen('filename', 'permission string');
```

where *fid* is a variable that stores the file identifier (it can be named anything) and the ***permission strings*** include:

r reading (this is the default)
w writing
a appending

After the **fopen** is attempted, the value returned should be tested to make sure that the file was opened successfully. For example, if attempting to open for reading and the file does not exist, the **fopen** will not be successful. As the **fopen** function returns −1 if the file was not found, this can be tested to decide whether to print an error message or to carry on and use the file. For example, if it is desired to read from a file "samp.dat":

```
fid = fopen('samp.dat');
if fid == -1
   disp('File open not successful')
else
   % Carry on and use the file!
end
```

The **fopen** function also returns an error message if it is not successful; this can be stored and displayed. Also, when the first file is opened in a MATLAB session, it will have a file identifier value of 3, because MATLAB assigns three default identifiers (0, 1, and 2) for the ***standard input***, ***standard output***, and ***standard error***. This can be seen if the output from an **fopen** is not suppressed.

```
>> [fid, msg] = fopen('sample.dat')
fid =
    -1
msg =
No such file or directory
>> if fid == -1
        disp(msg)
    else
        % Carry on and use the file!
    end
```

```
No such file or directory
>> [fid, msg] = fopen('samp.dat')
fid =
     3
msg =
     ''
```

In this case there was no error message so the message string is empty.

Files should be closed when the program has finished reading from or writing or appending to them. The function that accomplishes this is the **fclose** function, which returns 0 if the file close was successful or −1 if not. Individual files can be closed by specifying the file identifier or, if more than one file is open, all open files can be closed by passing the string 'all' to the **fclose** function. The general forms are:

```
closeresult = fclose(fid);

closeresult = fclose('all');
```

The result from the **fclose** function should also be checked with an **if-else** statement to make sure it was successful, and a message should be printed (if the close was not successful, that might mean the file was corrupted and the user would want to know that). So, the outline of the code will be:

```
fid = fopen('filename', 'permission string');
if fid == -1
   disp('File open not successful')
else

   % do something with the file!

   closeresult = fclose(fid);

   if closeresult == 0
      disp('File close successful')
   else
      disp('File close not successful')
   end
end
```

9.3.2 Reading From Files Using fgetl

There are several lower-level functions that read from files. The **fgetl** and **fgets** functions read strings from a file one line at a time; the difference is that the **fgets** keeps the newline character if there is one at the end of the line, whereas the **fgetl** function gets rid of it. Both of these functions require first opening the file, and then closing it when finished. As the **fgetl** and **fgets** functions read one line at a time, these functions are typically inside some form of a loop.

We will concentrate on the **fgetl** function, which reads strings from a file one line at a time. Other input functions will be covered in Section 9.3.4. The **fgetl** function affords more control over how the data are read than other input functions. The **fgetl** function reads one line of data from a file into a string; string functions can then be used to manipulate the data. As **fgetl** only reads one line, it is normally placed in a loop that keeps going until the ***end of the file*** is reached. The function **feof** returns **logical true** if the end of the file has been reached. The function call **feof(fid)** would return **logical true** if the end of the file has been reached for the file identified by *fid*, or **logical false** if not. A general algorithm for reading from a file into strings would be:

- Attempt to open the file
  - Check to ensure the file open was successful.
- If opened, loop until the end of the file is reached
  - For each line in the file:
    - read it into a string
    - manipulate the data
- Attempt to close the file
  - Check to make sure the file close was successful.

The following is the generic code to accomplish these tasks:

```
fid = fopen('filename');
if fid == -1
   disp('File open not successful')
else
   while feof(fid) == 0
       % Read one line into a string variable
       aline = fgetl(fid);
       % Use string functions to extract numbers, strings,
       %    etc. from the line
       % Do something with the data!
   end
   closeresult = fclose(fid);
   if closeresult == 0
       disp('File close successful')
   else
       disp('File close not successful')
   end
end
```

The permission string could be included in the call to the **fopen** function. For example:

```
fid = fopen('filename', 'r');
```

but the 'r' is not necessary as reading is the default. The condition on the **while** loop can be interpreted as saying "while the file end-of-file is false". Another way to write this is

```
while ~feof(fid)
```

which can be interpreted similarly as "while we're not at the end of the file".

For example, assume that there is a data file "subjexp.dat", which has on each line a number followed by a space followed by a character code. The **type** function can be used to display the contents of this file (as the file does not have the default extension .m, the extension on the file name must be included).

```
>> type subjexp.dat
5.3 a
2.2 b
3.3 a
4.4 a
1.1 b
```

The **load** function would not be able to read this into a matrix as it contains both numbers and text. Instead, the **fgetl** function can be used to read each line as a string, and then string functions are used to separate the numbers and characters. For example, the following just reads each line and prints the number with two decimal places and then the rest of the string:

fileex.m

```
% Reads from a file one line at a time using fgetl
% Each line has a number and a character
% The script separates and prints them

% Open the file and check for success
fid = fopen('subjexp.dat');
if fid == -1
    disp('File open not successful')
else
    while feof(fid) == 0
        aline = fgetl(fid);
        % Separate each line into the number and character
        % code and convert to a number before printing
        [num, charcode] = strtok(aline);
        fprintf('%.2f %s\n', str2double(num), charcode)
    end

    % Check the file close for success
    closeresult = fclose(fid);
    if closeresult == 0
        disp('File close successful')
    else
        disp('File close not successful')
    end
end
```

The following is an example of executing this script:

```
>> fileex
5.30  a
2.20  b
3.30  a
4.40  a
1.10  b
File close successful
```

In this example, every time the loop action is executed, the **fgetl** function reads one line into a string variable. The string function **strtok** is then used to store the number and the character in separate variables, both of which are string variables (the second variable actually stores the blank space and the letter). If it is desired to perform calculations using the number, the function **str2double** would be used to convert the number stored in the string variable into a **double** value.

PRACTICE 9.1

Modify the script *fileex* to sum the numbers from the file. Create your own file in this format first.

9.3.3 Writing and Appending to Files

There are several lower-level functions that can write to files. Like the other low-level functions, the file must be opened first for writing (or appending) and should be closed once the writing has been completed.

We will concentrate on the **fprintf** function, which can be used to write to a file and also to append to a file. To write one line at a time to a file, the **fprintf** function can be used. We have, of course, been using **fprintf** to write to the screen. The screen is the default output device, so if a file identifier is not specified, the output goes to the screen; otherwise, it goes to the specified file. The default file identifier number is 1 for the screen. The general form is:

```
fprintf(fid, 'format', variable(s));
```

The **fprintf** function actually returns the number of bytes that was written to the file, so if it is not desired to see that number, the output should be suppressed with a semicolon as shown here.

The following is an example of writing to a file named "tryit.txt":

Note
When writing to the screen, the value returned by **fprintf** is not seen, but could be stored in a variable.

```
>> fid = fopen('tryit.txt', 'w');

>> for i = 1:3
       fprintf(fid, 'The loop variable is %d\n', i);
   end

>> fclose(fid);
```

The permission string in the call to the **fopen** function specifies that the file is opened for writing to it. Just like when reading from a file, the results from **fopen** and **fclose** should really be checked to make sure they were successful. The **fopen** function attempts to open the file for writing. If the file already exists, the contents are erased so it is as if the file had not existed. If the file does not currently exist (which would be the norm), a new file is created. The **fopen** could fail, for example, if there isn't space to create this new file.

To see what was written to the file we could then open it (for reading) and loop to read each line using **fgetl**:

```
>> fid = fopen('tryit.txt');

>> while ~feof(fid)
         aline = fgetl(fid);
         disp(aline)
     end
The loop variable is 1
The loop variable is 2
The loop variable is 3
>> fclose(fid);
```

Of course, we could also just display the contents using **type**.

Here is another example in which a matrix is written to a file. First, a 2×4 matrix is created and then it is written to a file using the format string `'%d %d\n'`, which means that each column from the matrix will be written as a separate line in the file.

```
>> mat = [20 14 19 12; 8 12 17 5]
mat =

    20  14  19  12
     8  12  17   5

>> fid = fopen('randmat.dat','w');
>> fprintf(fid,'%d %d\n',mat);
>> fclose(fid);
```

As this is a matrix, the **load** function can be used to read it in.

```
>> load randmat.dat
>> randmat
randmat =
      20     8
      14    12
      19    17
      12     5

>> randmat'
ans =
    20    14    19    12
     8    12    17     5
```

Transposing the matrix will display it in the form of the original matrix. If this is desired to begin with, the matrix variable *mat* can be transposed before using

fprintf to write to the file. (Of course, it would be much simpler in this case to just use **save** instead!)

PRACTICE 9.2

Create a 3×5 matrix of random integers, each in the range from 1 to 100. Write the sum of each row to a file called "myrandsums.dat" using **fprintf**. Confirm that the file was created correctly.

The **fprintf** function can also be used to append to an existing file. The permission string is 'a', so the general form of the **fopen** would be:

```
fid = fopen('filename', 'a');
```

Then, using **fprintf** (typically in a loop), we would write to the file starting at the end of the file. The file would then be closed using **fclose**. What is written to the end of the file doesn't have to be in the same format as what is already in the file when appending.

9.3.4 Alternate File Input Functions

The function **fscanf** reads formatted data into a matrix, using conversion formats such as %d for integers, %s for strings, and %f for floats (**double** values). The **textscan** function reads text data from a file and stores the data in a cell array; it also uses conversion formats. The **fscanf** and **textscan** functions can read the entire data file into one data structure. In terms of level, these two functions are somewhat in between the **load** function and the lower-level functions, such as **fgetl**. The file must be opened using **fopen** first, and should be closed using **fclose** after the data has been read. However, no loop is required; they will read in the entire file automatically into a data structure.

Instead of using the **fgetl** function to read one line at a time, once a file has been opened the **fscanf** function can be used to read from this file directly into a matrix. However, the matrix must be manipulated somewhat to get it back into the original form from the file. The format of using the function is:

```
mat = fscanf(fid, 'format', [dimensions])
```

The **fscanf** reads into the matrix variable *mat* columnwise from the file identified by *fid*. The 'format' includes conversion characters much like those used in the **fprintf** function. The 'format' specifies the format of every line in the file, which means that the lines must be formatted consistently. The dimensions specify the desired dimensions of *mat*; if the number of values in the file is not known, **inf** can be used for the second dimension.

For example, the following would read in from the file subjexp.dat; each line contains a number, followed by a space, and then a character.

```
>> type subjexp.dat
5.3 a
2.2 b
3.3 a
4.4 a
1.1 b

>> fid = fopen('subjexp.dat');

>> mat = fscanf(fid,'%f %c',[2, inf])
mat =
    5.3000    2.2000    3.3000    4.4000    1.1000
   97.0000   98.0000   97.0000   97.0000   98.0000

>> fclose(fid);
```

The **fopen** opens the file for reading. The **fscanf** then reads from each line one double and one character, and places each pair in separate columns in the matrix (in other words every line in the file becomes a column in the matrix). Note that the space in the format string is important: `'%f %c'` specifies that there is a float, a space, and a character. The dimensions specify that the matrix is to have two rows by however many columns are necessary (equal to the number of lines in the file). As matrices store values that are all the same type, the characters are stored as their ASCII equivalents in the character encoding (e.g., 'a' is 97).

Once this matrix has been created, it may be more useful to separate the rows into vector variables and to convert the second back to characters, which can be accomplished as follows:

```
>> nums = mat(1,:);

>> charcodes = char(mat(2,:))
charcodes =
abaab
```

Of course, the results from **fopen** and **fclose** should be checked but were omitted here for simplicity.

PRACTICE 9.3

Write a script to read in this file using **fscanf**, and sum the numbers.

QUICK QUESTION!

Instead of using the dimensions [2, **inf**] in the **fscanf** function, could we use [**inf** ,2]?

Answer: No, [**inf**, 2] would not work. Because **fscanf** reads each row from the file into a column in the matrix, the number of rows in the resulting matrix is known but the number of columns is not.

QUICK QUESTION!

Why is the space in the conversion string '%f %c' important? Would the following also work?

```
>> mat = fscanf(fid,'%f%c', [2, inf])
```

Answer: No, that would not work. The conversion string '%f %c' specifies that there is a real number, then a space, then a character. Without the space in the conversion string, it would specify a real number immediately followed by a character (which would be the space in the file). Then, the next time it would be attempting to read the next real number, but the file position indicator would be pointing to the character on the first line; the error would cause the **fscanf** function to halt. The end result follows:

```
>> fid = fopen('subjexp.dat');
>> mat = fscanf(fid,'%f%c', [2, inf])
mat =
     5.3000
    32.0000
```

The 32 is the numerical equivalent of the space character ' ', as seen here.

```
>> double(' ')
ans =
    32
```

Another option for reading from a file is to use the **textscan** function. The **textscan** function reads text data from a file and stores the data in column vectors in a cell array. The **textscan** function is called, in its simplest form, as

```
cellarray = textscan(fid, 'format');
```

where the 'format' includes conversion characters much like those used in the **fprintf** function. The 'format' essentially describes the format of columns in the data file, which will then be read into column vectors. For example, to read the file 'subjexp.dat' we could do the following (again, for simplicity, omitting the error-check of **fopen** and **fclose**):

```
>> fid = fopen('subjexp.dat');
>> subjdata = textscan(fid,'%f %c');
>> fclose(fid)
```

The format string '%f %c' specifies that on each line there is a **double** value followed by a space followed by a character. This creates a 1×2 cell array variable called *subjdata*. The first element in this cell array is a column vector of doubles (the first column from the file); the second element is a column vector of characters (the second column from the file), as shown here:

```
>> subjdata
subjdata =
    [5x1 double]  [5x1 char]
>> subjdata{1}
ans =
    5.3000
    2.2000
    3.3000
    4.4000
    1.1000
```

```
>> subjdata{2}
ans =
a
b
a
a
b
```

To refer to individual values from the vector, it is necessary to index into the cell array using curly braces and then index into the vector using parentheses. For example, to refer to the third number in the first element of the cell array:

```
>> subjdata{1}(3)
ans =
    3.3000
```

A script that reads in these data and echo prints it is shown here:

textscanex.m

```
% Reads data from a file using textscan
fid = fopen('subjexp.dat');
if fid == -1
    disp('File open not successful')
else
   % Reads numbers and characters into separate elements
   % in a cell array
   subjdata = textscan(fid,'%f %c');
   len = length(subjdata{1});
   for i = 1:len
       fprintf('%.1f %c\n',subjdata{1}(i),subjdata{2}(i))
   end

   closeresult = fclose(fid);
   if closeresult == 0
       disp('File close successful')
   else
       disp('File close not successful')
   end
end
```

Executing this script produces the following results:

```
>> textscanex
5.3 a
2.2 b
3.3 a
4.4 a
1.1 b
File close successful
```

PRACTICE 9.4.

Modify the script *textscanex* to calculate the average of the column of numbers.

9.3.4.1 Comparison of Input File Functions

To compare the use of these input file functions, consider the example of a file called "xypoints.dat" that stores the x and y coordinates of some data points in the following format:

```
>> type xypoints.dat
x2.3y4.56
x7.7y11.11
x12.5y5.5
```

What we want is to be able to store the x and y coordinates in vectors so that we can plot the points. The lines in this file store combinations of characters and numbers, so the **load** function cannot be used. It is necessary to separate the characters from the numbers so that we can create the vectors. The following is the outline of the script to accomplish this:

fileInpCompare.m

```
fid = fopen('xypoints.dat');

if fid == -1
    disp('File open not successful')
else
   % Create x and y vectors for the data points
   % This part will be filled in using different methods

   % Plot the points
   plot(x,y,'k*')
   xlabel('x')
   ylabel('y')

   % Close the file
   closeresult = fclose(fid);
   if closeresult == 0
       disp('File close successful')
   else
       disp('File close not successful')
   end
end
```

We will now complete the middle part of this script using four different methods: **fgetl**, **fscanf** (two ways), and **textscan**.

To use the **fgetl** function, it is necessary to loop until the end-of-file is reached, reading each line as a string, and parsing the string into the various components and converting the strings containing the actual x and y coordinates to numbers. This would be accomplished as follows:

```
% using fgetl
x = [];
y = [];
while feof(fid) == 0
   aline = fgetl(fid);
   aline = aline(2:end);
   [xstr, rest] = strtok(aline,'y');
   x = [x str2double(xstr)];
   ystr = rest(2:end);
   y = [y str2double(ystr)];
end
```

Instead, to use the **fscanf** function, we need to specify the format of every line in the file as a character, a number, a character, a number, and the newline character. As the matrix that will be created will store every line from the file in a separate column, the dimensions will be *4 x n*, where *n* is the number of lines in the file (and as we do not know that, **inf** is specified instead). The x characters will be in the first row of the matrix (the ASCII equivalent of 'x' in each element), the x coordinates will be in the second row, the ASCII equivalent of 'y' will be in the third row, and the fourth row will store the y coordinates. The code would be:

```
% using fscanf

mat = fscanf(fid, '%c%f%c%f\n', [4, inf]);
x = mat(2,:);
y = mat(4,:);
```

Note that the newline character in the format string is necessary. The data file itself was created by typing in the MATLAB Editor/Debugger, and to move down to the next line, the Enter key was used, which is equivalent to the newline character. It is an actual character that is at the end of every line in the file. It is important to note that if the **fscanf** function is looking for a number, it will skip over whitespace characters including blank spaces and newline characters. However, if it is looking for a character, it would read a whitespace character including the newline.

In this case, after reading in 'x2.3y4.56' from the first line of the file, if we had as the format string '%c%f%c%f' (without the '\n'), it would then attempt to read again using '%c%f%c%f', but the next character it would read for the first '%c' would be the newline character, and then it would find the 'x' on the second line for the '%f'—not what is intended! (The difference between this and the previous example is that previously we read a number followed by a character on each line. Thus, when looking for the next number it would skip over the newline character.)

Since we know that every line in the file contains the letter 'x' and 'y', not just any random characters, we can build that into the format string:

```
% using fscanf method 2

mat = fscanf(fid, 'x%fy%f\n', [2, inf]);
x = mat(1,:);
y = mat(2,:);
```

In this case the characters 'x' and 'y' are not read into the matrix, so the matrix only has the x coordinates (in the first row) and the y coordinates (in the second row).

Finally, to use the **textscan** function, we could put '%c' in the format string for the 'x' and 'y' characters, or build those in as with **fscanf**. If we build those in, the format string essentially specifies that there are four columns in the file, but it will only read the columns with the numbers into column vectors in the cell array *xydat*. The reason that the newline character is not necessary is that with **textscan**, the format string specifies what the columns look like in the file, whereas, with **fscanf**, it specifies the format of every line in the file. Thus, it is a slightly different way of viewing the file format.

```
% using textscan

xydat = textscan(fid,'x%fy%f');
x = xydat{1};
y = xydat{2};
```

To summarize, we have now seen four methods of reading from a file. The function **load** will work only if the values in the file are all of the same type and there is the same number on every line in the file, so that they can be read into a matrix. If this is not the case, lower-level functions must be used. To use these, the file must be opened first and then closed when the reading has been completed.

The **fscanf** function will read into a matrix, converting the characters to their ASCII equivalents. The **textscan** function will instead read into a cell array that stores each column from the file into separate column vectors of the cell array. Finally, the **fgetl** function can be used in a loop to read each line from the file as a separate string; string manipulating functions must then be used to break the string into pieces and convert to numbers.

QUICK QUESTION!

If a data file is in the following format, which file input function(s) could be used to read it in?

```
48    25    23    23
12    45     1    31
31    39    42    40
```

Answer: Any of the file input functions could be used, but as the file consists of only numbers and four on each line, the **load** function would be the easiest.

■ Explore Other Interesting Features

Reading from and writing to binary files, using the functions **fread**, **fwrite**, **fseek**, and **frewind**. Note that to open a file to both read from it and write to it, the plus sign must be added to the permission string (e.g., 'r+').

Use **help load** to find some example MAT-files in MATLAB.

The **dlmread** function reads from an ASCII-delimited file into a matrix; also investigate the **dlmwrite** function.

The Import Tool to import files from a variety of file formats.

In the MATLAB Product Help, enter "Supported File Formats" to find a table of the file formats that are supported, and the functions that read from them and write to them. ■

SUMMARY

COMMON PITFALLS

- Misspelling a file name, which causes a file open to be unsuccessful.
- Using a lower-level file I/O function, when **load** or **save** could be used.
- Forgetting that **fscanf** reads columnwise into a matrix, so every line in the file is read into a column in the resulting matrix.
- Forgetting that **fscanf** converts characters to their ASCII equivalents.
- Forgetting that **textscan** reads into a cell array (so curly braces are necessary to index).
- Forgetting to use the permission string 'a' for appending to a file (which means the data already in the file would be lost if 'w' was used!).

PROGRAMMING STYLE GUIDELINES

- Use **load** when the file contains the same kind of data on every line and in the same format on every line.
- Always close files that were opened.
- Always check to make sure that files were opened and closed successfully.
- Make sure that all data are read from a file; e.g., use a conditional loop to loop until the end of the file is reached rather than using a <u>**for**</u> loop.
- Be careful to use the correct formatting string when using **fscanf** or **textscan**.
- Store groups of related variables in separate MAT-files.

MATLAB Functions and Commands			
xlswrite	fclose	feof	textscan
xlsread	fgetl	fprintf	
fopen	fgets	fscanf	

Exercises

1. Create a spreadsheet that has on each line an integer student identification number followed by three quiz grades for that student. Read that information from the spreadsheet into a matrix, and print the average quiz score for each student.
2. The **xlswrite** function can write the contents of a cell array to a spreadsheet. A manufacturer stores information on the weights of some parts in a cell array. Each row stores the part identifier code followed by weights of some sample parts. To simulate this, create the following cell array:

```
>> parts = {'A22', 4.41 4.44 4.39 4.39
            'Z29', 8.88 8.95 8.84 8.92}
```

   Then, write this to a spreadsheet file.
3. A spreadsheet *popdata.xls* stores the population every 20 years for a small town that underwent a boom and then decline. Create this spreadsheet (include the header row) and then read the headers into a cell array and the numbers into a matrix. Plot the data using the header strings on the axis labels.

Year	Population
1920	4021
1940	8053
1960	14,994
1980	9942
2000	3385

4. Create a multiplication table and write it to a spreadsheet.
5. Read numbers from any spreadsheet file, and write the variable to a MAT-file.
6. Clear out any variables that you have in your Command Window. Create a matrix variable and two vector variables.
   - Make sure that you have your Current Folder set.
   - Store all variables to a MAT-file
   - Store just the two vector variables in a different MAT-file
   - Verify the contents of your files using **who**.
7. Create a set of random matrix variables with descriptive names (e.g., *ran2by2int*, *ran3by3double*, etc.) for use when testing matrix functions. Store all of these in a MAT-file.
8. What is the difference between a data file and a MAT-file?
9. Write a script that will prompt the user for the name of a file from which to read. Loop to error-check until the user enters a valid filename that can be opened. (Note: this would be part of a longer program that would actually do something with the file, but for this problem all you have to do is to error-check until the user enters a valid filename that can be read from.)

10. A file "potfilenames.dat" stores potential file names, one per line. The names do not have any extension. Write a script that will print the names of the valid files, once the extension ".dat" has been added. "Valid" means that the file exists in the Current Directory, so it could be opened for reading. The script will also print how many of the file names were valid.
11. A set of data files named "exfile1.dat", "exfile2.dat", etc. has been created by a series of experiments. It is not known exactly how many there are, but the files are numbered sequentially with integers beginning with 1. The files all store combinations of numbers and characters, and are not in the same format. Write a script that will count how many lines in total are in the files. Note that you do not have to process the data in the files in any way; just count the number of lines.
12. Write a script that will read from a file x and y data points in the following format:

```
x 0 y 1
x 1.3 y 2.2
```

The format of every line in the file is the letter 'x', a space, the x value, space, the letter 'y', space, and the y value. First, create the data file with 10 lines in this format. Do this by using the Editor/Debugger, then File Save As *xypts.dat*. The script will attempt to open the data file and error-check to make sure it was opened. If so, it uses a **for** loop and **fgetl** to read each line as a string. In the loop, it creates x and y vectors for the data points. After the loop, it plots these points and attempts to close the file. The script should print whether or not the file was successfully closed.
13. Modify the script from the previous problem. Assume that the data file is in exactly that format, but do not assume that the number of lines in the file is known. Instead of using a **for** loop, loop until the end of the file is reached. The number of points, however, should be in the plot title.

Medical organizations store a lot of very personal information of their patients. There is an acute need for improved methods of storing, sharing, and encrypting all of these medical records. Being able to read from and write to the data files is just the first step.

14. For a biomedical experiment, the names and weights of some patients have been stored in a file *patwts.dat*. For example, the file might look like this:

```
Darby George   166.2
Helen Dee      143.5
Giovanni Lupa  192.4
Cat Donovan    215.1
```

Create this data file first. Then, write a script *readpatwts* that will first attempt to open the file. If the file open is not successful, an error message should be printed. If it is successful, the script will read the data into strings, one line at a time. Print for each person the name in the form 'last,first' followed by the weight. Also, calculate and print the average weight. Finally, print whether or not the file close was successful. For example, the result of running the script would look like this:

```
>> readpatwts
George,Darby 166.2
Dee,Helen 143.5
Lupa,Giovanni 192.4
Donovan,Cat 215.1
The ave weight is 179.30
File close successful
```

15. Create a data file to store blood donor information for a biomedical research company. For every donor, store the person's name, blood type, Rh factor, and blood pressure information. The Blood type is either A, B, AB, or O. The Rh factor is + or −. The blood pressure consists of two readings: systolic and diastolic (both are **double** numbers). Write a script to read from your file into a data structure and print the information from the file.
16. A data file called "mathfile.dat" stores three characters on each line: an operand (a single digit number), an operator (a one character operator, such as +, −, /, \, *, ^), and then another operand (a single digit number). For example, it might look like this:

```
>> type mathfile.dat
5+2
8-1
3+3
```

You are to write a script that will use **fgetl** to read from the file, one line at a time, perform the specified operation, and print the result.

17. Assume that a file named *testread.dat* stores the following:

```
110x0.123y5.67z8.45
120x0.543y6.77z11.56
```

Assume that the following are typed SEQUENTIALLY. What would the values be?

```
tstid = fopen('testread.dat')

fileline = fgetl(tstid)

[beg, endline] = strtok(fileline,'y')

length(beg)

feof(tstid)
```

18. Create a data file to store information on hurricanes. Each line in the file should have the name of the hurricane, its speed in miles per hour, and the diameter of its eye in miles. Then, write a script to read this information from the file and create a vector of structures to store it. Print the name and area of the eye for each hurricane.

19. Create a file "parts_inv.dat" that stores on each line a part number, cost, and quantity in inventory, in the following format:

```
123 5.99 52
```

Use **fscanf** to read this information, and print the total dollar amount of inventory (the sum of the cost multiplied by the quantity for each part).

20. Students from a class took an exam for which there were 2 versions, marked either A or B on the front cover (½ of the students had version A, ½ Version B). The exam results are stored in a file called "exams.dat", which has on each line the version of the exam (the letter 'A' or 'B') followed by a space followed by the integer exam grade. Write a script that will read this information from the file using **fscanf**, and separate the exam scores into two separate vectors: one for Version A, and one for Version B. Then, the grades from the vectors will be printed in the following format (using **disp**).

```
A exam grades:
   99    80    76

B exam grades:
   85    82   100
```

Note: no loops or selection statements are necessary!

21. Create a file which stores on each line a letter, a space, and a real number. For example, it might look like this:

```
e 5.4
f 3.3
c 2.2
```

Write a script that uses **textscan** to read from this file. It will print the sum of the numbers in the file. The script should error-check the file open and close, and print error messages as necessary.

22. Write a script to read in division codes and sales for a company from a file that has the following format:

```
A    4.2
B    3.9
```

Print the division with the highest sales.

23. A data file is created as a **char** matrix and then saved to a file; for example,

```
>> cmat = char('hello', 'ciao', 'goodbye')
cmat =
hello
ciao
goodbye
>> save stringsfile.dat cmat -ascii
```

Can the **load** function be used to read this in? What about **textscan**?

24. Create a file of strings as in the previous exercise, but create the file by opening a new file, type in strings, and then save it as a data file. Can the **load** function be used to read this in? What about **textscan**?

25. Write a script that creates a cell array of strings, each of which is a two-word phrase. The script is to write the first word of each phrase to a file "examstrings.dat" in the format shown below. You do not have to error-check on the file open or file close. The script should be general and should work for any cell array containing two-word phrases.
26. The Wind Chill Factor (WCF) measures how cold it feels with a given air temperature (T, in degrees Fahrenheit) and wind speed (V, in miles per hour). One formula for the WCF follows:

    $$WCF = 35.7 + 0.6\,T - 35.7\left(V^{0.16}\right) + 0.43\,T\left(V^{0.16}\right)$$

    Create a table showing WCFs for temperatures ranging from −20 to 55 in steps of 5, and wind speeds ranging from 0 to 55 in steps of 5. Write this to a file *wcftable.dat*. If you have version R2016a or later, write the script as a live script.
27. Write a script that will loop to prompt the user for n circle radii. The script will call a function to calculate the area of each circle, and will write the results in sentence form to a file.
28. Create a file that has some college department names and enrollments. For example, it might look like this:

    ```
    Aerospace 201
    Mechanical 66
    ```

    Write a script that will read the information from this file and create a new file that has just the first four characters from the department names, followed by the enrollments. The new file will be in this form:

    ```
    Aero 201
    Mech 66
    ```

29. An engineering corporation has a data file "vendorcust.dat" which has names of its vendors and customers for various products, along with a title line. The format is that every line has the vendor name and then the customer name, separated by one space. For example, it might look like this (although you cannot assume the length):

    ```
    >> type vendorcust.dat
    Vendor Customer
    Acme XYZ
    Tulip2you Flowers4me
    Flowers4me Acme
    XYZ Cartesian
    ```

    The "Acme" company wants a little more zing in their name, however, so they've changed it to "Zowie"; now this data file has to be modified. Write a script that will read in from the "vendorcust.dat" file and replace all occurrences of "Acme" with "Zowie", writing this to a new file called "newvc.dat".

30. Environmental engineers are trying to determine whether the underground aquifers in a region are being drained by a new spring water company in the area. Well depth data has been collected every year at several locations in the area. Create a data file that stores on each line the year, an alphanumeric code representing the location, and the measured well depth that year. Write a script that will read the data from the file and determine whether or not the average well depth has been lowered.
31. Write a menu-driven program that will read in an employee data base for a company from a file, and do specified operations on the data. The file stores the following information for each employee:
    - Name
    - Department
    - Birth Date
    - Date Hired
    - Annual Salary
    - Office Phone Extension

    You are to decide exactly how this information is to be stored in the file. Design the layout of the file, and then create a sample data file in this format to use when testing your program. The format of the file is up to you. However, space is critical. Do not use any more characters in your file than you have to! Your program is to read the information from the file into a data structure, and then display a menu of options for operations to be done on the data. You may not assume in your program that you know the length of the data file. The menu options are:
    1. Print all of the information in an easy-to-read format to a new file.
    2. Print the information for a particular department.
    3. Calculate the total payroll for the company (the sum of the salaries).
    4. Find out how many employees have been with the company for N years (N might be 10, for example).
    5. Exit the program.

CHAPTER 10

Advanced Functions

KEY TERMS

variable number of arguments
nested functions
anonymous functions
function handle
function function
recursive functions
outer function
inner function
recursion
general (inductive) case
base case
infinite recursion

CONTENTS

Functions were introduced in Chapter 3 and then expanded on in Chapter 6. In this chapter, several advanced features of functions and types of functions will be described. All of the functions that we have seen so far have had a well-defined number of input and output arguments, but we will see that it is possible to have a ***variable number of arguments***. ***Nested functions*** are also introduced, which are functions contained within other functions. ***Anonymous functions*** are simple one-line functions that are called using their ***function handle***. Other uses of function handles will also be demonstrated, including ***function functions*** and built-in function functions in the MATLAB® software. Finally, ***recursive functions*** are functions that call themselves. A recursive function can return a value, or may simply accomplish a task such as printing.

10.1 VARIABLE NUMBERS OF ARGUMENTS

The functions that we've written thus far have contained a fixed number of input arguments and a fixed number of output arguments. For example, in the following function that we have defined previously, there is one input argument and two output arguments:

MATLAB®. http://dx.doi.org/10.1016/B978-0-12-804525-1.00010-6

areacirc.m

```
function [area, circum] = areacirc(rad)
%areacirc returns the area and
% the circumference of a circle
% Format: areacirc(radius)

area = pi * rad .* rad;
circum = 2 * pi * rad;
end
```

However, this is not always the case. It is possible to have a ***variable number of arguments***, both input and output arguments. A built-in cell array **varargin** can be used to store a variable number of input arguments and a built-in cell array **varargout** can be used to store a variable number of output arguments. These are cell arrays because the arguments could be different types, and only cell arrays can store different kinds of values in the various elements. The function **nargin** returns the number of input arguments that were passed to the function, and the function **nargout** determines how many output arguments are expected to be returned from a function.

10.1.1 Variable Number of Input Arguments

For example, the following function *areafori* has a variable number of input arguments, either 1 or 2. The name of the function stands for "area, feet or inches." If only one argument is passed to the function, it represents the radius in feet. If two arguments are passed, the second can be a character 'i' indicating that the result should be in inches (for any other character, the default of feet is assumed). One foot = 12 in. The function uses the built-in cell array **varargin**, which stores any number of input arguments. The function **nargin** returns the number of input arguments that were passed to the function. In this case, the radius is the first argument passed and so it is stored in the first element in **varargin**. If a second argument is passed (if **nargin** is 2), it is a character that specifies the units.

Note
Curly braces are used to refer to the elements in the cell array **varargin**.

areafori.m

```
function area = areafori(varargin)
% areafori returns the area of a circle in feet
% The radius is passed, and potentially the unit of
% inches is also passed, in which case the result will be
% given in inches instead of feet
% Format: areafori(radius) or areafori(radius,'i')

n = nargin; % number of input arguments
radius = varargin{1}; % Given in feet by default
if n == 2
```

```
        unit = varargin{2};
        % if inches is specified, convert the radius
        if unit=='i'
            radius = radius*12;
        end
    end
    area=pi*radius .^ 2;
    end
```

Some examples of calling this function follow:

```
>> areafori(3)
ans =
   28.2743

>> areafori(1,'i')
ans =
   452.3893
```

In this case, it was assumed that the radius will always be passed to the function. The function header can therefore be modified to indicate that the radius will be passed, and then a variable number of remaining input arguments (either none or 1):

areafori2.m

```
function area = areafori2(radius, varargin)
% areafori2 returns the area of a circle in feet
% The radius is passed, and potentially the unit of
% inches is also passed, in which case the result will be
% given in inches instead of feet
% Format: areafori2(radius) or areafori2(radius,'i')

n=nargin; %number of input arguments

if n==2
    unit = varargin{1};
    % if inches is specified, convert the radius
    if unit=='i'
        radius = radius *12;
    end
end
area = pi *radius .^ 2;
end
```

```
>> areafori2(3)
ans =
   28.2743

>> areafori2(1,'i')
ans =
   452.3893
```

Note that **nargin** returns the total number of input arguments, not just the number of arguments in the cell array **varargin**.

There are basically two formats for the function header with a variable number of input arguments. For a function with one output argument, the options are:

```
function outarg = fnname(varargin)

function outarg = fnname(input arguments, varargin)
```

Either some input arguments are built into the function header and **varargin** stores anything else that is passed, or all of the input arguments go into **varargin**.

PRACTICE 10.1

The sum of a geometric series is given by

$$1 + r + r^2 + r^3 + r^4 + \ldots + r^n$$

Write a function called *geomser* that will receive a value for *r* and calculate and return the sum of the geometric series. If a second argument is passed to the function, it is the value of *n*; otherwise, the function generates a random integer for *n* (ranging from 5 to 30). Note that loops are not necessary to accomplish this. The following examples of calls to this function illustrate what the result should be:

```
>> g = geomser(2,4)  % 1+2^1+2^2+2^3+2^4
g =
    31

>> geomser(1)     % 1 + 1^1 + 1^2 + 1^3 + ... ?
ans =
    12
```

Note that in the last example, a random integer was generated for *n* (which must have been 11). Use the following header for the function, and fill in the rest:

```
function sgs = geomser(r, varargin)
```

10.1.2 Variable Number of Output Arguments

A variable number of output arguments can also be specified. For example, one input argument is passed to the following function *typesize*. The function will always return a character specifying whether the input argument was a scalar ('s'), vector ('v'), or matrix ('m'). This character is returned through the output argument *arrtype*.

Additionally, if the input argument was a vector, the function returns the length of the vector, and if the input argument was a matrix, the function returns the number of rows and the number of columns of the matrix. The output argument **varargout** is used, which is a cell array. So, for a vector the length is returned through **varargout**, and for a matrix both the number of rows and columns are returned through **varargout**.

typesize.m

```
function [arrtype, varargout] = typesize(inputval)
% typesize returns a character 's' for scalar, 'v'
% for vector, or 'm' for matrix input argument
% also returns length of a vector or dimensions of matrix
% Format: typesize(inputArgument)

[r, c ] = size(inputval);

if r==1 && c==1
    arrtype = 's';
elseif r==1 || c==1
    arrtype = 'v';
    varargout{1} = length(inputval);
else
    arrtype = 'm';
    varargout{1} = r;
    varargout{2} = c;
end
end
```

```
>> typesize(5)
ans =
s

>> [arrtype, len] = typesize(4:6)
arrtype =
v
len =
     3

>> [arrtype, r, c] = typesize([4:6;3:5])
arrtype =
m

r =
     2
c =
     3
```

In the examples shown here, the user must actually know the type of the argument in order to determine how many variables to have on the left-hand side of the assignment statement. An error will result if there are too many variables.

```
>> [arrtype, r, c] = typesize(4:6)
Error in typesize (line 7)
[r, c ] = size(inputval);

Output argument "varargout{2}" (and maybe others) not assigned during call
to "\path\typesize.m>typesize".
```

The function **nargout** can be called to determine how many output arguments were used to call a function. For example, in the function *mysize* below, a matrix is passed to the function. The function behaves like the built-in function **size** in that it returns the number of rows and columns. However, if three variables are used to store the result of calling this function, it also returns the total number of elements:

Note

The function **nargout** does not return the number of output arguments in the function header, but the number of output arguments expected from the function (meaning that the number of variables in the vector on the left side of the assignment statement when calling the function).

mysize.m

```
function [row, col, varargout] = mysize(mat)
% mysize returns dimensions of input argument
% and possibly also total # of elements
% Format: mysize(inputArgument)

[row, col] = size(mat);

if nargout == 3
    varargout{1} = row*col;
end
end
```

```
>> [r, c] = mysize(eye(3))
r =
     3
c =
     3

>> [r, c, elem] = mysize(eye(3))
r =
     3
c =
     3
elem =
     9
```

In the first call to the *mysize* function, the value of **nargout** was 2, so the function only returned the output arguments *row* and *col*. In the second call, as there were three variables on the left of the assignment statement, the value of **nargout** was 3; thus, the function also returned the total number of elements.

There are basically two formats for the function header with a variable number of output arguments:

```
function varargout = fnname(input args)

function [output args, varargout] = fnname(input args)
```

Either some output arguments are built into the function header, and **varargout** stores anything else that is returned or all go into **varargout**. The function is called as follows:

```
[variables] = fnname(input args);
```

QUICK QUESTION!

A temperature in degrees Celsius is passed to a function called *converttemp*. How could we write this function so that it converts this temperature to degrees Fahrenheit, and possibly also to degrees Kelvin, depending on the number of output arguments? The conversions are:

$$F = \frac{9}{5}C + 32$$

$$K = C + 273.15$$

Here are possible calls to the function:

```
>> df = converttemp(17)
df =
   62.6000
>> [df, dk] = converttemp(17)
df =
   62.6000
dk =
  290.1500
```

Answer: We could write the function in two different ways: one with only **varargout** in the function header and another with an output argument for the degrees F and also **varargout** in the function header.

converttemp.m

```
function [degreesF, varargout] = converttemp(degreesC)
% converttemp converts temperature in degrees C
% to degrees F and maybe also K
% Format: converttemp(C temperature)

degreesF = 9/5*degreesC + 32;
n = nargout;
if n == 2
   varargout{1} = degreesC + 273.15;
end
end
```

converttempii.m

```
function varargout = converttempii(degreesC)
% converttempii converts temperature in degrees C
% to degrees F and maybe also K
% Format: converttempii(C temperature)

varargout{1} = 9/5*degreesC + 32;
n = nargout;
if n == 2
    varargout{2} = degreesC + 273.15;
end
end
```

10.2 NESTED FUNCTIONS

Just as loops can be nested, meaning one inside of another, functions can be nested. The terminology for ***nested functions*** is that an ***outer function*** can have within it ***inner functions***. When functions are nested, every function must have an **end** statement (much like loops). The general format of a nested function is as follows:

```
outer function header

      body of outer function

      inner function header
            body of inner function
      end % inner function

      more body of outer function

end % outer function
```

The inner function can be in any part of the body of the outer function, so there may be parts of the body of the outer function before and after the inner function. There can be multiple inner functions.

The scope of any variable is the workspace of the outermost function in which it is defined and used. That means that a variable defined in the outer function could be used in an inner function (without passing it).

For example, the following function calculates and returns the volume of a cube. Three arguments are passed to it, for the length and width of the base of the cube, and also the height. The outer function calls a nested function that calculates and returns the area of the base of the cube.

Note
It is not necessary to pass the length and width to the inner function, as the scope of these variables includes the inner function.

nestedvolume.m

```
function outvol = nestedvolume(len, wid, ht)
% nestedvolume receives the length, width, and
% height of a cube and returns the volume; it calls
% a nested function that returns the area of the base
% Format: nestedvolume(length,width,height)

outvol = base * ht;

    function outbase = base
    % returns the area of the base
    outbase = len * wid;
    end % base function

end % nestedvolume function
```

An example of calling this function follows:

```
>> v = nestedvolume(3,5,7)
v =
   105
```

Output arguments are different from variables. The scope of an output argument is just the nested function; it cannot be used in the outer function. In this example, *outbase* can only be used in the *base* function; its value, for example, could not be printed from *nestedvolume*.

A variable defined in the inner function ***could*** be used in the outer function, but if it is not used in the outer function the scope is just the inner function.

Examples of nested functions will be used in the section on Graphical User Interfaces.

10.3 ANONYMOUS FUNCTIONS AND FUNCTION HANDLES

An anonymous function is a very simple, one-line function. The advantage of an anonymous function is that it does not have to be stored in a separate file. This can greatly simplify programs, as often calculations are very simple and the use of anonymous functions reduces the number of code files necessary for a program. Anonymous functions can be created in the Command Window or in any script or user-defined function. The syntax for an anonymous function follows:

```
fnhandlevar = @ (arguments) functionbody
```

where *fnhandlevar* stores the ***function handle***; it is essentially a way of referring to the function. The handle is returned by the @ operator and then this handle is assigned to the variable *fnhandlevar* on the left. The arguments, in parentheses, correspond to the argument(s) that are passed to the function, just like any other kind of function. The functionbody is the body of the function, which is any valid MATLAB expression. For example, here is an anonymous function that calculates and returns the area of a circle:

```
>> cirarea = @ (radius) pi * radius .^ 2;
```

The function handle variable name is *cirarea*. There is one input argument, *radius*. The body of the function is the expression `pi * radius .^ 2`. The `.^` array operator is used so that a vector of radii can be passed to the function.

The function is then called using the handle and passing argument(s) to it; in this case, the radius or vector of radii. The function call using the function handle looks just like a function call using a function name:

```
>> cirarea(4)
ans =
   50.2655
```

```
>> areas = cirarea(1:4)
areas =
    3.1416   12.5664   28.2743   50.2655
```

The type of *cirarea* can be found using the **class** function:

```
>> class(cirarea)
ans =
function_handle
```

Unlike functions stored in code files, if no argument is passed to an anonymous function, the parentheses must still be in the function definition and in the function call. For example, the following is an anonymous function that prints a random real number with two decimal places, as well as a call to this function:

```
>> prtran = @ () fprintf('%.2f\n',rand);
>> prtran()
0.95
```

Typing just the name of the function handle will display its contents, which is the function definition.

```
>> prtran
prtran =
    @ () fprintf('%.2f\n',rand)
```

This is why parentheses must be used to call the function, even though no arguments are passed.

An anonymous function can be saved to a MAT-file and can then be loaded when needed.

```
>> cirarea = @ (radius) pi * radius .^ 2;
>> save anonfns cirarea
>> clear
>> load anonfns
>> who
Your variables are:
cirarea
>> cirarea
cirarea =
    @ (radius) pi * radius .^ 2
```

Other anonymous functions could be appended to this MAT-file. Even though an advantage of anonymous functions is that they do not have to be saved in individual code files, it is frequently useful to save groups of related anonymous functions in a single MAT-file. Anonymous functions that are used frequently can be saved in a MAT-file and then loaded from this MAT-file in every MATLAB Command Window.

PRACTICE 10.2

Create your own anonymous functions to perform some temperature conversions. Store these anonymous functions in a file called "tempconverters.mat."

10.4 USES OF FUNCTION HANDLES

Function handles can also be created for functions other than anonymous functions, both built-in and user-defined functions. For example, the following would create a function handle for the built-in **factorial** function:

```
>> facth = @factorial;
```

The @ operator gets the handle of the function, which is then stored in a variable *facth*.

The handle could then be used to call the function, just like the handle for the anonymous functions, as in:

```
>> facth(5)
ans =
   120
```

Using the function handle to call the function instead of using the name of the function does not in itself demonstrate why this is useful, so an obvious question would be why function handles are necessary for functions other than anonymous functions.

10.4.1 Function Functions

One reason for using function handles is to be able to pass functions to other functions—these are called ***function functions***. For example, let's say we have a function that creates an *x* vector. The *y* vector is created by evaluating a function at each of the *x* points, and then these points are plotted.

fnfnexamp.m

```
function fnfnexamp(funh)
% fnfnexamp receives the handle of a function
% and plots that function of x (which is 1:.25:6)
% Format: fnfnexamp(function handle)

x = 1:.25:6;
y = funh(x);
plot(x,y,'ko')
xlabel('x')
ylabel('fn(x)')
title(func2str(funh))
end
```

What we want to do is pass a function to be the value of the input argument *funh*, such as **sin**, **cos**, or **tan**. Simply passing the name of the function does not work:

```
>> fnfnexamp(sin)
Error using sin
Not enough input arguments.
```

Instead, we have to pass the handle of the function:

```
>> fnfnexamp(@sin)
```

which creates the *y* vector as **sin(x)** and then brings up the plot as seen in Fig. 10.1. The function **func2str** converts a function handle to a string; this is used for the title.

Passing the handle to the **cos** function instead would graph cosine instead of sine:

```
>> fnfnexamp(@cos)
```

We could also pass the handle of any user-defined or anonymous function to the *fnfnexamp* function. Note that if a variable stores a function handle, just the name of the variable would be passed (not the @ operator). For example, for our anonymous function defined previously,

```
>> fnfnexamp(cirarea)
```

The function **func2str** will return the definition of an anonymous function as a string that could also be used as a title. For example:

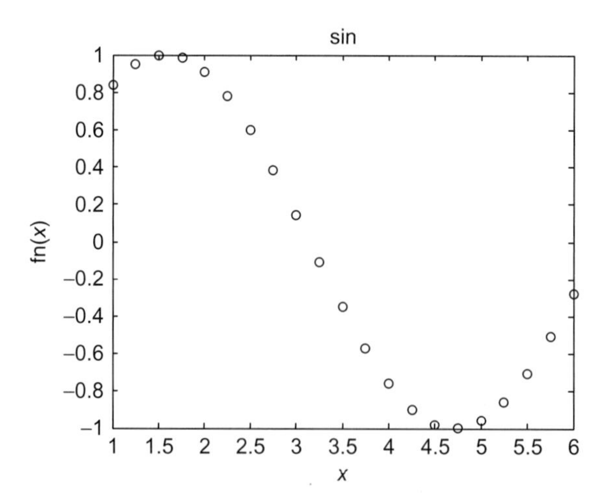

FIGURE 10.1
Plot of **sin** created by passing handle of function to plot.

```
>> cirarea = @ (radius) pi * radius .^ 2;
>> fnname = func2str(cirarea)
fnname =
@(radius)pi*radius.^2
```

There is also a built-in function **str2func** that will convert a string to a function handle. A string containing the name of a function could be passed as an input argument, and then converted to a function handle.

fnstrfn.m

```
function fnstrfn(funstr)
% fnstrfn receives the name of a function as a string
% it converts this to a function handle and
% then plots the function of x (which is 1:.25:6)
% Format: fnstrfn(function name as string)
x = 1:.25:6;
funh = str2func(funstr);
y = funh(x);
plot(x,y,'ko')
xlabel('x')
ylabel('fn(x)')
title(funstr)
end
```

This would be called by passing a string to the function, and would create the same plot as in Fig. 10.1:

```
>> fnstrfn('sin')
```

PRACTICE 10.3

Write a function that will receive as input arguments an *x* vector and a function handle, and will create a vector *y* that is the function of *x* (whichever function handle is passed) and will also plot the data from the *x* and *y* vectors with the function name in the title.

MATLAB has some built-in function functions. One built-in function function is **fplot**, which plots a function between limits that are specified. The form of the call to **fplot** is:

```
fplot(fnhandle, [xmin, xmax])
```

For example, to pass the **sin** function to **fplot,** one would pass its handle (see Fig. 10.2 for the result).

```
>> fplot(@sin, [-pi, pi])
```

The **fplot** function is a nice shortcut—it is not necessary to create *x* and *y* vectors, and it plots a continuous curve rather than discrete points.

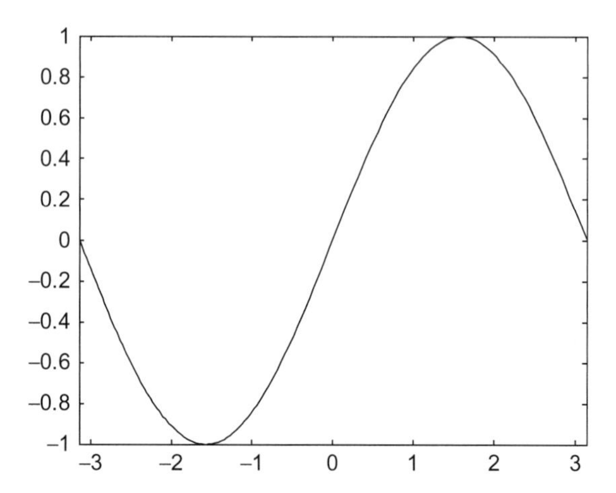

FIGURE 10.2
Plot of **sin** created using **fplot**.

QUICK QUESTION!

Could you pass an anonymous function to the **fplot** function?

Answer: Yes, as in:

```
>> cirarea = @ (radius) pi*radius .^2;
>> fplot(cirarea, [1, 5])
>> title(func2str(cirarea))
```

Note that in this case the @ operator is not used in the call to **fplot**, as *cirarea* already stores the function handle.

The function function **feval** will evaluate a function handle and execute the function for the specified argument. For example, the following is equivalent to **sin(3.2)**:

```
>> feval(@sin, 3.2)
ans =
   -0.0584
```

Another built-in function function is **fzero**, which finds a zero of a function near a specified value. For example:

```
>> fzero(@cos,4)
ans =
    4.7124
```

10.4.2 Timing Functions

The function **timeit** (introduced in R2013b) can be used to time functions, and is more robust than using **tic** and **toc**. The **timeit** function takes one input argument, which is a function handle; this can be the handle of any type of function. The time is returned in seconds.

```
>> fh = @ () prod(1:10000000);
>> timeit(fh)
ans =
    0.0308
```

A warning message is thrown if the function is too fast.

```
>> fh = @ () prod(1:100);
>> timeit(fh)
Warning: The measured time for F may be inaccurate because it is running too
fast. Try measuring something that takes longer.
> In timeit (line 158)
ans =
    1.3993e-06
```

10.5 RECURSIVE FUNCTIONS

Recursion is when something is defined in terms of itself. In programming, a ***recursive function*** is a function that calls itself. Recursion is used very commonly in programming, although many simple examples (including some shown in this section) are actually not very efficient and can be replaced by iterative methods (loops or vectorized code in MATLAB). Nontrivial examples go beyond the scope of this book, so the concept of recursion is simply introduced here.

The first example will be of a factorial. Normally, the factorial of an integer n is defined iteratively:

```
n! = 1 * 2 * 3 * ... * n
```

For example, `4! = 1 * 2 * 3 * 4, or 24.`

Another, recursive, definition is:

```
n! = n * (n - 1)!        general case
1! = 1                   base case
```

This definition is recursive because a factorial is defined in terms of another factorial. There are two parts to any recursive definition: the ***general (or inductive) case***, and the ***base case***. We say that, in general, the factorial of n is defined as n multiplied by the factorial of $(n-1)$, but the base case is that the factorial of 1 is just 1. The base case stops the recursion.

For example:

```
3! = 3 * 2!
          2! = 2 * 1!
                     1! = 1
              = 2
   = 6
```

The way this works is that 3! is defined in terms of another factorial, as 3 * 2!. This expression cannot as yet be evaluated because first we have to find out the value of 2!. So, in trying to evaluate the expression 3 * 2!, we are interrupted by the recursive definition. According to the definition, 2! is 2 * 1!. Again, the expression 2 * 1! cannot yet be evaluated because first we have to find the value of 1!. According to the definition, 1! is 1. As we now know what 1! is, we can continue with the expression that was just being evaluated; now we know that 2 * 1! is 2 * 1, or 2. Thus, we can now finish the previous expression that was being evaluated; as now we know that 3 * 2! is 3 * 2, or 6.

This is the way that recursion always works. With recursion, the expressions are put on hold with the interruption of the general case of the recursive definition. This keeps happening until the base case of the recursive definition applies. This finally stops the recursion, and then the expressions that were put on hold are evaluated in the reverse order. In this case, first the evaluation of 2 * 1! was completed, and then 3 * 2!.

There must always be a base case to end the recursion, and the base case must be reached at some point. Otherwise, ***infinite recursion*** would occur (theoretically, although MATLAB will stop the recursion eventually).

We have already seen the built-in function **factorial** in MATLAB to calculate factorials, and we have seen how to implement the iterative definition using a running product. Now we will instead write a recursive function called *fact*. The function will receive an integer *n*, which we will for simplicity assume is a positive integer, and will calculate *n!* using the recursive definition given previously.

fact.m

```
function facn = fact(n)
% fact recursively finds n!
% Format: fact(n)
if n == 1
    facn = 1;
else
    facn = n * fact(n-1);
end
end
```

The function calculates one value, using an **if-else** statement to choose between the base and general cases. If the value passed to the function is 1, the function returns 1 as 1! is equal to 1. Otherwise, the general case applies. According to the definition, the factorial of n, which is what this function is calculating, is defined as n multiplied by the factorial of $(n-1)$. So, the function assigns `n * fact(n-1)` to the output argument.

How does this work? Exactly the way the example was sketched previously for 3!. Let's trace what would happen if the integer 3 is passed to the function:

```
fact(3) tries to assign 3 * fact(2)
                          fact(2) tries to assign 2 * fact(1)
                                                      fact(1) assigns 1
                          fact(2) assigns 2
fact(3) assigns 6
```

When the function is first called, 3 is not equal to 1, so the statement

```
facn = n * fact(n-1);
```

is executed. This will attempt to assign the value of 3 * fact(2) to *facn*, but this expression cannot be evaluated yet and therefore a value cannot be assigned yet because first the value of fact(2) must be found.

Thus, the assignment statement has been interrupted by a recursive call to the *fact* function. The call to the function fact(2) results in an attempt to assign 2 * fact(1), but again, this expression cannot yet be evaluated. Next, the call to the function fact(1) results in a complete execution of an assignment statement as it assigns just 1. Once the base case has been reached, the assignment statements that were interrupted can be evaluated, in the reverse order.

Calling this function yields the same result as the built-in **factorial** function, as follows:

```
>> fact(5)
ans =
    120

>> factorial(5)
ans =
    120
```

The recursive factorial function is a very common example of a recursive function. It is somewhat of a lame example, however, as recursion is not necessary to find a factorial; a **for** loop can be used just as well in programming (or, of course, the built-in function in MATLAB).

Another better example is of a recursive function that does not return anything, but simply prints. The following function *prtwords* receives a sentence and prints the words in the sentence in reverse order. The algorithm for the *prtwords* function follows:

- Receive a sentence as an input argument.
- Use **strtok** to break the sentence into the first word and the rest of the sentence.

- If the rest of the sentence is not empty (in other words, if there is more to it), recursively call the *prtwords* function and pass to it the rest of the sentence.
- Print the word.

The function definition follows:

prtwords.m

```
function prtwords(sent)
% prtwords recusively prints the words in a string
% in reverse order
% Format: prtwords(string)

[word, rest] = strtok(sent);
if ~isempty(rest)
   prtwords(rest);
end
disp(word)
end
```

Here is an example of calling the function, passing the sentence "what does this do":

```
>> prtwords('what does this do')
do
this
does
what
```

An outline of what happens when the function is called follows:

```
The function receives 'what does this do'
It breaks it into word = 'what', rest = 'does this do'
Since "rest" is not empty, calls prtwords, passing "rest"
    The function receives 'does this do'
    It breaks it into word = 'does', rest = 'this do'
    Since "rest" is not empty, calls prtwords, passing "rest"
        The function receives 'this do'
        It breaks it into word = 'this', rest = 'do'
        Since "rest" is not empty, calls prtwords, passing "rest"
            The function receives 'do'
            It breaks it into word = 'do', rest = ''
            "rest" is empty so no recursive call
            Print 'do'
        Print 'this'
    Print 'does'
Print 'what'
```

In this example, the base case is when the rest of the string is empty, in other words, the end of the original sentence has been reached. Every time the function is called, the execution of the function is interrupted by a recursive call to the function until the base case is reached. When the base case is reached, the entire function can be executed, including printing the word (in the base case, the word 'do').

Once that execution of the function is completed, the program returns to the previous version of the function in which the word was 'this,' and finishes the execution by printing the word 'this.' This continues; the versions of the function are finished in the reverse order, so the program ends up printing the words from the sentence in the reverse order.

PRACTICE 10.4

For the following function,

recurfn.m

```
function outvar = recurfn(num)
% Format: recurfn(number)

if num < 0
   outvar = 2;
else
   outvar = 4 +recurfn(num-1);
end
end
```

what would be returned by the call to the function `recurfn(3.5)`? Think about it, and then type in the function and test it.

■ Explore Other Interesting Features

Other function functions and Ordinary Differential Equation (ODE) solvers can be found using **help funfun**.

The function function **bsxfun**. Look at the example in the documentation page of subtracting the column mean from every element in each column of a matrix.

The ODE solvers include **ode45** (which is used most often), **ode23**, and several others. Error tolerances can be set with the **odeset** function.

Investigate the use of the functions **narginchk** and **nargoutchk**.

The function **nargin** can be used not just when using **varargin**, but also for error-checking for the correct number of input arguments into a function. Explore examples of this. ■

SUMMARY

COMMON PITFALLS

- Thinking that **nargin** is the number of elements in **varargin** (it may be, but not necessarily; **nargin** is the total number of input arguments).
- Trying to pass just the name of a function to a function function; instead, the function handle must be passed.
- Forgetting the base case for a recursive function.

PROGRAMMING STYLE GUIDELINES

- If some inputs and/or outputs will always be passed to/from a function, use standard input arguments/output arguments for them. Use **varargin** and **varargout** only when it is not known ahead of time whether other input/output arguments will be needed.
- Use anonymous functions whenever the function body consists of just a simple expression.
- Store related anonymous functions together in one MAT-file.
- Use iteration instead of recursion when possible.

MATLAB Reserved Words
end (for functions)

MATLAB Functions and Commands	
varargin	str2func
varargout	fplot
nargin	feval
nargout	fzero
func2str	timeit

MATLAB Operators
handle of functions @

Exercises

1. Write a function that will print a random integer. If no arguments are passed to the function, it will print an integer in the inclusive range from 1 to 100. If one argument is passed, it is the max and the integer will be in the inclusive range from 1 to max. If two arguments are passed, they represent the min and max and it will print an integer in the inclusive range from min to max.
2. Write a function *numbers* that will create a matrix in which every element stores the same number num. Either two or three arguments will be passed to the function. The first argument will always be the number num. If there are two arguments, the second will be the size of the resulting square ($n \times n$) matrix. If there are three arguments, the second and third will be the number of rows and columns of the resulting matrix.
3. The overall electrical resistance of n resistors in parallel is given as:

$$R_T = \left(\frac{1}{R_1} + \frac{1}{R_2} + \frac{1}{R_3} + \cdots + \frac{1}{R_n}\right)^{-1}$$

   Write a function *Req* that will receive a variable number of resistance values and will return the equivalent electrical resistance of the resistor network.
4. Write a function that will receive the radius r of a sphere. It will calculate and return the volume of the sphere ($4/3\ \pi r^3$). If the function call expects two output arguments, the function will also return the surface area of the sphere ($4\ \pi r^2$).
5. Most lap swimming pools have lanes that are either 25 yd long or 25 m long; there's not much of a difference. A function "convyards" is to be written to help swimmers calculate how far they swam. The function receives as input the number of yards. It calculates and returns the equivalent number of meters, and, if (and only if) two output arguments are expected, it also returns the equivalent number of miles. The relevant conversion factors are:

```
1 meter = 1.0936133 yards
1 mile = 1760 yards
```

6. Write a function *unwind* that will receive a matrix as an input argument. It will return a row vector created columnwise from the elements in the matrix. If the number of expected output arguments is 2, it will also return this as a column vector.
7. Write a function "cylcalcs" that will receive the radius and height of a cylinder and will return the area and volume of the cylinder. If the function is called as an expression or in an assignment statement with one variable on the left, it will return the area and volume together in one vector. On the other hand, if the function is called in an assignment statement with two variables on the left, the function will return the area and volume separately (in that order). The function will call a subfunction to calculate the area. The formulas are:

```
Area = pi * radius²
Volume = Area * height
```

8. Information on some hurricanes is stored in a vector of structures; the name of the vector variable is *hurricanes*. For example, one of the structures might be initialized as follows:

```
struct('Name','Bettylou', 'Avespeed',18,...
        'Size', struct('Width',333,'Eyewidth',22));
```

   Write a function *printHurr* that will receive a vector of structures in this format as an input argument. It will print, for every hurricane, its *Name* and *Width* in a sentence format to the screen. If a second argument is passed to the function, it is a file identifier for an output file (which means that the file has already been opened), and the function will print in the same format to this file (and does not close it).
9. The built-in function **date** returns a string containing the day, month, and year. Write a function (using the **date** function) that will always return the current day. If the function call expects two output arguments, it will also return the month. If the function call expects three output arguments, it will also return the year.
10. List some built-in functions to which you pass a variable number of input arguments. (Note: This is not asking for **varargin**, which is a built-in cell array, or **nargin**.)
11. List some built-in functions that have a variable number of output arguments (or, at least one!).
12. Write a function that will receive a variable number of input arguments: the length and width of a rectangle, and possibly also the height of a box that has this rectangle as its base. The function should return the rectangle area if just the length and width are passed, or also the volume if the height is also passed.
13. Write a function to calculate the volume of a cone. The volume V is $V = AH$ where A is the area of the circular base ($A = \pi r^2$ where r is the radius) and H is the height. Use a nested function to calculate A.
14. The two real roots of a quadratic equation $ax^2 + bx + c = 0$ (where a is nonzero) are given by

$$\frac{-b \pm \sqrt{D}}{2a}$$

   where the discriminant $D = b^2 - 4 \times a \times c$. Write a function to calculate and return the roots of a quadratic equation. Pass the values of a, b, and c to the function. Use a nested function to calculate the discriminant.
15. The velocity of sound in air is $49.02\sqrt{T}$ ft/s where T is the air temperature in degrees Rankine. Write an anonymous function that will calculate this. One argument, the air temperature in degrees R, will be passed to the function and it will return the velocity of sound.
16. Create a set of anonymous functions to do length conversions and store them in a file named *lenconv.mat*. Call each with a descriptive name, such as *cmtoinch* to convert from centimeters to inches.

17. Write an anonymous function to convert from fluid ounces to milliliters. The conversion is one fluid ounce is equivalent to 29.57 mL.
18. Why would you want to use an anonymous function?
19. Write an anonymous function to implement the following quadratic: `3x^2-2x+5`. Then, use **fplot** to plot the function in the range from −6 to 6.
20. Write a function that will receive data in the form of *x* and *y* vectors, and a handle to a plot function and will produce the plot. For example, a call to the function would look like `wsfn(x,y,@bar)`.
21. Write a function *plot2fnhand* that will receive two function handles as input arguments, and will display in two Figure Windows, plots of these functions with the function names in the titles. The function will create an *x* vector that ranges from 1 to *n* (where *n* is a random integer in the inclusive range from 4 to 10). For example, if the function is called as follows

```
>> plot2fnhand(@sqrt, @exp)
```

    and the random integer is 5, the first Figure Window would display the **sqrt** function of $x = 1:5$, and the second Figure Window would display **exp(x)** for $x = 1:5$.
22. Use **feval** as an alternative way to accomplish the following function calls:

    `abs(-4)`
    `size(zeros(4))` Use **feval** twice for this one!

23. There is a built-in function function called **cellfun** that evaluates a function for every element of a cell array. Create a cell array, then call the **cellfun** function, passing the handle of the **length** function and the cell array to determine the length of every element in the cell array.
24. A recursive definition of a^n where *a* is an integer and *n* is a nonnegative integer follows:

```
a^n = 1             if n == 0
    = a * a^(n-1)   if n > 0
```

    Write a recursive function called *mypower*, which receives *a* and *n* and returns the value of a^n by implementing the previous definition. Note: The program should NOT use ^ operator anywhere; this is to be done recursively instead! Test the function.
25. What does this function do:

```
function outvar = mystery(x,y)
if y==1
   outvar = x;
else
   outvar = x +mystery(x,y-1);
end
```

    Give one word to describe what this function does with its two arguments.

    The Fibonacci numbers is a sequence of numbers 0, 1, 1, 2, 3, 5, 8, 13, 21, 34... The sequence starts with 0 and 1. All other Fibonacci numbers are obtained by adding the previous two Fibonacci numbers. The higher up in the sequence that you go, the closer the fraction of one Fibonacci number divided by the previous is to the golden ratio. The Fibonacci numbers can be seen in an astonishing number of examples in nature, for example, the arrangement of petals on a sunflower.

26. The Fibonacci numbers is a sequence of numbers F_i:

    ```
    0  1  1  2  3  5  8  13  21  34 ...
    ```

    where F_0 is 0, F_1 is 1, F_2 is 1, F_3 is 2, and so on. A recursive definition is:

    $F_0 = 0$
    $F_1 = 1$
    $F_n = F_{n-2} + F_{n-1} \quad \text{if } n > 1$

    Write a recursive function to implement this definition. The function will receive one integer argument *n*, and it will return one integer value that is the *n*th Fibonacci number. Note that in this definition there is one general case but two base cases. Then, test the function by printing the first 20 Fibonacci numbers.
27. Use **fgets** to read strings from a file and recursively print them backwards.

Introduction to Object-Oriented Programming and Graphics

KEY TERMS

procedural languages
object-oriented languages
classes
hybrid languages
abstract data types
objects
properties
methods
class definition
instances
instantiation
superclass/subclass
value classes
handle classes
object handle
graphics primitives
root object
constructor function
ordinary method
overloading
attributes
copy constructor
destructor function
events
event-driven programming
listeners
callback

CONTENTS

Most programming languages are either ***procedural*** or ***object-oriented***. Procedural programs are comprised of ***functions***, each of which performs a task. Object-oriented programs use ***classes***, which contain both data and functions to manipulate the data. ***Hybrid languages*** can utilize both of these programming paradigms. All of our programs so far have been procedural, but the MATLAB® software uses objects in its graphics, and thus has object-oriented programming (OOP) capabilities.

In this chapter, we will first introduce some of the concepts and terminologies of OOP using graphics objects and will show how to manipulate plot properties using this system. Later we will show how user-defined classes can be created.

11.1 OBJECT-ORIENTED PROGRAMMING

This short section is intended to introduce the very basic ideas behind OOP as well as some of the terminology that is used. This section is very dense in terms

MATLAB®. http://dx.doi.org/10.1016/B978-0-12-804525-1.00011-8

of the terminology. It is hoped that by introducing the terms here, and then giving examples in the next two sections, the terms will be easier to understand.

Built-in data types have certain capabilities associated with them. For example, we can perform mathematical operations such as adding and dividing with number types such as **double**. When a variable is created that stores a value of a particular type, operations can then be performed that are suitable for that type.

Similarly, ***abstract data types*** are data types that are defined by both data and operational capabilities. In MATLAB, these are called ***classes***. Classes define both the data and the functions that manipulate the data. Once a class has been defined, ***objects*** can be created of the class.

To define a class, both the data and the functions that manipulate the data must be defined. The data are called ***properties***, and are similar to variables in that they store the values. The functions are called ***methods***. A ***class definition*** consists of the definition of the properties and the definition of the methods.

Once a class has been defined, ***objects*** can be created from the class. The objects are called ***instances*** of the class and an object that is created is an ***instantiation*** of the class. The properties and methods of the object can be referenced using the object name and the dot operator.

Inheritance is when one class is derived from another. The initial class is called the ***base***, ***parent***, or ***superclass***, and the derived class is called the ***derived***, ***child***, or ***subclass***. A subclass is a new class that has the properties and methods of the superclass (this is what is called ***inheritance***), plus it can have its own properties and methods. The methods in a subclass can override methods from the superclass if they have the same name.

MATLAB has built-in classes and also allows for user-defined classes. There are two types of classes in MATLAB: ***value classes*** and ***handle classes***. The differences will be explained in Section 11.3. MATLAB uses handle classes to implement graphical objects used in plots.

11.2 USING OBJECTS WITH GRAPHICS AND PLOT PROPERTIES

MATLAB uses graphics in all of its figures. All figures consist of ***objects***, each of which is referenced using an ***object handle***. In versions of MATLAB prior to version R2014b, the object handles were unique real numbers that were used to refer to the object. With the new graphics system that was initiated in MATLAB R2014b, the object handles actually store objects, which are derived from a superclass called **handle**.

Plot objects include ***graphics primitives***, which are basic building blocks of plots, as well as the axes used to orient the objects. The graphics primitives include objects such as lines and text. For example, a plot that uses straight line segments uses the **line** graphics primitive. More of the graphics primitives will be discussed in Section 12.3. The objects are organized hierarchically, and there are properties associated with each object.

The computer screen is called the ***root object*** and is referenced using the function **groot** (which is short for "graphics root"). When plots are made, they appear in the Figure Window; the Figure Window itself appears on the computer screen. The hierarchy in MATLAB can be summarized as follows:

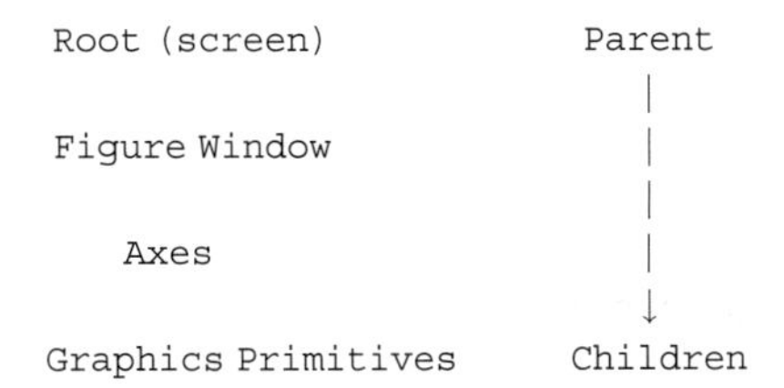

In other words, the Figure Window is in the screen; it includes Axes, which are used to orient graphics primitives, which are the building blocks of plots.

11.2.1 Objects and Properties

A Figure Window is an object; the data in objects are stored in properties. Just calling the **figure** function will bring up a blank Figure Window and return its handle; by assigning the handle of this Figure Window to an object variable, the properties can then be referenced. For example, if no other figures have been created yet, the following will create Figure (1).

```
>> f = figure
f =
 Figure (1) with properties:

    Number: 1
      Name: ''
     Color: [0.9400 0.9400 0.9400]
  Position: [440 378 560 420]
     Units: 'pixels'

Show all properties
```

By default only a few of the properties are listed; what is shown are the names of the properties and the values for this instance of the object; these include:

- the figure number: 1
- the name of the figure: none was given so this is an empty string

- the color: given as a vector storing the values of the red, green, and blue components of the color
- the position of the Figure Window within the screen, specified in the units of pixels (which is shown next); this is a vector consisting of four values: the number of pixels that the lower left corner of the Figure Window is from the left of the screen, the number of pixels that the lower left corner of the Figure Window is from the bottom of the screen, the length in pixels and the height in pixels
- the units: pixels

For the Color property, the three numbers in the vector are real numbers in the range from 0 to 1. Zero for a color component means none of that color, whereas one is the brightest possible hue. All zeros represent black, and all ones represent the color white. The default Color property value of `[0.94 0.94 0.94]` is a very light gray.

By clicking on the "all properties" link, all of the properties can be seen. As long as the Figure Window is not closed, the handle *f* can be used to refer to the Figure Window; but when the Figure Window is closed, the handle is deleted. Another method of viewing the properties is to pass the handle to the **get** function as follows.

```
>> get(f)
              Alphamap: [1x64 double]
          BeingDeleted: 'off'
            BusyAction: 'queue'
         ButtonDownFcn: ''
              Children: [0x0 GraphicsPlaceholder]
              Clipping: 'on'
       CloseRequestFcn: 'closereq'
                 Color: [0.9400 0.9400 0.9400]
              Colormap: [64x3 double]
             CreateFcn: ''
           CurrentAxes: [0x0 GraphicsPlaceholder]
      CurrentCharacter: ''
         CurrentObject: [0x0 GraphicsPlaceholder]
          CurrentPoint: [0 0]
             DeleteFcn: ''
          DockControls: 'on'
              FileName: ''
     GraphicsSmoothing: 'on'
      HandleVisibility: 'on'
         IntegerHandle: 'on'
         Interruptible: 'on'
        InvertHardcopy: 'on'
           KeyPressFcn: ''
         KeyReleaseFcn: ''
```

```
                MenuBar: 'figure'
                   Name: ''
               NextPlot: 'add'
                 Number: 1
            NumberTitle: 'on'
       PaperOrientation: 'portrait'
          PaperPosition: [0.2500 2.5000 8 6]
      PaperPositionMode: 'manual'
              PaperSize: [8.5000 11]
              PaperType: 'usletter'
             PaperUnits: 'inches'
                 Parent: [1x1 Root]
                Pointer: 'arrow'
      PointerShapeCData: [16x16 double]
    PointerShapeHotSpot: [1 1]
               Position: [440 378 560 420]
               Renderer: 'opengl'
           RendererMode: 'auto'
                 Resize: 'on'
          SelectionType: 'normal'
         SizeChangedFcn: ''
                    Tag: ''
                ToolBar: 'auto'
                   Type: 'figure'
          UIContextMenu: [0x0 GraphicsPlaceholder]
                  Units: 'pixels'
               UserData: []
                Visible: 'on'
    WindowButtonDownFcn: ''
  WindowButtonMotionFcn: ''
      WindowButtonUpFcn: ''
      WindowKeyPressFcn: ''
    WindowKeyReleaseFcn: ''
   WindowScrollWheelFcn: ''
            WindowStyle: 'normal'
               XDisplay: 'Quartz'
```

You may not understand most of these properties; do not worry about it! Notice, however, that the Parent of this figure is the Root object and that there are no Children since there is nothing in this Figure Window.

QUICK QUESTION!

What would the following display:

```
>> f
```

Answer: It would display the same abbreviated list of properties as was shown when the handle was first created.

The **get** function returns a structure consisting of the names and values of the properties; once stored in a structure variable, the dot operator can then be used to refer to the value of a particular property.

```
>> fstruct = get(f);
>> fstruct.Color
ans =
    0.9400    0.9400    0.9400
```

The **get** function can also be used to retrieve just one particular property; e.g., the Color property as follows.

```
>> get(f,'Color')
ans =
    0.9400    0.9400    0.9400
```

PRACTICE 11.1

Call the **groot** function and store the resulting handle in an object variable. What are the dimensions of your screen?

(Note that as of R2015b, pixels are now a fixed size and do not necessarily correspond exactly to the actual number of pixels on the screen.)

The function **set** can be used to change property values. The **set** function is called in the format:

```
set(objhandle, 'PropertyName', property value)
```

For example, the position of the Figure Window could be modified as follows.

```
>> set(f,'Position',[400 450 600 550])
```

Another method for referencing or changing a property of an object is to use the dot notation. The format for this is:

```
objecthandle.PropertyName
```

This method is new as of R2014b and is an alternative to using **get** and **set**. Using the dot notation is preferable to using **get** and **set** since the handles are now objects and this is the standard syntax for referencing object properties.

For example, the following stores the current value of the Color property in a variable and then modifies the Color property to a darker gray.

```
>> c = f.Color
c =
    0.9400    0.9400    0.9400

>> f.Color = [0.5 0.5 0.5]
```

Note that this is the same as the notation to refer to a field in a structure, but it is not a structure; this is directly referencing a property within an object. This can be confusing, so here is a review of the variables that have been created:

- The variable *f* is the handle of the Figure Window object. Objects have properties, which contain the data. Object properties can be referenced using the dot operator, e.g., *f.Name*.
- The variable *fstruct* is a structure that contains the names and values of the properties. Fields in structures can be referenced using the dot operator, e.g., *fstruct.Name*.

QUICK QUESTION!

So, is *f* the same as *fstruct*? Is *f.Name* the same as *fstruct.Name*?

Answer: The variables *f* and *fstruct* are not the same. They are different types; *f* is the handle of an object, whereas *fstruct* is a structure variable. However, *f.Name* and *fstruct.Name* are equivalent; they both refer to the name of the Figure Window.

Note, however, that changing a property of *f* will not change the value in the structure variable. The Name property was an empty string by default; changing *f.Name* will not automatically change *fstruct.Name*.

```
>> f.Name
ans =
    ''
>> fstruct.Name
ans =
    ''
>> f.Name = 'FW Title'
f =
  Figure (2: FW Title) with properties:

      Number: 2
        Name: 'FW Title'
       Color: [0.9400 0.9400 0.9400]
    Position: [440 378 560 420]
       Units: 'pixels'

  Show all properties
>> fstruct.Name
ans =
    ''
```

For the figure object stored in *f*, its built-in class is matlab.ui.Figure; "ui" is the abbreviation for "user interface" and is used in many graphics names. This can be seen using the **class** function.

```
>> class(f)
ans =
matlab.ui.Figure
```

The class of *fstruct*, on the other hand, is **struct**:

```
>> class(fstruct)
ans =
struct
```

Recall that a class definition consists of the data (properties) and functions to manipulate the data (methods). There are built-in functions **properties** and **methods** that will display the properties and methods for a particular class. For example, for the figure referenced by the handle *f*, we can find the properties; note that they are not listed in alphabetical order as with **get** and that only the names of the properties are returned (not the values).

```
>> properties(f)
Properties for class matlab.ui.Figure:
    Position
    Units
    Renderer
    RendererMode
    Visible
    Color

          etc.
```

The methods for the figure *f* are as follows.

```
>> methods(f)

Methods for class matlab.ui.Figure:

Figure       clo       double    horzcat   reset
addlistener  details   eq        isprop    set
addprop      disp      findobj   java      vertcat
cat          display   get       ne

Static methods:

loadobj

Methods of matlab.ui.Figure inherited from handle.
```

Again, much of this will not make sense but notice that the methods are derived from the superclass **handle**. The methods, or functions, that can be used with the object *f* include **get** and **set**. Also, the methods **eq** and **ne** are defined; these are equivalent to the equality (==) and inequality (~=) operators. That means that the equality and inequality operators can be used to determine whether two figure handles are equal to each other or not.

QUICK QUESTION!

Given two figure objects

```
>> f = figure(1);
>> g = figure(2);
```

Assuming that both Figure Windows are still open, what would be the value of the following expression?

```
>> f == g
```

Answer: This would be false. However, consider the following, which demonstrates that all of the color components of the two figures are the same, and that if one figure handle variable is assigned to another, they are equivalent.

```
>> f.Color == g.Color
ans =
     1     1     1
>> h = f;
>> f == h
ans =
     1
```

The various plot functions return a handle for the plot object, which can then be stored in a variable. In the following, the **plot** function plots a **sin** function in a Figure Window and returns the object handle. This handle will remain valid as long as the object exists.

```
>> x = -2*pi: 1/5 : 2*pi;
>> y = sin(x);
>> hl = plot(x,y)
hl =
  Line with properties:

              Color: [0 0.4470 0.7410]
          LineStyle: '-'
          LineWidth: 0.5000
             Marker: 'none'
         MarkerSize: 6
    MarkerFaceColor: 'none'
              XData: [1x63 double]
              YData: [1x63 double]
              ZData: [1x0 double]

  Show all properties
```

Notice that the plot is generated using the **line** graphics primitive. As with the Figure Window, the properties can be viewed and modified using either the dot notation or using **get** and **set**. For example, we can find that the parent of the plot is an Axes object.

```
>> axhan = hl.Parent
ans =
  Axes with properties:

             XLim: [-8 8]
             YLim: [-1 1]
               etc.
```

QUICK QUESTION!

How could you change the *x*-axis limit to `[-10 10]`?

Answer: There are two ways:

```
>> axhan.XLim = [-10 10]

>> set(axhan, 'XLim', [-10 10])
```

Note that both modify the axes in the Figure Window. Using the dot notation (the first example) in an assignment statement will show the new value of *axhan* whereas the **set** method does not.

The objects, their properties, what the properties mean, and valid values can be found in the MATLAB Help Documentation. Search for Graphics Object Properties to see a list of the property names and a brief explanation of each.

For example, the Color property is a vector that stores the color of the line as three separate values for the Red, Green, and Blue intensities, in that order. Each value is in the range from 0 (which means none of that color) to 1 (which is the brightest possible hue of that color). In the previous plot example, the Color was [0 0.4470 0.7410], which means no red, some green, and a lot of blue; in other words the line drawn for the **sin** function was a royal blue hue. More examples of possible values for the Color vector include:

[1 0 0] is red
[0 1 0] is green
[0 0 1] is blue
[1 1 1] is white
[0 0 0] is black
[0.5 0.5 0.5] is a shade of gray

(Note that in earlier versions of MATLAB the default color was [0 0 1], or blue.)

Changing the line width in the figure makes it easier to see the line and its color, as shown in Fig. 11.1. Also, tab completion is available for class properties and methods; e.g., if you are not sure of the exact property name for the line width, typing "hl.Li" and then hitting the tab key would display the options.

```
>> hl.LineWidth = 4;
```

PRACTICE 11.2

Create *x* and *y* vectors, and use the **plot** function to plot the data points represented by these vectors. Store the handle in a variable and do not close the Figure Window! Inspect the properties and then change the line width and color. Next, put markers for the points and change the marker size and edge color.

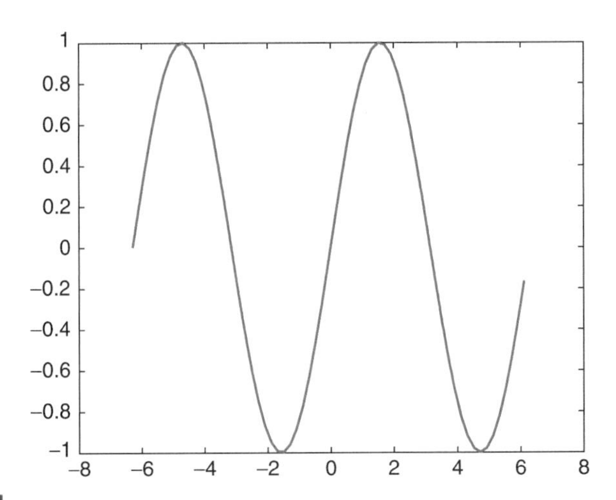

FIGURE 11.1
Thicker line.

In addition to handles for objects, the built-in functions **gca** and **gcf** return the handles for the current axes and figure, respectively (the function names stand for "get current axes" and "get current figure"). In the following example, two Figure Windows are opened. The current figure and the current axes are the most recently created.

```
>> x = -pi: 0.1: pi;
>> f1 = figure(1);
>> p1 = plot(x,sin(x));
>> f2 = figure(2);
>> p2 = plot(x,cos(x),'Color',[0 1 1]);
>> curfig = gcf
curfig =
  Figure (2) with properties:

      Number: 2
        Name: ''
       Color: [0.9400 0.9400 0.9400]
    Position: [440 378 560 420]
       Units: 'pixels'

  Show all properties
```

This Children property of the current figure stores the axes that orient the plot; these axes are also returned by the **gca** function.

```
>> curfig.Children
ans =
  Axes with properties:

             XLim: [-4 4]
             YLim: [-1 1]
           XScale: 'linear'
           YScale: 'linear'
    GridLineStyle: '-'
         Position: [0.1300 0.1100 0.7750 0.8150]
            Units: 'normalized'

  Show all properties
>> curfig.Children == gca
ans =
     1
```

Within the axes, the Line primitive was used to create the plot. This can be referenced using the dot operator twice. The variable *curfig* stores the handle of the current figure; its Children property stores the current axes, and the Children property of the axes is the Line primitive.

```
>> curfig.Children.Children
ans =
  Line with properties:

              Color: [0 1 1]
          LineStyle: '-'
          LineWidth: 0.5000
             Marker: 'none'
         MarkerSize: 6
    MarkerFaceColor: 'none'
              XData: [1x63 double]
              YData: [1x63 double]
              ZData: [1x0 double]

  Show all properties
```

Thus, the hierarchy is: Figure Window - > Axes - > Line.

11.3 USER-DEFINED CLASSES AND OBJECTS

There are many examples of built-in classes in MATLAB, including the **handle** class used by plot functions. It is also possible for users to define classes, and then create or instantiate objects of those classes.

11.3.1 Class Definitions

Classes are defined using the keyword **classdef**. The class definition is organized by blocks, and typically at a minimum contains properties (the data) using the keyword **properties** and methods (the functions that act on the data) using the keyword **methods**. One special case of a method is a ***constructor function*** that initializes the properties. Class definitions are stored in code files with the same name as the class; the constructor function also is given the same name.

Every block begins with the keyword and ends with **end**. The organization of a basic class definition, in which there are two properties and one method, which is a constructor function, follows.

MyClassName.m

```
classdef MyClassName

   properties
       prop1
       prop2
   end

   methods

     % Constructor function
     function obj = MyClassName(val1, val2)
                obj.prop1 = val1;
                obj.prop2 = val2;
     end

     % Other methods that operate on properties

   end
end
```

The class definition is stored in a code file with the same name as the class. Within the **classdef** block, there are blocks for **properties** and **methods**. In the **properties** block, the names of all properties are given. It is also possible to assign default values for the properties using the assignment operator; if this is not done, MATLAB initializes each to the empty vector.

The constructor function must have the same name as the class. It only returns one output argument, which is the initialized object. If no constructor function is defined, MATLAB creates one that uses the default values given in the **properties** definition, if any, or empty vectors if not. It is best to write the constructor function to allow for the case in which no input arguments are passed, using **nargin** to check to determine how many arguments were passed to the function.

The following is a simple class definition in which there are two properties: x, which is not initialized so the default value is the empty vector and y, which is initialized to 33. The class has one constructor function; if two arguments are passed to it, they are stored in the two properties, if not, the default values are used.

SimpleClass.m

```
classdef SimpleClass

    properties
        x
        y = 33;
    end

    methods

        function obj = SimpleClass(val1, val2)
            if nargin == 2
                obj.x = val1;
                obj.y = val2;
            end
        end
    end
end
```

Once the class has been defined, objects can be created or instantiated by assigning the name of the class. For example, the following instantiates an object named *myobject*; since the output is not suppressed, the property names and their values are shown:

```
>> myobject = SimpleClass
myobject =
  SimpleClass with properties:

    x: []
    y: 33
```

Instantiating this object automatically calls the constructor function; since no arguments were passed, the default values were used for the properties.

In the following example input arguments are passed to the constructor.

```
>> newobject = SimpleClass(4, 22)
newobject =
  SimpleClass with properties:

    x: 4
    y: 22
```

The properties and methods can be seen using the **properties** and **methods** functions:

```
>> properties(myobject)
Properties for class SimpleClass:
    x
    y
>> methods(myobject)

Methods for class SimpleClass:

SimpleClass
```

The properties can be accessed using the dot operator to either display or modify their values.

```
>> myobject.x = 11
myobject =
  SimpleClass with properties:

    x: 11
    y: 33
```

We will now modify the class definition, making the constructor function more general, and adding a new method.

SimpleClassii.m

```
classdef SimpleClassii

    properties
        x
        y = 33;
    end

    methods
        function obj = SimpleClassii(varargin)
            if nargin == 0
                obj.x = 0;
            elseif nargin == 1
                obj.x = varargin{1};
            else
                obj.x = varargin{1};
                obj.y = varargin{2};
            end
        end

        function outarg = ordMethod(obj, arg1)
            outarg = obj.x + obj.y + arg1;
     end
  end
end
```

In the class *SimpleClassii*, there are two methods. The constructor, which has the same name as the class, is general in that it accepts any number of input arguments. If no arguments are passed, the property *x* is initialized to 0 (*y* is not initialized since a default value was already assigned to it in the **properties** block). If only one input argument is passed, it is assumed to be the value of the property *x*, and is assigned to *obj.x*. If two or more input arguments are passed, the first is the value of *x* and the second is the value to be stored in *y*. Although it is best to use **nargin** and **varargin** to allow for any number of input arguments, future examples will assume the correct number of input arguments for simplicity. If the properties are to be a certain type, the constructor function should also check and ensure that the input arguments are of the correct type and either typecast them or change them if not.

The following examples demonstrate instantiating two objects of the class *SimpleClassii*.

```
>> objA = SimpleClassii
objA =
  SimpleClassii with properties:

     x: 0
     y: 33

>> objB = SimpleClassii(4, 9)
objB =
  SimpleClassii with properties:

     x: 4
     y: 9
```

QUICK QUESTION!

What would the value of the properties be for the following:

```
>> ob = SimpleClassii(1, -6, 7, 200)
```

Answer:

```
x: 1
y: -6
```

The last two arguments to the constructor were ignored.

Every time an object is instantiated, the constructor function is automatically called. So, there are two ways of initializing properties: in the property definition block, and by passing values to the constructor method.

The second method in SimpleClassii is an example of an ***ordinary method***. The method *ordMethod* adds the values of the input argument, the *x* property, and the *y* property together and returns the result. When calling this method, the object to be used must always be passed to the method, which is why there are two input arguments in the function header: the object and the value to be summed with the properties.

There are two ways in which the method *ordMethod* can be called. One way is by explicitly passing the object to be used:

```
>> resultA = ordMethod(objA, 5)
resultA =
    38
```

The other way is by using the dot operator with the object, as follows:

```
>> resultB = objB.ordMethod(11)
resultB =
    24
```

Both of these methods are identical in their effect; they both pass the object to be used and the value to be added. Notice that regardless of how the method is called, there are still two input arguments in the function header: one for the object (whether it is passed through the argument list or by using the dot operator) and one for the value to be added. Although it is common for the object to be the first input argument, it is not necessary to do so.

11.3.2 Overloading Functions and Operators

By default MATLAB creates an assignment operator for classes, which allows one object of the class to be assigned to another. This performs *memberwise* assignment, which means it assigns each property individually. Thus, one object can be assigned to another using the assignment operator. Other operators, however, are not defined automatically for classes. Instead, the programmer has the power to define operators. For example, what would it mean for one object to be less than another? The programmer has the power to define "<" any way he or she wants! Of course, it makes sense for the operator to be defined in a way that is consistent with the definition for MATLAB classes. For example, it makes sense to define the equality operator to determine whether two objects are equal to each other or not (and that would typically be memberwise). An error message will be thrown if an operator is used that has not been defined.

Recall that all operators have a functional form. For example, the expression `a+b` can also be written as **plus(a,b)**. When defining an operator for a user-defined class, a member function is defined with the function name for the operator, e.g., **plus**. This is called ***overloading***, as it gives another definition for an existing function. Which function is used (the built-in or user-defined) depends on the context, which means the types of the arguments that are used in the expression.

In addition to the operator functions, it is also possible to overload other functions for a class. For example, one function that is frequently overloaded is the function **disp**. By creating a class member function **disp**, one can customize the way in which object properties are displayed. One aspect of overloading the **disp** function is that when the assignment operator is used to assign a value to an object or an object property, and the semicolon is not used to suppress the output, the **disp** function is automatically called to display the properties—so, the format of the output that is created in the overloaded **disp** function will be seen with every unsuppressed assignment.

To illustrate some of these concepts, a class to represent a rectangle, *Rectangle*, will be developed. There are two properties, for the length and width of the rectangle. There are four methods:

- a constructor function, or method, *Rectangle*
- an ordinary method *rectarea* that calculates the area of a Rectangle object
- two overloaded functions:
  - *disp*, which displays the properties in a formatted sentence
  - *lt*, which is the function for the less than operator

What does it mean for one Rectangle object to be less than another? In the following definition, the *lt* function is defined as **true** if the area of one Rectangle object is less than another. However, this is our choice. Depending on the application, it may make more sense to define it using just the length, just the width, or perhaps based on the perimeters of the Rectangle objects. This is a cool thing about classes; the programmer can define these operator functions in any way desired.

Rectangle.m

```
classdef Rectangle

    properties
        len = 0;
        width = 0;
    end

    methods

        function obj = Rectangle(l, w)
            if nargin == 2
                obj.len = l;
                obj.width = w;
            end
        end

        function outarg = rectarea(obj)
            outarg = obj.len * obj.width;
        end
```

```
            function disp(obj)
                fprintf('The rectangle has length %.2f', obj.len)
                fprintf(' and width %.2f\n', obj.width)
            end

            function out = lt(obja, objb)
                out = rectarea(obja) < rectarea(objb);
            end
        end
end
```

For simplicity, the constructor only checks for two input arguments; it does not check for a variable number of arguments, nor does it verify the types of the input arguments. If **nargin** is not 2, the default values from the properties block are used.

Here are examples of instantiating *Rectangle* objects, both using the constructor function and using the assignment operator that MATLAB provides for classes:

```
>> rect1 = Rectangle(3,5)
rect1 =
The rectangle has length 3.00 and width 5.00
>> rect2 = rect1;
>> rect2.width = 11
rect2 =
The rectangle has length 3.00 and width 11.00
```

Notice that the overloaded *disp* function in the class definition is used for displaying the objects when the output is not suppressed. It can also be called explicitly.

```
>> rect1.disp
The rectangle has length 3.00 and width 5.00
```

As the *lt* operator was overloaded, it can be used to compare *Rectangle* objects.

```
>> rect1 < rect2
ans =
     1
```

Other operators, e.g., **gt** (greater than), however, have not been defined within the class so they cannot be used, and MATLAB will throw an error message.

```
>> rect1 > rect2
Undefined operator '>' for input arguments of type 'Rectangle'.
```

Care must be taken when overloading operator functions. The function in the class definition takes precedence over the built-in function, when objects of the class are used in the expression.

QUICK QUESTION!

Could we mix types in the expression? For example, what if we wanted to know whether the area of *rect1* was less than 20, could we use the expression

```
rect1 < 20 ?
```

Answer: No, not with the overloaded *lt* function as written, which assumes that both arguments are *Rectangle* objects. The following error message would be generated:

```
>> rect1 < 20
Undefined function 'rectarea' for input arguments of type 'double'.
Error in < (line 30)
    out = rectarea(obja) < rectarea(objb);
```

However, it is possible to rewrite the function to handle this case. In the following modified version, the type of each of the input arguments is checked. If the argument is not a *Rectangle* object, the type is checked to see whether it is the type **double**. If it is, then the input argument is modified to be a rectangle with the number specified as the length and a width of 1 (so the area will be calculated correctly). Otherwise, the argument is simply typecast to be a *Rectangle* object so that no error is thrown (another option would be to print an error message).

```
function out = lt(inp1, inp2)
    if ~isa(inp1,'Rectangle')
       if isa(inp1, 'double')
          inp1 = Rectangle(inp1,1);
       else
          inp1 = Rectangle;
       end
    end
    if ~isa(inp2,'Rectangle')
       if isa(inp2, 'double')
          inp2 = Rectangle(inp2,1);
       else
          inp2 = Rectangle;
       end
    end
    out = rectarea(inp1) < rectarea(inp2);
end
```

With the modified function, expressions mixing Rectangle objects and double values can now be used:

```
>> rect1 < 20
ans =
     1
```

11.3.3 Inheritance and the Handle Class

Inheritance is when one class is derived from another. The initial class is called the *superclass* and the derived class is called the *subclass*. A subclass is a new class that has the properties and methods of the superclass, plus it can have its own properties and methods. The methods in a subclass can override methods from the superclass if they have the same name.

11.3.3.1 Subclasses

The syntax for the subclass definition includes the "<" operator followed by the name of the superclass in the first line of the code file. (Note: this is not the less than operator!) The subclass inherits all of the properties and methods of the superclass, and then its own properties and methods can be added. For example, a subclass might inherit two properties from the superclass and then also define one of its own. The constructor function would initialize all three properties, as seen in the example that follows.

MySubclass.m

```
classdef MySubclass < Superclass

    properties
        prop3
    end

    methods

      % Constructor function
      function obj = MySubclass(val1, val2,val3)
                  obj@Superclass(val1,val2)
                  obj.prop3 = val3;
      end

        % Other methods that operate on properties
      end
  end
```

The first line in the constructor uses the syntax obj@Superclass in order to call the constructor method of the super class to initialize the two properties defined in the super class.

For example, our class *Rectangle* can be a superclass for a subclass *Box*. The subclass *Box* inherits the *len* and *width* properties, and has its own property *height*. In the following class definition for *Box*, there is also a constructor function named *Box* and an ordinary method to calculate the volume of a *Box* object.

Box.m

```
classdef Box < Rectangle
    properties
        height = 0;
    end
    methods
        function obj = Box(l,w,h)
            if nargin < 3
                l = 0;
                w = 0;
                h = 0;
            end
            obj@Rectangle(l,w)
            obj.height = h;
        end

        function out = calcvol(obj)
            out = obj.len * obj.width * obj.height;
        end
    end
end
```

The values of all three properties should be passed to the constructor function. If not, the input arguments are all assigned default values. Next, the *Box* constructor calls the *Rectangle* constructor to initialize the *len* and *width* properties. The syntax for the call to the *Rectangle* constructor is:

```
obj@Rectangle(l,w)
```

Note that this call to the Rectangle constructor must be executed first, before other references to the object properties. Finally, the constructor initializes the *height* property.

The following is an example of instantiating a *Box* object. Notice that since the result of the assignment is not suppressed, the *disp* function from the *Rectangle* class is called. All three of the properties were initialized, but only the length and width were displayed.

```
>> mybox = Box(2,5,8)
mybox =
The rectangle has length 2.00 and width 5.00
>> mybox.height
ans =
     8
```

To remedy this, we would have to overload the *disp* function again within the *Box* class.

```
function disp(obj)
    fprintf('The box has a length of %.2f,',obj.len)
    fprintf(' width %.2f\nand height %.2f\n',...
                 obj.width,obj.height)
end
```

```
>> mybox = Box(2,5,8)
mybox =
The box has a length of 2.00, width 5.00
and height 8.00
```

11.3.3.2 Value and Handle Classes

There are two types of classes in MATLAB: ***value classes*** and ***handle classes***. Value classes are the default; so far, the classes that have been demonstrated have all been value classes. Handle classes are subclasses that are derived from the abstract class **handle**, which is a built-in class. The class definition for a **handle** class begins with:

```
classdef MyHandclass < handle
```

There is a very fundamental difference between value classes and handle classes. With value classes, if one object is copied to another, they are completely independent; changing one does not affect the other. With handle classes, on the other hand, if one handle object is copied to another it does not copy the data; instead, it creates a reference to the same data. All objects refer to the same data (the properties).

User-defined classes can be either value classes or handle classes. Built-in classes are also either value classes or handle classes. For example, built-in numeric types such as **double** are value classes, whereas plot objects are handle objects.

Since **double** is a value class, we can assign one **double** variable to another—and then changing the value of one does not affect the other.

```
>> num = 33;
>> value = num;
>> value = value + 4
value =
    37
>> num
num =
    33
```

On the other hand, plot object handles are handle objects. When assigning one plot handle variable to another, they both refer to the same plot.

```
>> x = 0: 0.1 : pi;
>> plothan = plot(x,sin(x));
>> handleb = plothan;
```

Both variables *plothan* and *handleb* refer to the same plot; they are not different plots. As a result, a property such as the line width could be changed by either

```
>> set(plothan,'LineWidth',3)
```

or

```
>> set(handleb,'LineWidth',3)
```

Either of these would accomplish the same thing, changing the line width in the one plot to 3.

As an example of a user-defined **handle** class, let us modify the value class *Rectangle* to be a handle class called *HandleRect* (and simplify a bit by not overloading the **lt** function).

HandleRect.m

```
classdef HandleRect < handle
    properties
        len = 0;
        width = 0;
    end

    methods
```

```
        function obj = HandleRect(l, w)
            if nargin == 2
               obj.len = l;
               obj.width = w;
            end
        end

        function outarg = rectarea(obj)
           outarg = obj.len * obj.width;
        end

        function disp(obj)
           fprintf('The rectangle has length %.2f', obj.len)
           fprintf('and width %.2f\n', obj.width)
        end
    end
end
```

By instantiating an object, we can find the properties and methods as follows.

```
>> HRectangle = HandleRect(3,5)
HRectangle =
The rectangle has length 3.00 and width 5.00
>> properties(HRectangle)
Properties for class HandleRect:
     len
     width
>> methods(HRectangle)

Methods for class HandleRect:

HandleRect  disp        rectarea

Methods of HandleRect inherited from handle.
```

By clicking on the underlined Methods link, the inherited methods can be seen.

```
Methods for class handle:

addlistener  findobj   gt       lt
delete       findprop  isvalid  ne
eq           ge        le       notify
```

Notice that these inherited methods include overloaded operator functions for the operators `>`, `<`, `>=`, `<=`, `==`, and `~=`. Since the assignment operator is defined automatically for all classes, and the equality operator is overloaded for handle classes, we can assign one *HandleRect* object to another, and then verify that they are identical.

```
>> HRecA = HandleRect(2,7.5)
HRecA =
The rectangle has length 2.00 and width 7.50
```

```
>> HRecB = HRecA
HRecB =
The rectangle has length 2.00 and width 7.50
>> HRecA == HRecB
ans =
     1
```

QUICK QUESTION!

What would happen if the value of *HRecA.len* was changed to 6?

Answer:

```
>> HRecA.len = 6;
>> HRecA
HRecA =
The rectangle has length 6.00 and width 7.50
>> HRecB
HRecB =
The rectangle has length 6.00 and width 7.50
```

However, if we then create another object HRecC with the same properties as HRecA and HRecB, HRecC is not equal to either HRecA or HRecB.

```
>> HRecC = HandleRect(2,7.5)
HRecC =
The rectangle has length 2.00 and width 7.50
>> HRecA == HRecC
ans =
     0
```

This illustrates one of the important concepts about handle classes: assigning one object to another does not make a new copy; instead, it creates another reference to the same object. However, instantiating an object by calling the constructor function does create a new object, even if it happens to have the same properties as other object(s). The following is an illustration of the three *HandleRect* objects that have been created:

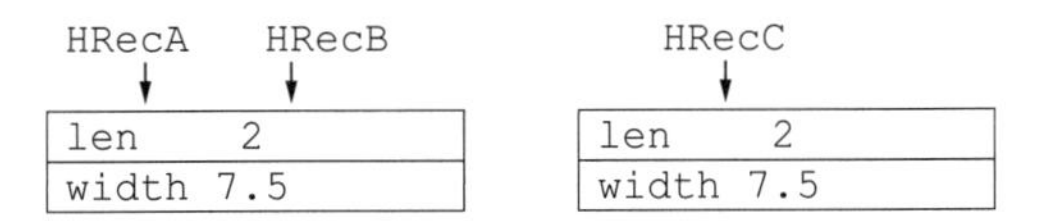

Since both *HRecA* and *HRecB* refer to the same object, changing a property using one of the instances will change that property for both *HRecA* and *HRecB*, but will not affect *HRecC*.

```
>> HRecA.len = 11;
>> HRecB.len
ans =
     11
>> HRecC.len
ans =
     2
```

PRACTICE PROBLEM 11.3

The **gt**, **lt**, **le**, and **ge** functions are overloaded in **handle** classes for the operators >, <, <=, and >= respectively. Create a **handle** class with multiple properties, instantiate at least two objects of this class, and design experiments with the objects to determine whether these overloaded operators are implemented memberwise or not.

Because handle class objects are references to the locations in which the objects are stored, there are differences between handle and value classes in the way that objects are passed to functions, and in the manner in which functions can change objects. We will create two simple classes, a value class *valClass*, and a handle class *hanClass*, to illustrate the differences. Both will have just one double property *x*. There will be four methods; for simplicity, none of them error-check:

- a simple constructor
- a function *add* that receives two objects, adds the *x* properties together, and returns an object in which the *x* property is the sum of the two inputs
- a function *timestwo* that receives one object and returns an object in which the property is the property of the input argument multiplied by two
- a function *timesthree* that receives one object, multiplies its property by three, but does not return anything

The constructor and *add* functions behave similarly. In both classes, the *add* function receives two input arguments which are objects, and returns an object which is distinct from the input objects. The other two functions, however, behave differently in the *valClass* and *hanClass*. We will first examine the value class.

valClass.m

```
classdef valClass
 properties
    x = 0;
 end

 methods
    function obj = valClass(in)
        if nargin == 1
           obj.x = in;
        end
    end
```

```
        function out = add(obja, objb)
            out = valClass(obja.x + objb.x);
        end

        function outobj = timestwo(inobj)
            outobj = valClass(inobj.x * 2);
        end

        function timesthree(obj)
            % Note: this function does not
            % accomplish anything; MATLAB
            % flags the following line
            obj.x = obj.x*3;
        end
    end
end
```

Note that the *add* and *timestwo* functions call the *valClass* constructor in order to make the output a *valClass* object.

Instantiating two objects and calling the *timestwo* function creates the following result.

```
>> va = valClass(3);
>> vb = valClass(5);
>> vmult2 = timestwo(vb)
vmult2 =
  valClass with properties:

      x: 10
>> vb
vb =
  valClass with properties:

      x: 5
```

Note that in the base workspace, initially there are two objects *va* and *vb*. While the function is executing, the function's workspace has the input argument *inobj* and output argument *outobj*. These are all separate objects. The value of the object *va* was passed to the input object *inobj*, and a separate output argument is created in the function, which is then returned to the object *vmult2* in the assignment statement. At this point the function's workspace would no longer exist but the base workspace would now have *va*, *vb*, and *vmult*.

Before the function call to *timestwo* we have:

Base workspace:

va	vb
x: 3	x: 5

While the *timestwo* function is executing we have:

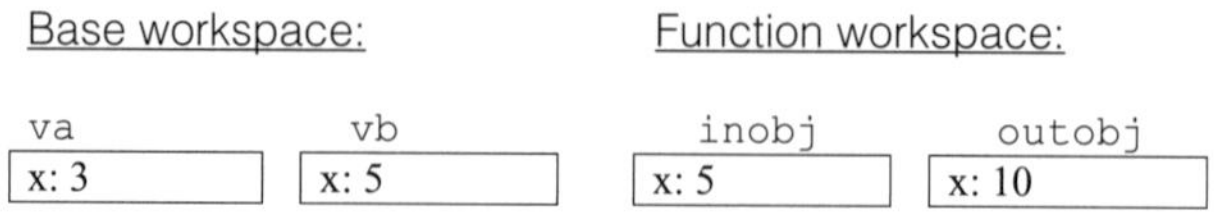

Base workspace: | | Function workspace: |
---|---|---|---
va | vb | inobj | outobj
x: 3 | x: 5 | x: 5 | x: 10

After *timestwo* has stopped executing and has returned the object we have:

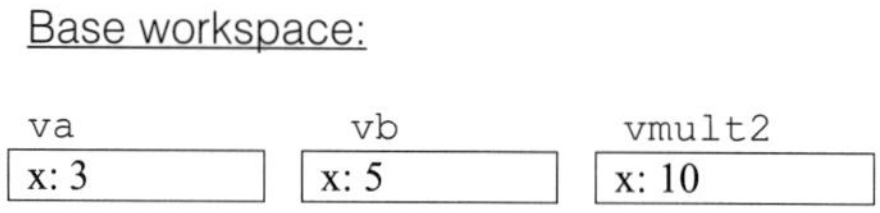

Base workspace: | |
---|---|---
va | vb | vmult2
x: 3 | x: 5 | x: 10

Now let us examine the behavior of the *timesthree* function, which does not return any output argument, so the call cannot occur in an assignment statement.

```
>> clear
>> va = valClass(3);
>> vb = valClass(5);
>> timesthree(vb)
>> va
va =
  valClass with properties:

    x: 3
>> vb
vb =
  valClass with properties:

    x: 5
```

Before the function call to *timesthree* we have:

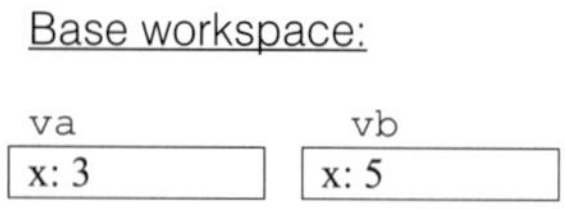

Base workspace: |
---|---
va | vb
x: 3 | x: 5

Initially in the *timesthree* function we have:

Base workspace:		Function workspace:
va	vb	obj
x: 3	x: 5	x: 5

Once the assignment statement in the *timesthree* function has executed we have:

Base workspace:		Function workspace:
va	vb	obj
x: 3	x: 5	x: 15

After *timesthree* has stopped executing we have:

Within the body of the function, the value of the input argument was modified. However, that value was not returned. It was also a separate object from the two objects in the base workspace. Therefore, the function accomplished nothing, which is why MATLAB flags it. It appears to behave as though it is a handle class method, as we will see next.

The next example is similar, but is a **handle** class rather than a value class. Let us examine the *timestwo* and *timesthree* functions in the following **handle** class.

hanClass.m

```
classdef hanClass < handle
    properties
        x = 0;
    end

    methods
        function obj = hanClass(in)
            if nargin == 1
              obj.x = in;
            end
        end

        function out = add(obja, objb)
            out = hanClass(obja.x + objb.x);
        end

    function outobj = timestwo(inobj)
        outobj = hanClass(inobj.x * 2);
    end

     function timesthree(obj)
         obj.x = obj.x * 3;
     end
    end
end
```

Instantiating two objects and calling the *timestwo* function creates the following result.

```
>> ha = hanClass(12);
>> hb = hanClass(7);
>> hmult2 = timestwo(hb)
hmult2 =
```

```
  hanClass with properties:

     x: 14
>> hb
hb =
  hanClass with properties:

     x: 7
>> ha
ha =
  hanClass with properties:

     x: 12
```

Note that in the base workspace, there are two objects, *ha* and *hb*. While the function is executing, there are also the input argument *inobj* and output argument *outobj*. The value of the object *hb*, which is a reference to the object, was passed to the input object *inobj*, which means that *inobj* refers to the same object. Within the body of the function, a separate object *outobj* is created.

Before the function call to *timestwo* we have:

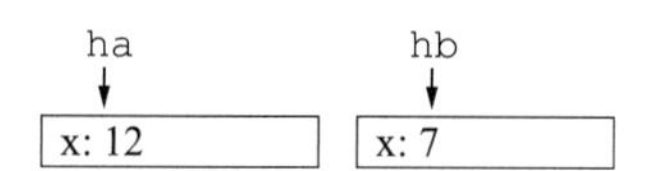

While the *timestwo* function is executing we have:

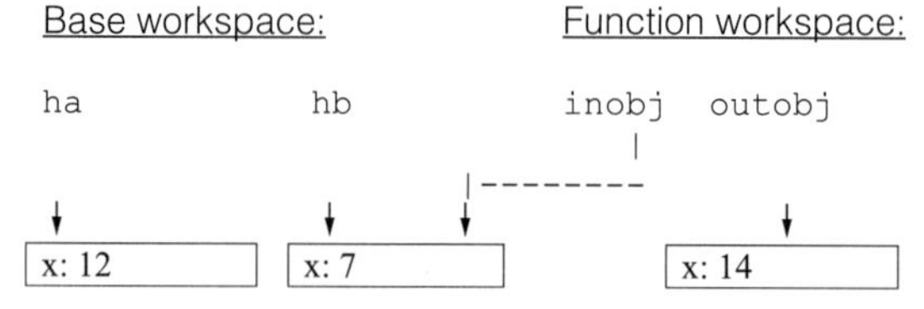

After *timestwo* has stopped executing and has returned the object we have:

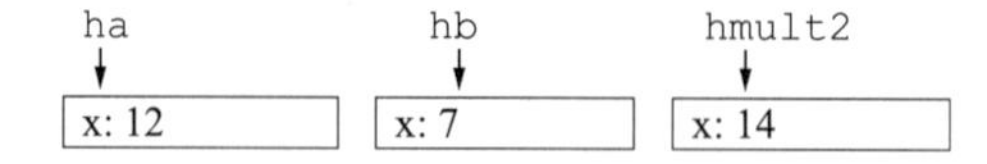

Now let us examine the behavior of the *timesthree* function. As with the value class, this function does not return any output argument, so calls to it cannot occur in an assignment statement. However, this function does accomplish something; it modifies the object passed as an input argument.

```
>> clear
>> ha = hanClass(12);
```

```
>> hb = hanClass(7);
>> timesthree(hb)
>> ha
ha =
  hanClass with properties:

      x: 12
>> hb
hb =
  hanClass with properties:

      x: 21
```

Before the function call to *timesthree* we have:

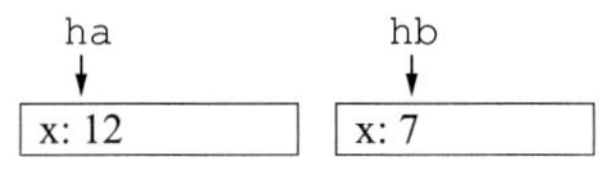

While the *timesthree* function begins we have:

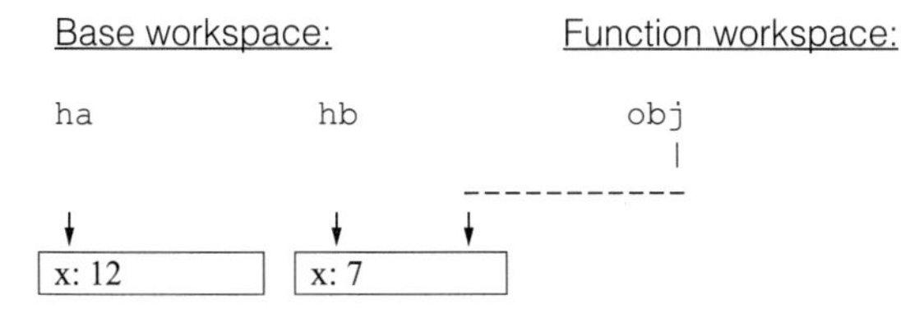

Once the *timesthree* function executes its assignment statement we have:

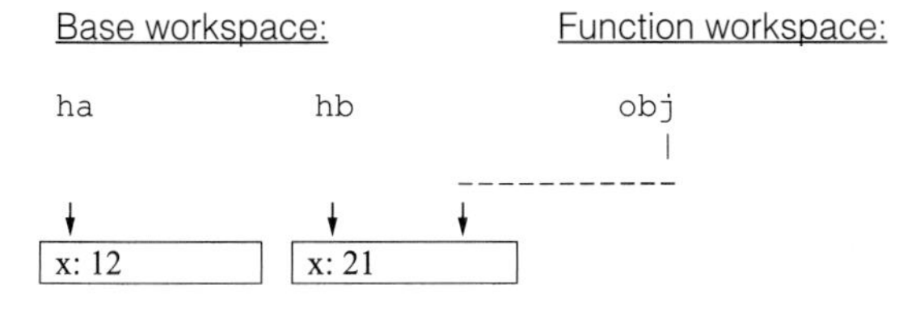

After *timesthree* has stopped executing and has returned the object we have:

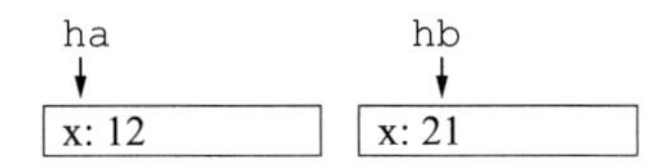

Thus, because handle class objects store references, passing a handle class object to a function can allow the function to modify its properties (without returning anything). This cannot happen with value class objects.

Notice that errors will occur for both the value and handle classes if the *timesthree* functions are called in assignment statements, since these functions do not return any values.

```
>> vmult3 = timesthree(vb)
Error using valClass/timesthree
Too many output arguments.

>> hmult3 = timesthree(hb)
Error using hanClass/timesthree
Too many output arguments.
```

11.3.4 Property Definition Blocks

The behaviors of, and access to, classes, properties, and methods can be controlled by what are called ***attributes***. The attributes are specified in parentheses in the first line of any block.

There are several attributes that relate to properties. One of the most important attributes is access to the properties. There are three types of access to properties:

- **public**: access is possible from anywhere; this is the default
- **private**: access is possible only by methods in this class
- **protected**: access is possible only by methods in this class or any of its subclasses

The attributes that control the access to properties are:

- **GetAccess**: read access, which means the ability to determine the values of properties
- **SetAccess**: write access, which means the ability to initialize or modify values of properties
- **Access**: both read and write access

In the case in which different properties are to have different attributes, multiple property definition blocks can be defined. So far, since we have not specified, both the read and write access to all properties in our class definitions have been public. This means that we have been able to see and to modify, the values of the properties for example from the Command Window.

The following is an example of a value class in which one property *num*, is public, and the other property *word*, has public GetAccess but protected SetAccess.

valAttributes.m

```
classdef valAttributes
    properties
        num = 0;
    end

    properties (SetAccess = protected)
        word = 'hello';
    end

    methods
        function obj = valAttributes(n,w)
            if nargin == 2
                obj.num = n;
                obj.word = w;
            end
        end

    end
end
```

Once an object has been instantiated using the constructor, the property *num* can be queried and modified (because Access is public by default).

```
>> valA = valAttributes(3, 'hi')
valA =
  valAttributes with properties:

     num: 3
    word: 'hi'
>> valA.num
ans =
     3
>> valA.num = 14
valA =
  valAttributes with properties:

     num: 14
    word: 'hi'
```

The property *word*, however, can be queried (because GetAccess is public by default) but it cannot be changed (because SetAccess was set to protected). Only methods within the class (or subclasses if defined) can change the word property.

```
>> valA.word
ans =
hi
```

```
>> valA.word = 'ciao'
You cannot set the read-only property 'word' of valAttributes.
```

This is a very useful aspect of objects. Protecting objects by only allowing class methods to modify them is a very powerful tool. One example of this is to error-check property values.

Although the access attributes are the most commonly used, there are other attributes that can be set (e.g., Constant for values that are to be constant for all objects). A table of all attributes can be found by searching the documentation for Property Attributes.

11.3.5 Method Types

There are different types of methods that can be defined in class definitions; we have already seen constructor functions, ordinary methods, and overloading functions. As with properties, there are also attributes that control the behavior of methods.

11.3.5.1 Constructor Functions

As we have seen, all classes should have constructor functions, which initialize the properties in an object. If a class definition does not include a constructor, MATLAB will initialize the properties to either default values provided in the property definition block, or to empty vectors. Constructor functions always have the same name as the class, and return only the initialized object; this is true for both value and handle classes. Constructor functions should be general and should allow for the case in which no input arguments are passed by creating default property values. To be truly general, the types of the input arguments should also be checked to make sure that they are the correct types for the properties. Overloading the set.PropertyName functions for all properties allows for even more control over all functions that set the properties, including the constructor function.

If the class being defined is a subclass, the constructor function of the superclass must be called to initialize its properties. MATLAB automatically makes implicit calls to the superclass constructors using no arguments when the class is defined as a subclass in the class definition line. It is also possible to explicitly call the constructors and pass arguments to be the property values; this is necessary if the superclass constructors require that input arguments be passed.

Some languages have what are called ***copy constructor*** functions, which allow an object to be constructed by copying all properties of one object into another. MATLAB does not have a copy constructor, but it would be possible to write a constructor function so that it checks to see whether the input argument is an object of the class type, and if so, copy the properties. The beginning of a simplified version of such a constructor for a class *MyClass* might be:

```
function obj = MyClass(varargin)
if nargin == 1
    val = varargin{1};
    if isa(val, 'MyClass')
        % Copy all properties
        obj.Prop = val.Prop;
    else
        % etc.


    end
end
end
```

11.3.5.2 Access Methods

MATLAB has special access methods that allow you to query a property and to assign a value to a property. These methods have special names that include the name of the property:

```
get.PropertyName
set.PropertyName
```

These methods cannot be called directly and they do not show in the list of functions returned by the **methods** function. Instead, they are automatically called whenever a property is queried or an attempt is made to assign a value to a property. They can, however, be overloaded.

One reason to overload the set.PropertyName method is to be able to error-check to make sure that only correct values are being assigned to a property. For example, the following is a simple value class in which the property is a grade that should be in the inclusive range from 0 to 100; the *set.grade* function ensures this.

valSet.m

```
classdef valSet
    properties
        grade = 0;
    end

    methods

        function obj = valSet(in)
            if nargin == 1
                obj.grade = in;
            end
        end
```

```
        function obj = set.grade(obj,val)
            if val >= 0 && val <= 100
                obj.grade = val;
            else
                error('Grade not in range')
            end
        end
    end
end
```

The *set.grade* function restricts values for the *grade* property to be in the correct range, both when instantiating objects and when attempting to modify an object. Note that the input and output argument names for the object must be the same. The **error** function throws an error message.

```
>> valobj = valSet(98)
valobj =
  valSet with properties:

    grade: 98
>> badobj = valSet(105)
Error using valSet/set.grade (line 18)
Grade not in range
Error in valSet (line 10)
               obj.grade = in;
>> valobj.grade = 99
valobj =
  valSet with properties:

    grade: 99
>> valobj.grade = -5
Error using valSet/set.grade (line 18)
Grade not in range
```

The *set.grade* function would be slightly different in a handle class, since the function can modify properties of an object without returning the object. An equivalent example for a handle class follows; note that the function does not return any output argument.

hanSet.m

```
classdef hanSet < handle
    properties
        grade = 0;
    end

    methods
```

```
            function obj = hanSet(in)
                if nargin == 1
                    obj.grade = in;
                end
            end

            function set.grade(obj,val)
                if val >= 0 && val <= 100
                    obj.grade = val;
                else
                    error('Grade not in range')
                end
            end
        end
    end
```

Note that the **get** and **set** method blocks cannot have any attributes.

11.3.5.3 Method Attributes

As with property attributes, method attributes are defined in parentheses in the first line of the method block. In the case in which different methods are to have different attributes, multiple method definition blocks can be defined.

There are several attributes that relate to methods. These include **Access**, which controls from where the method can be called. There are three types of access to methods:

- **public**: access is possible from anywhere; this is the default
- **private**: access is possible only by methods in this class
- **protected**: access is possible only by methods in this class or any of its subclasses

All of the methods in the examples shown thus far have been public. This means, for example, that we have been able to call our class methods from the Command Window. It is very common, however, to restrict access so that only methods within the class itself (or, any subclasses) can call other methods.

Besides **Access**, other method attributes include the following, all of which are the type **logical** and have a default value of **false**.

- **Abstract**: If true, there is no implementation of the method
- **Hidden**: If true, the method is not seen in lists of methods
- **Sealed**: If true, the method cannot be redefined in a subclass

- **Static**: If true, the method is a static method which means that it is not called by any particular object; static methods are the same for all objects in the class

Static methods are not associated with any object. One reason to have a static method is to perform calculations that are typical for the class, e.g., conversion of units. The following is a simple example.

```
classdef StatClass
    methods (Static)
        function out = statex(in)
            out = in * 10;
        end
    end
end
```

As static methods are not associated with any instantiation of the class, they are therefore called with the name of the class, not by any object.

```
>> StatClass.statex(4)
ans =
   40
```

11.3.5.4 Destructor Functions

Just as constructor functions create objects, destructor functions destroy objects. In MATLAB, destructor functions are optional. If a destructor function is defined within a class, this is accomplished by overloading the **delete** function. There are specific rules that make an overloaded **delete** function a destructor function: the function must have just one input argument, which is an object of the class, and it must not have any output arguments. Also, it cannot have the value **true** for the attributes **Sealed**, **Static**, or **Abstract**. The reason for having a class destructor function is to be able to "clean up." For example, if the class opens a file, the destructor function might make sure that the file is closed properly.

11.3.6 Events and Listeners

In addition to the properties and methods blocks that are normally in a class definition, handle classes (but not value classes) can also have an **events** block. ***Events*** are some type of action that takes place; ***event-driven programming*** is when an event triggers an action. In handle classes in MATLAB, there are three main concepts:

- ***event***: an action, such as changing the value of a property or the user clicking the mouse
- ***listener***: something that detects an event and initiates an action based on it
- ***callback***: a function that is called by the listener as a result of the event

There are two methods defined in the handle class that are used with events:

- **notify**: notifies listener(s) when an event has occurred
- **addlistener**: adds a listener to an object, so that it will know when an event has occurred

Examples of these concepts will be provided later in the section on Graphical User Interfaces (GUI's). GUI's are graphical objects that the user manipulates, e.g., a push button. When the user pushes a button, for example, that is an event that can cause a callback function to be called to perform a specified operation.

11.3.7 Advantages of Classes

There are many advantages to using OOP and instantiating objects versus using procedural programming and data structures such as structs. Creating one's own classes offers more control over behaviors. Since operators and other functions can be overloaded, programmers can define them precisely as needed. Also, by redefining the set.PropertyName methods, the range of values that can be assigned is strictly controlled.

Another advantage of objects is that when using struct.field, if a fieldname is not spelled correctly in an assignment statement, e.g.,

```
struct.feild = value;
```

this would just create a new field with the incorrect fieldname, and add it to the structure! If this was attempted with a class object, however, it would throw an error.

■ Explore Other Interesting Features

- Creating a directory for a class so that not all methods have to be in the same file as the class definition
- The Constant attribute
- Map containers
- Enumeration class ■

SUMMARY

COMMON PITFALLS

- Confusing value and handle classes
- Not realizing that constructor functions are called automatically when objects are instantiated

PROGRAMMING STYLE GUIDELINES

- Use the dot notation to reference properties instead of **get** and **set**
- Use **nargin** to check the number of input arguments to a constructor function
- Write constructor functions to allow for no input arguments
- Call an ordinary method by using the dot operator with the object rather than explicitly passing the object

MATLAB Keyword
classdef

MATLAB Functions and Commands		
line	set	gca
groot	properties	gcf
get	methods	

MATLAB Operators
dot operator for object properties and methods.

Exercises

1. Create a **double** variable. Use the functions **methods** and **properties** to see what are available for the class **double**.
2. Create a simple **plot** and store the handle in a variable. Use the three different methods (dot notation, **set**, and structure) to change the Color property.
3. Create a **bar** chart and store the handle in a variable. Change the EdgeColor property to red.
4. Create a class *circleClass* that has a property for the radius of the circle and a constructor function. Make sure that there is a default value for the radius, either in the properties block or in the constructor. Instantiate an object of your class and use the **methods** and **properties** functions.
5. Add ordinary methods to *circleClass* to calculate the area and circumference of the circle.
6. Create a class that will store the price of an item in a store, as well as the sales tax rate. Write an ordinary method to calculate the total price of the item, including the tax.
7. Create a class designed to store and view information on software packages for a particular software superstore. For every software package, the information needed includes the item number, the cost to the store, the price passed

on to the customer, and a code indicating what kind of software package it is (e.g., 'c' for a compiler, 'g' for a game, etc.). Include an ordinary method *profit* that calculates the profit on a particular software product.

8. Create the *Rectangle* class from this chapter. Add a function to overload the **gt** (greater than) operator. Instantiate at least two objects and make sure that your function works.
9. Create a class *MyCourse* that has properties for a course number, number of credits, and grade. Overload the **disp** function to display this information.
10. Construct a class named *Money* that has five properties for dollars, quarters, dimes, nickels, and pennies. Include an ordinary function *equivtotal* that will calculate and return the equivalent total of the properties in an object (e.g., 5 dollars, 7 quarters, 3 dimes, 0 nickels, and 6 pennies is equivalent to 7.11). Overload the **disp** function to display the properties.
11. Write a program that creates a class for complex numbers. A complex number is a number of the form $a+bi$, where a is the real part, b is the imaginary part, and $i=\sqrt{-1}$. The class *Complex* should have properties for the real and imaginary parts. Overload the **disp** function to print a complex number.
12. Create a base class *Square* and then a derived class *Cube*, similar to the Rectangle/Box example from the chapter. Include methods to calculate the area of a square and volume of a cube.
13. Create a base class named *Point* that has properties for *x* and *y* coordinates. From this class derive a class named *Circle* having an additional property named *radius*. For this derived class, the *x* and *y* properties represent the center coordinates of a circle. The methods of the base class should consist of a constructor, an *area* function that returns 0, and a distance function that returns the distance between two points ($\text{sqrt}((x_2-x_1)^2+(y_2-y_1)^2)$). The derived class should have a constructor and an overloaded function named *area* that returns the area of a circle. Write a script that has two objects of each class and calls all of the methods.
14. Take any value class (e.g., *MyCourse* or *Square*) and make it into a handle class. What are the differences?
15. Create a class that stores information on a company's employees. The class will have properties to store the employee's name, a 10-digit ID, their department, and a rating from 0 to 5. Overwrite the *set.propertyname* function to check that each property is the correct class and that:
    - The employee ID has 10 digits
    - The department has one of the following codes: HR (Human Resources), IT (Information Technology), MK (Marketing), AC (Accounting), or RD (research and Development)
    - The rating is a number from 0 to 5.

    The rating should not be accessible to anyone without a password. Overwrite the *set.rating* and *get.rating* functions to prompt the user for a password. Then, write a function that returns the rating.
16. Create a handle class that logs the times a company's employees arrive and leave at work. The class must have the following characteristics. As the

employer, you do not want your employees to access the information stored. The class will store date, hour, minute, second, and total time as properties. The constructor function will input the data from the *clock* function, which returns a vector with format [year month day hour minute second]. Each time an employee arrives or leaves, they must call a method *LogTime* that will store the new times with the old times. For example, property hour will be

Hour 1
Hour 2
Hour 3

Include a method *calcPay* that calculates the money owed if it is assumed that the employees are paid $15 per hour. In order to do this, call a separate method that calculates the time elapsed between the last two time entries. Use the function **etime**. This method should only be available to call by *calcPay*, and the existence of *calcPay* should only be known to the coder.

17. You head a team developing a small satellite in competition for a NASA contract. Your design calls for a central satellite that will deploy sensor nodes. These nodes must remain within 30 km of the satellite to allow for data transmission. If they pass out of range, they will use an impulse of thrust propulsion to move back towards the satellite. Make a *Satellite* class with the following properties:

- *location*: An [*X Y Z*] vector of coordinates, with the satellite as the origin.
- *magnetData*: A vector storing magnetic readings.
- *nodeAlerts*: An empty string to begin with, stores alerts when nodes go out of range.

Satellite also has the following methods:

- *Satellite*: The constructor, which sets location to [0 0 0] and magnetData to 0.
- *retrieveData*: Takes data from a node, extends the magnetData vector.

Then, make the *sensorNode* class as a subclass of Satellite. It will have the following properties:

- *distance*: The magnitude of the distance from the satellite. Presume that a node's location comes from on-board, real-time updating GPS (i.e., do not worry about updating node.location).
- *fuel*: Sensor nodes begin with 100 kg of fuel.

sensorNode also has the following methods:

- *sensorNode*: The constructor.
- *useThrust*: Assume this propels node towards satellite. Each usage consumes 2 kg of fuel. If the fuel is below 5 kg, send an alert message to the satellite.
- *checkDistance*: Check the magnitude of the distance between
- useMagnetometer: Write this as a stub. Have the "magnetometer reading" be a randomized number in the range 0 to 100.
- *sendAlert*: set the "nodeAlerts" Satellite property to the string 'Low fuel.'

First, treat both classes as value classes. Then, adjust your code so that both are handle classes. Which code is simpler?

CHAPTER 12

Advanced Plotting Techniques

KEY TERMS

histogram	area plot	plot properties
stem plot	bin	core objects
pie chart	animation	text box

CONTENTS

In Chapter 3, we introduced the use of the function **plot** in the MATLAB® software to get simple, two-dimensional (2D) plots of *x* and *y* points represented by two vectors *x* and *y*. We have also seen some functions that allow customization of these plots. In Chapter 11, we introduced handle classes, object handles and methods of examining and modifying plot object properties. In this chapter we will explore other types of plots, ways of customizing plots, and some applications that combine plotting with functions and file input. Additionally, animation, three-dimensional (3D) plots, and core graphics primitives will be introduced. Note that in R2014b the default colors were modified, so in versions prior to that the colors in the plots may differ from what is shown.

In the latest versions of MATLAB, the PLOTS tab can be used to very easily create advanced plots. The method is to create the variables in which the data are stored, and then select the PLOTS tab. The plot functions that can be used are then highlighted; simply clicking the mouse on one will plot the data using that function and open up the Figure Window with that plot. For example, by creating *x* and *y* variables, and highlighting them in the Workspace Window, the 2D plot types will become visible. If, instead, *x*, *y*, and *z* variables are highlighted, the 3D plot types will become available. These are extremely fast methods for users to create plots in MATLAB. However, as this text focuses on programming concepts, the programmatic methodologies will be explained in this chapter.

MATLAB®. http://dx.doi.org/10.1016/B978-0-12-804525-1.00012-X

12.1 PLOT FUNCTIONS AND CUSTOMIZING PLOTS

So far, we have used **plot** to create 2D plots and **bar** to create bar charts. We have seen how to clear the Figure Window using **clf**, and how to create and number Figure Windows using **figure**. Labeling plots has been accomplished using **xlabel**, **ylabel**, **title**, and **legend**, and we have also seen how to customize the strings passed to these functions using **sprintf**. The **axis** function changes the axes from the defaults that would be taken from the data in the x and y vectors to the values specified. Finally, the **grid** and **hold** toggle functions print grids or not, or lock the current graph in the Figure Window so that the next plot will be superimposed.

Another function that is very useful with all types of plots is **subplot**, which creates a matrix of plots in the current Figure Window, as we have seen in Chapter 5. The **sprintf** function is frequently used to create customized axis labels and titles within the matrix of plots.

Besides **plot** and **bar**, there are many other plot types, such as ***histograms***, ***stem plots***, ***pie charts***, and ***area plots***, as well as other functions that customize graphs. Described in this section are some of the other plotting functions.

12.1.1 Bar, Barh, Area, and Stem Functions

The functions **bar**, **barh**, **area**, and **stem** essentially display the same data as the **plot** function, but in different forms. The **bar** function draws a bar chart (as we have seen before), **barh** draws a horizontal bar chart, **area** draws the plot as a continuous curve and fills in under the curve that is created, and **stem** draws a stem plot.

For example, the following script creates a Figure Window that uses a 2×2 **subplot** to demonstrate four plot types using the same data points (see Fig. 12.1). Notice how the axes are set by default.

subplottypes.m

```
% Subplot to show plot types

year = 2016:2020;
pop = [0.9 1.4 1.7 1.3 1.8];
subplot(2,2,1)
plot(year,pop)
title('plot')
xlabel('Year')
ylabel('Population (mil)')
subplot(2,2,2)
bar(year,pop)
```

```
title('bar')
xlabel('Year')
ylabel('Population (mil)')
subplot(2,2,3)
area(year,pop)
title('area')
xlabel('Year')
ylabel('Population (mil)')
subplot(2,2,4)
stem(year,pop)
title('stem')
xlabel('Year')
ylabel('Population (mil)')
```

Note

The third argument in the call to the **subplot** function is a single index into the matrix created in the Figure Window; the numbering is row wise (in contrast to the normal column wise unwinding that MATLAB uses for matrices).

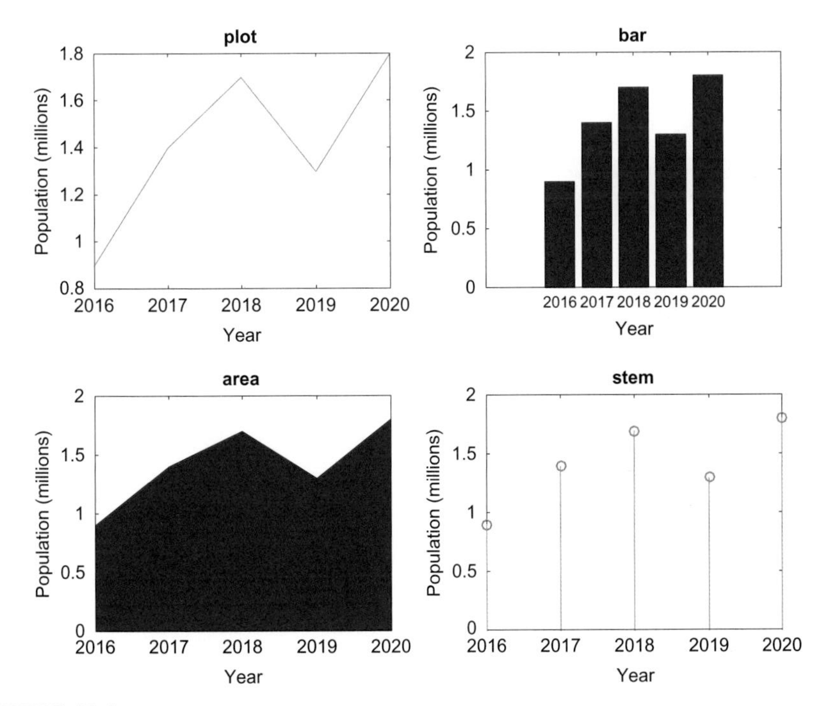

FIGURE 12.1
Subplot to display **plot**, **bar**, **area**, and **stem** plots.

QUICK QUESTION!

Could we produce this **subplot** using a loop?

Answer: Yes, we can store the names of the plots in a cell array. These names are put in the titles, and also concatenated with the string '(x,y)' and passed to the **eval** function to evaluate the function.

loopsubplot.m

```
% Demonstrates evaluating plot type names in order to
% use the plot functions and put the names in titles

year = 2016:2020;
pop = [0.9 1.4 1.7 1.3 1.8];
titles = {'plot', 'bar', 'area', 'stem'};
for i = 1:4
    subplot(2,2,i)
    eval([titles{i} '(year,pop)'])
    title(titles{i})
    xlabel('Year')
    ylabel('Population (mil)')
end
```

QUICK QUESTION!

What are some different options for plotting more than one graph?

Answer: There are several methods, depending on whether you want them in one Figure Window superimposed (using **hold on**), in a matrix in one Figure Window (using **subplot**), or in multiple Figure Windows (using **figure(n)**).

For a matrix, the **bar** and **barh** functions will group together the values in each row. For example:

```
>> groupages = [8 19 43 25; 35 44 30 45]
groupages =
     8    19    43    25
    35    44    30    45
```

produces the plot shown in Fig. 12.2.

```
>> bar(groupages)
>> xlabel('Group')
>> ylabel('Ages')
```

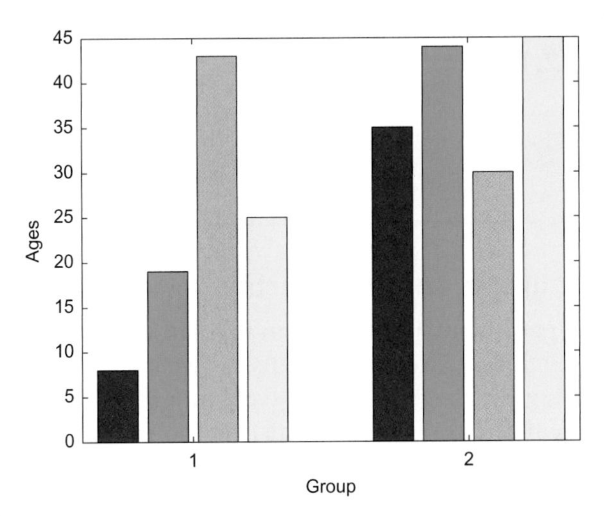

FIGURE 12.2
Data from a matrix in a bar chart.

Note that MATLAB groups together the values in the first row and then in the second row. It cycles through colors to distinguish the bars. The 'stacked' option will stack rather than group the values, so the "y" value represented by the top of the bar is the sum of the values from that row (shown in Fig. 12.3).

```
>> bar(groupages,'stacked')
>> xlabel('Group')
>> ylabel('Ages')
```

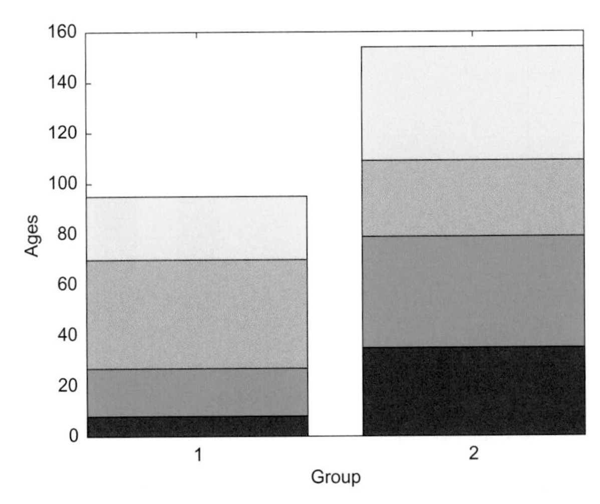

FIGURE 12.3
Stacked bar chart of matrix data.

PRACTICE 12.1

Create a file that has two lines with *n* numbers in each. Use **load** to read this into a matrix. Then, use **subplot** to show the **barh** and stacked **bar** charts side by side. Put labels 'Groups' for the two groups and 'Values' for the data values on the axes (note the difference between the *x* and *y* labels for these two plot types).

12.1.2 Histograms and Pie Charts

A ***histogram*** is a particular type of bar chart that shows the frequency of occurrence of values within a vector. Histograms use what are called ***bins*** to collect values that are in given ranges. MATLAB has a function **histogram** to create a histogram, (note that this replaced the **hist** function in R2014b). Calling the function with the form **histogram(vec)** by default takes the values in the vector *vec* and puts them into bins; the number of bins is determined by the **histogram** function (or **histogram(vec,n)** will put them into *n* bins) and plots this, as shown in Fig. 12.4.

```
>> quizzes = [10 8 5 10 10 6 9 7 8 10 1 8];
>> histogram(quizzes)
>> xlabel('Grade')
>> ylabel('#')
>> title('Quiz Grades')
```

In this example, the numbers range from 1 to 10 in the vector, and there are 10 bins in the range from 1 to 10. The heights of the bins represent the number of

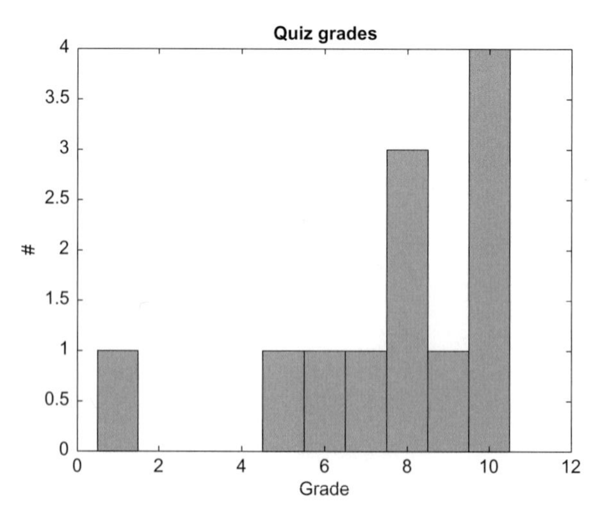

FIGURE 12.4
Histogram of data.

values that fall within that particular bin. The handle of a **histogram** can also be stored in an object variable; the properties can then be inspected and/or modified.

```
>> hhan = histogram(quizzes)
hhan =
  Histogram with properties:
            Data: [10 8 5 10 10 6 9 7 8 10 1 8]
          Values: [1 0 0 0 1 1 1 3 1 4]
         NumBins: 10
        BinEdges: [0.5000 1.5000 2.5000 3.5000 4.5000
                   5.5000 6.5000 7.5000 8.5000 9.5000 10.5000]
        BinWidth: 1
       BinLimits: [0.5000 10.5000]
   Normalization: 'count'
       FaceColor: 'auto'
       EdgeColor: [0 0 0]
  Show all properties
```

Histograms are used for statistical analyses of data; more statistics will be covered in Chapter 14.

MATLAB has a function, **pie,** that will create a pie chart. Calling the function with the form **pie(vec)** draws a pie chart using the percentage of each element of *vec* of the whole (the sum). It shows these starting from the top of the circle and going around counterclockwise. For example, the first value in the vector `[11 14 8 3 1]`, 11 is 30% of the sum, 14 is 38% of the sum, and so forth, as shown in Fig. 12.5.

```
>> pie([11 14 8 3 1])
```

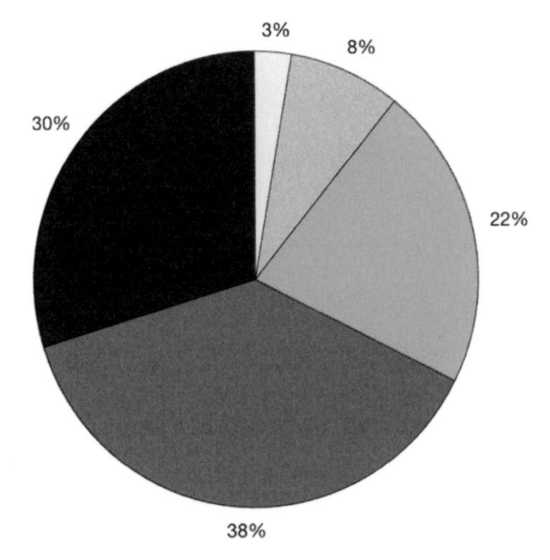

FIGURE 12.5
Pie chart showing percentages.

A cell array of labels can also be passed to the **pie** function; these labels will appear instead of the percentages (shown in Fig. 12.6).

```
>> pie([11 14 8 3 1], {'A','B','C','D', 'F'})
```

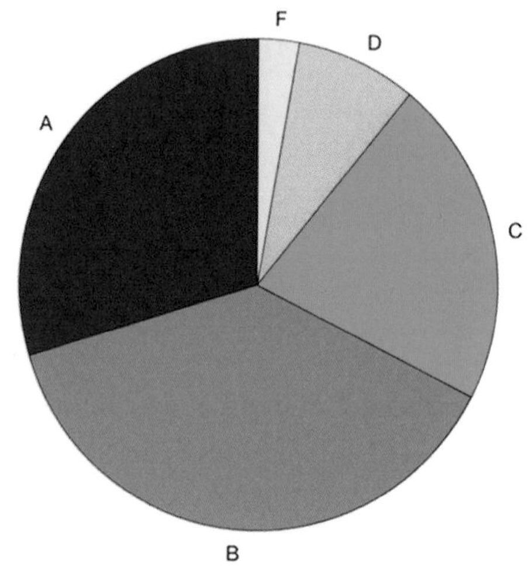

FIGURE 12.6
Pie chart with labels from a cell array.

PRACTICE 12.2

A chemistry professor teaches three classes. These are the course numbers and enrollments:

```
CH 101    111
CH 105     52
CH 555     12
```

Use **subplot** to show this information using **pie** charts: the **pie** chart on the right should show the percentage of students in each course, and on the left, the course numbers. Put appropriate titles on them.

12.1.3 Log Scales

The **plot** function uses linear scales for both the x and y axes. There are several functions that instead use logarithmic scales for one or both axes: the function **loglog** uses logarithmic scales for both the x and y axes, the function **semilogy** uses a linear scale for the x-axis and a logarithmic scale for the y-axis, and the function **semilogx** uses a logarithmic scale for the x-axis and a linear scale for the y-axis. The following example uses **subplot** to show the difference, for example, between using the **plot** and **semilogy** functions, as seen in Fig. 12.7.

```
>> subplot(1,2,1)
>> plot(logspace(1,10))
>> title('plot')
>> subplot(1,2,2)
>> semilogy(logspace(1,10))
>> title('semilogy')
```

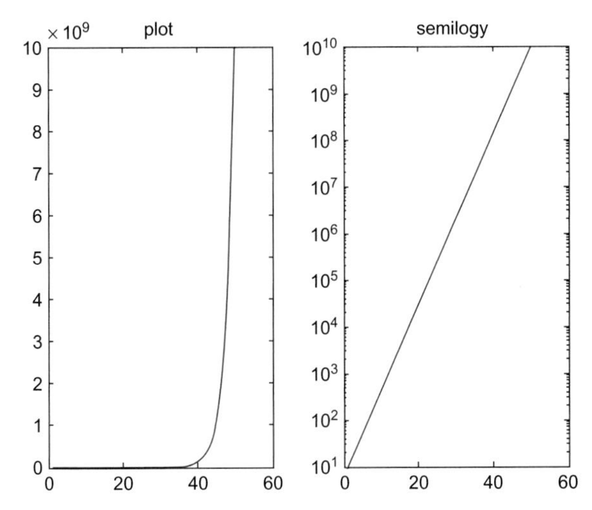

FIGURE 12.7
Plot versus semilogy.

12.1.4 Animation

There are several ways to ***animate*** a plot. These are visuals, so the results cannot really be shown here; it is necessary to type these into MATLAB to see the results.

We'll start by animating a plot of **sin(x)** with the vectors:

```
>> x = -2*pi : 1/100 : 2*pi;
>> y = sin(x);
```

This results in enough points that we'll be able to see the result using the built-in **comet** function, which shows the plot by first showing the point (x(1),y(1)), and then moving on to the point (x(2),y(2)), and so on, leaving a trail (like a comet!) of all of the previous points.

```
>> comet(x,y)
```

The end result looks similar to the result of **plot(x,y)**.

Another way of animating is to use the built-in function **movie**, which displays recorded movie frames. The frames are captured in a loop using the built-in function **getframe**, and are stored in a matrix. For example, the following script again animates the **sin** function. The **axis** function is used so that MATLAB will use the same set of axes for all frames, and using the **min** and **max** functions on the data vectors *x* and *y* will allow us to see all points. It displays the "movie" once in the **for** loop, and then again when the movie function is called. (Note that documentation is no longer available for the **movie** function, although it is still operational as of R2016a.)

sinmovie.m

```
% Shows a movie of the sin function
clear

x = -2*pi: 1/5 : 2*pi;
y = sin(x);
n = length(x);

for i = 1:n
    plot(x(i),y(i),'r*')
    axis([min(x)-1 max(x)+1 min(y)-1 max(y)+1])
    M(i) = getframe;
end
movie(M)
```

12.1.5 Customizing Plots

There are many ways to customize figures in the Figure Window. Clicking on the Plot Tools icon in the Figure Window itself will bring up the Property Editor and Plot Browser, with many options for modifying the current plot. Additionally, there are ***plot properties*** that can be modified from the defaults in the plot functions. Using the **help** facility with the function name will show all the options for that particular plot function.

For example, the **bar** and **barh** functions by default put a "width" of 0.8 between bars. When called as **bar(x,y)**, the width of 0.8 is used. If instead, consider a third argument is passed, it is the width, for example **barh(x,y,width)**. The following script uses **subplot** to show variations on the width. A width of 0.6 results in more space between the bars. A width of 1 makes the bars touch each other, and with a width greater than 1, the bars actually overlap. The results are shown in Fig. 12.8.

barwidths.m

```
% Subplot to show varying bar widths

year = 2016:2020;
pop = [0.9 1.4 1.7 1.3 1.8];

for i = 1:4
    subplot(1,4,i)
    % width will be 0.6, 0.8, 1, 1.2
    barh(year,pop,0.4+i*.2)
    title(sprintf('Width = %.1f',0.4+i*.2))
    xlabel('Population (mil)')
    ylabel('Year')
end
```

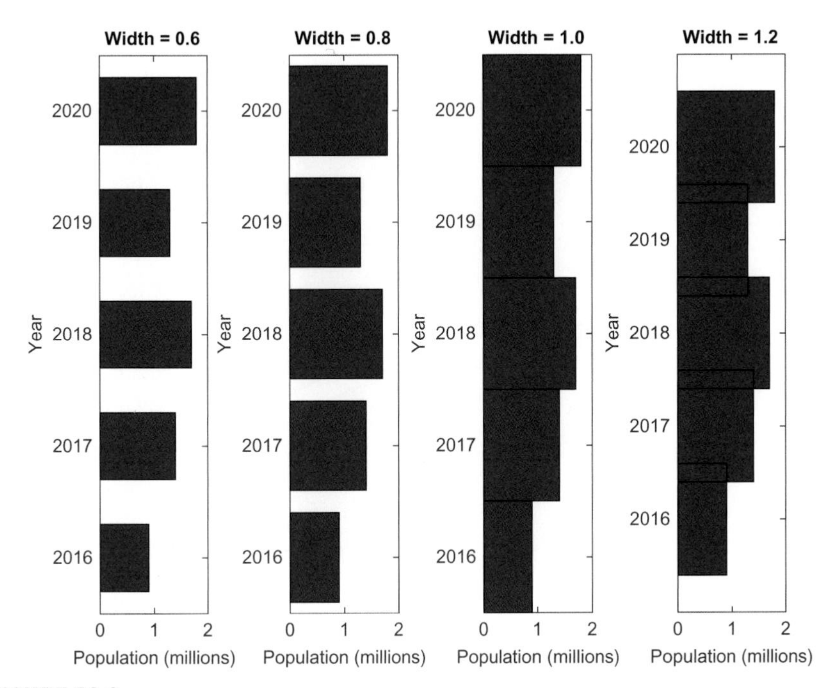

FIGURE 12.8
Subplot demonstrates varying widths in a bar chart.

PRACTICE 12.3

Use **help area** to find out how to change the base level on an **area** chart (from the default of 0).

As another example of customizing plots, pieces of a pie chart can be "exploded" from the rest. In this case, two vectors are passed to the **pie** function: first the data vector, then a **logical** vector; the elements for which the **logical** vector is **true** will be exploded from (separated from) the pie chart. A third argument, a cell array of labels, can also be passed. The result is seen in Fig. 12.9.

```
>> gradenums = [11 14 8 3 1];
>> letgrades = {'A','B','C','D','F'};
>> which = gradenums == max(gradenums)
which =
     0    1    0    0    0
>> pie(gradenums,which,letgrades)
>> title(strcat('Largest Fraction of Grades: ', ...
          letgrades(which)))
```

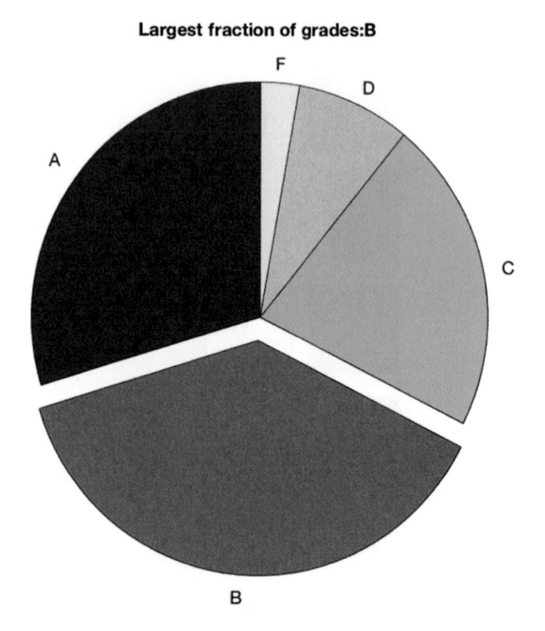

FIGURE 12.9
Exploding pie chart.

12.2 3D PLOTS

MATLAB has functions that will display 3D plots. Many of these functions have the same name as the corresponding 2D plot function with a '3' at the end. For example, the 3D line plot function is called **plot3**. Other functions include **bar3**, **bar3h**, **pie3**, **comet3**, and **stem3**.

Vectors representing *x*, *y*, and *z* coordinates are passed to the **plot3** and **stem3** functions. These functions show the points in 3D space. Clicking on the rotate 3D icon and then on the plot allows the user to rotate and see the plot from different angles. Also, using the **grid** function makes it easier to visualize, as shown in Fig. 12.10. The function **zlabel** is used to label the *z* axis.

```
>> x = 1:5;
>> y = [0 -2 4 11 3];
>> z = 2:2:10;
>> plot3(x,y,z,'k*')
>> grid
>> xlabel('x')
>> ylabel('y')
>> zlabel('z')
>> title('3D plot')
```

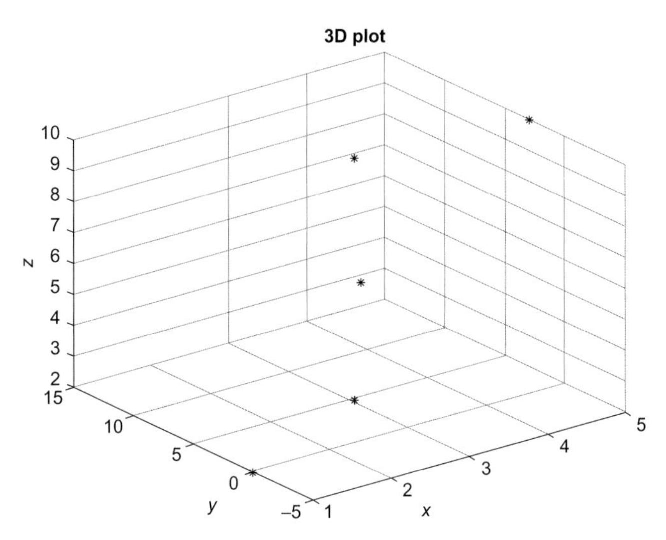

FIGURE 12.10
Three-dimensional plot with a grid.

For the **bar3** and **bar3h** functions, *y* and *z* vectors are passed and the function shows 3D bars as shown, for example, for **bar3** in Fig. 12.11.

```
>> y = 1:6;
>> z = [33 11 5 9 22 30];
>> bar3(y,z)
>> xlabel('x')
>> ylabel('y')
>> zlabel('z')
>> title('3D bar')
```

A matrix can also be passed, for example a 5×5 **spiral** matrix (which "spirals" the integers 1 to 25 or more generally from 1 to n^2 for **spiral(n)**) as shown in Fig. 12.12.

```
>> mat = spiral(5)
mat =
    21    22    23    24    25
    20     7     8     9    10
    19     6     1     2    11
    18     5     4     3    12
    17    16    15    14    13
>> bar3(mat)
>> title('3D spiral')
>> xlabel('x')
>> ylabel('y')
>> zlabel('z')
```

Similarly, the **pie3** function shows data from a vector as a 3D pie, as shown in Fig. 12.13.

```
>> pie3([3 10 5 2])
```

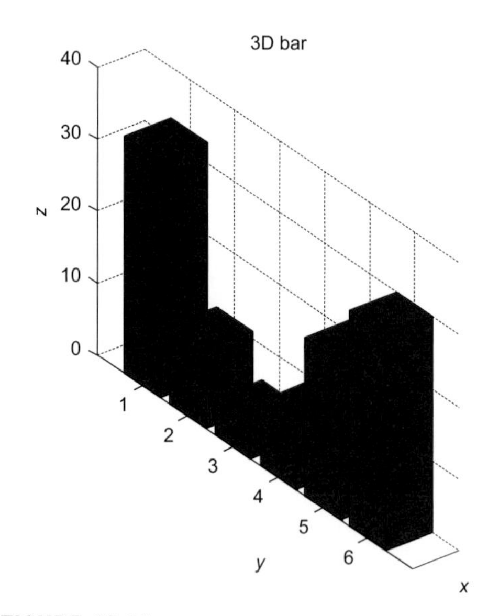

FIGURE 12.11
Three-dimensional bar chart.

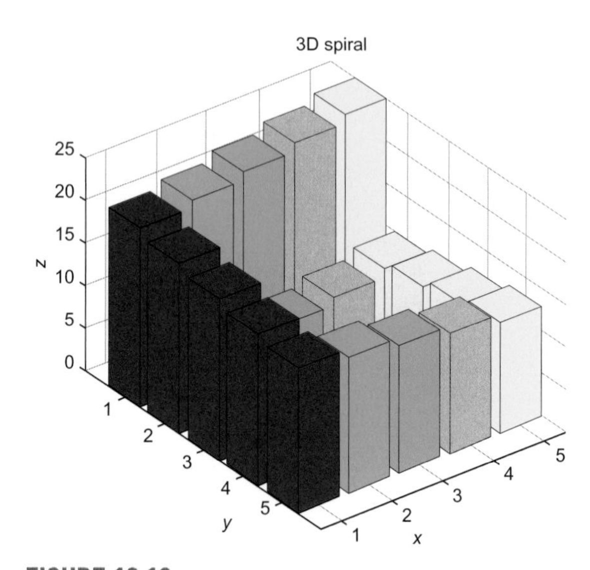

FIGURE 12.12
Three-dimensional bar chart of a spiral matrix.

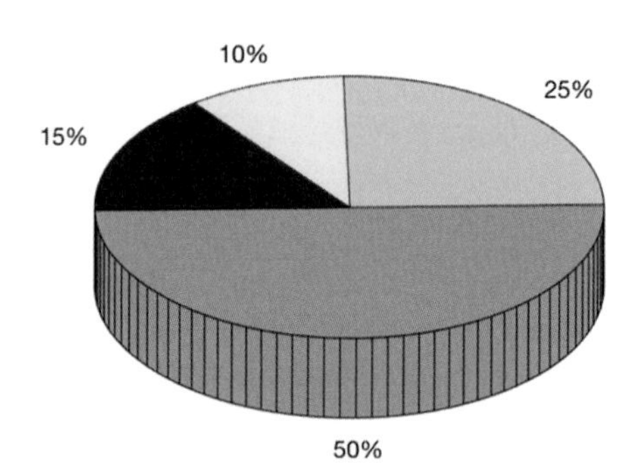

FIGURE 12.13
Three-dimensional pie chart.

Displaying the result of an animated plot in three dimensions is interesting. For example, try the following using the **comet3** function:

```
>> t = 0:0.001:12*pi;
>> comet3(cos(t), sin(t), t)
```

Other interesting 3D plot types include **mesh** and **surf**. The **mesh** function draws a wireframe mesh of 3D points, whereas the **surf** function creates a surface plot by using color to display the parametric surfaces defined by the points. MATLAB has several functions that will create the matrices used for the (*x*,*y*,*z*) coordinates for specified shapes (e.g., **sphere** and **cylinder**).

For example, passing an integer *n* to the **sphere** function creates $n+1 \times n+1$ *x*, *y*, and *z* matrices, which can then be passed to the **mesh** function (Fig. 12.14) or the **surf** function (Fig. 12.15).

```
>> [x,y,z] = sphere(15);
>> size(x)
ans =
    16    16
>> mesh(x,y,z)
>> title('Mesh of sphere')
```

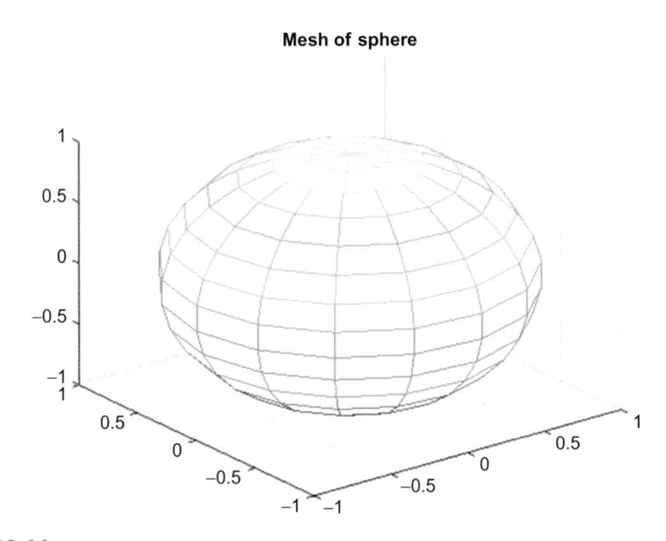

FIGURE 12.14
Mesh plot of sphere.

Additionally, the **colorbar** function displays a colorbar to the right of the plot, showing the range of colors.

```
>> [x,y,z] = sphere(15);
>> sh = surf(x,y,z);
>> title('Surf of sphere')
>> colorbar
```

Note that more options for colors will be described in Chapter 13.

One of the properties of the object stored in *sh* is the FaceAlpha property, which is a measure of the transparency. The result of modifying it to 0.5 is shown in Fig. 12.16.

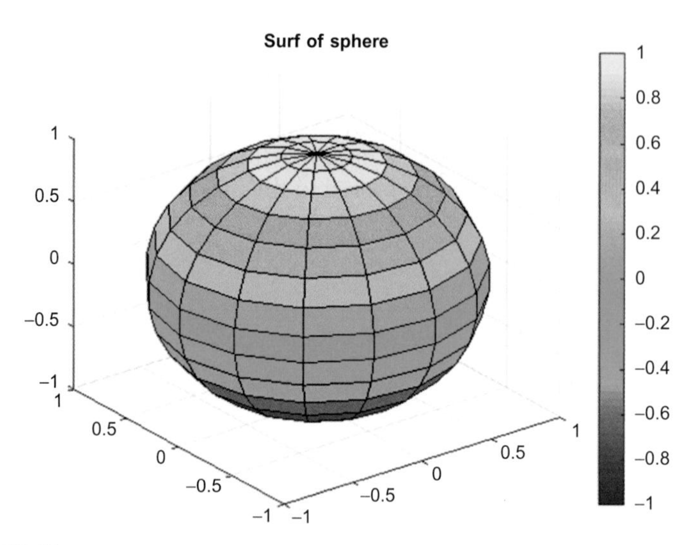

FIGURE 12.15
Surf plot of sphere.

```
>> sh
sh =
  Surface with properties:

       EdgeColor: [0 0 0]
       LineStyle: '-'
       FaceColor: 'flat'
    FaceLighting: 'flat'
       FaceAlpha: 0.5000
           XData: [16x16 double]
           YData: [16x16 double]
           ZData: [16x16 double]
           CData: [16x16 double]

  Show all properties
>> sh.FaceAlpha = 0.5;
```

The **meshgrid** function can be used to create (x,y) points for which $z=f(x,y)$; then the x, y, and z matrices can be passed to **mesh** or **surf**. For example, the following creates a surface plot of the function $\cos(x)+\sin(y)$, as seen in Fig. 12.17.

```
>> [x, y] = meshgrid(-2*pi: 0.1: 2*pi);
>> z = cos(x) + sin(y);
>> surf(x,y,z)
>> title('cos(x) + sin(y)')
>> xlabel('x')
>> ylabel('y')
>> zlabel('z')
```

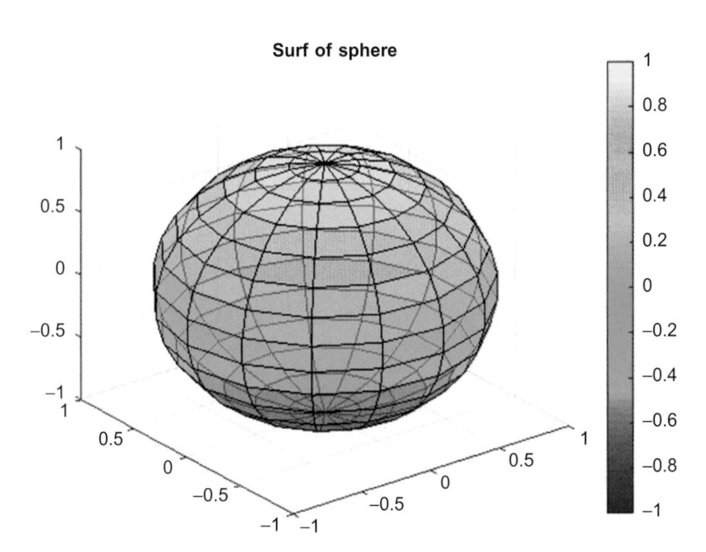

FIGURE 12.16
Surf plot of sphere with FaceAlpha modified.

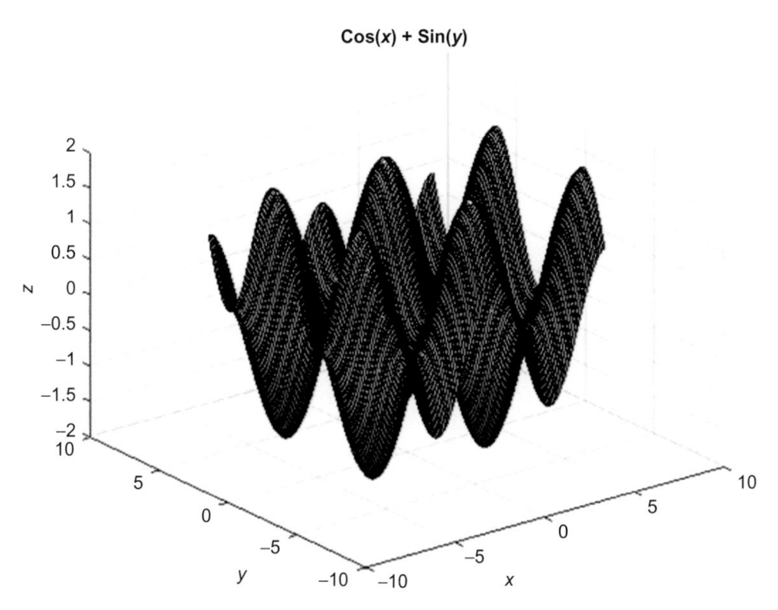

FIGURE 12.17
Use of **meshgrid** for f(*x*,*y*) points.

12.3 CORE GRAPHICS OBJECTS

Core objects in MATLAB are the very basic graphics primitives. A description can be found under the MATLAB help. In the documentation, search for Graphics and Graphics Objects. The core objects include:

- **line**
- **text**
- **rectangle**
- **patch**
- **image**

These are all built-in functions; **help** can be used to find out how each function is used. The first four of these core objects will be described in this section; images will be described in Section 13.1.

A **line** is a core graphics object, which is what is used by the **plot** function. The following is an example of creating a line object, setting some properties, and saving the handle in a variable *hl*:

```
>> x = -2*pi: 1/5 : 2*pi;
>> y = sin(x);
>> hl = line(x,y,'LineWidth', 6, 'Color', [0.5 0.5 0.5])
hl =
  Line with properties:

              Color: [0.5000 0.5000 0.5000]
          LineStyle: '-'
          LineWidth: 6
             Marker: 'none'
         MarkerSize: 6
    MarkerFaceColor: 'none'
              XData: [1x63 double]
              YData: [1x63 double]
              ZData: [1x0 double]

  Show all properties
```

As seen in Fig. 12.18, this draws a reasonably thick gray line for the **sin** function. As before, the handle will be valid as long as the Figure Window is not closed.

As another example, the following uses the **line** function to draw a circle. First, a white Figure Window is created. The x and y data points are generated, and then the **line** function is used, specifying a dotted red line with a line width of 4. The **axis** function is used to make the axes square, so the result looks like a circle, but then removes the axes from the Figure Window (using **axis square** and **axis off**, respectively). The result is shown in Fig. 12.19.

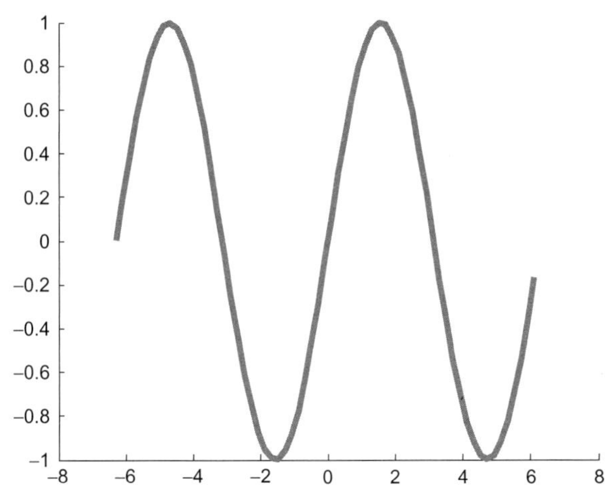

FIGURE 12.18
A **line** object with modified line width and color.

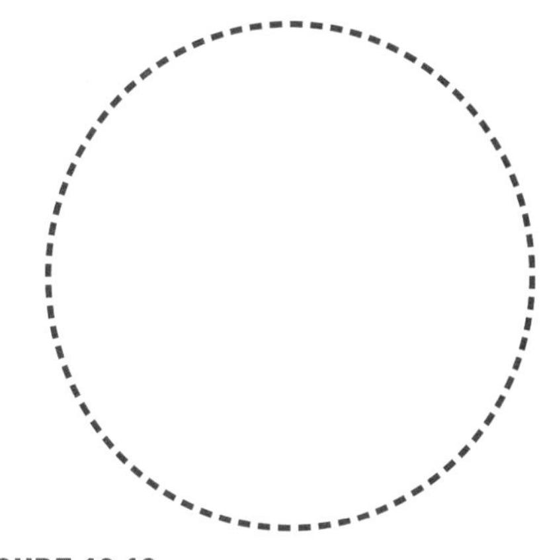

FIGURE 12.19
Use of line to draw a circle.

```
>> figure('Color', [1 1 1])
>> pts = 0:0.1:2*pi;
>> xcir = cos(pts);
>> ycir = sin(pts);
>> line(xcir, ycir, 'LineStyle',':', ...
        'LineWidth',4,'Color','r')
>> axis square
>> axis off
```

The **text** graphics function allows text to be printed in a Figure Window, including special characters that are printed using `\specchar`, where "specchar" is the actual name of the special character. The format of a call to the **text** function is

```
text(x,y,'text string')
```

where *x* and *y* are the coordinates on the graph of the lower left corner of the ***text box*** in which the text string appears. The special characters include letters of the Greek alphabet, arrows, and characters frequently used in equations. For example, Fig. 12.20 displays the Greek symbol for **pi** and a right arrow within the text box.

```
>> x = -4:0.2:4;
>> y = sin(x);
>> hp = line(x,y,'LineWidth',3);
>> thand = text(2,0,'Sin(\pi)\rightarrow')
thand =

  Text (Sin(\pi)\rightarrow) with properties:

                 String: 'Sin(\pi)\rightarrow'
```

```
               FontSize: 10
             FontWeight: 'normal'
               FontName: 'Helvetica'
                  Color: [0 0 0]
    HorizontalAlignment: 'left'
               Position: [2 0 0]
                  Units: 'data'

  Show all properties
```

Some of the other properties are shown here.

```
>> thand.Extent
ans =
    2.0000 -0.0321  0.6912  0.0613
>> thand.EdgeColor
ans =
none
>> thand.BackgroundColor
ans =
none
```

Although the Position specified was (2,0), the Extent is the actual extent of the text box, which cannot be seen as the BackgroundColor and EdgeColor are not specified. These can be seen using the dot notation or **get**, and can be changed using the dot notation or **set**. For example, the following produces the result seen in Fig. 12.21.

```
>> set(thand,'BackgroundColor',[0.8 0.8 0.8],...
         'EdgeColor',[1 0 0])
```

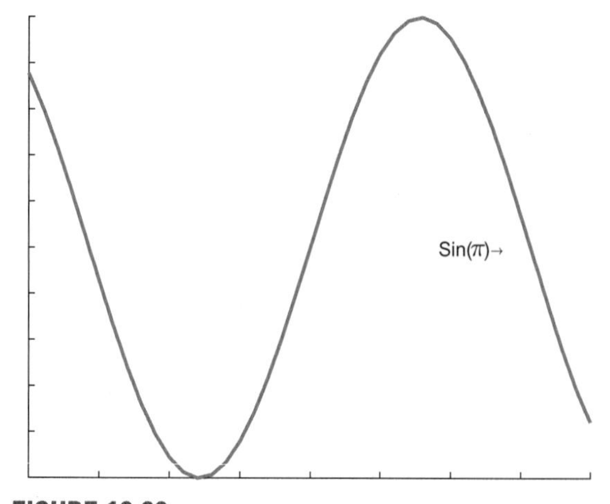

FIGURE 12.20
A **line** object with a text box.

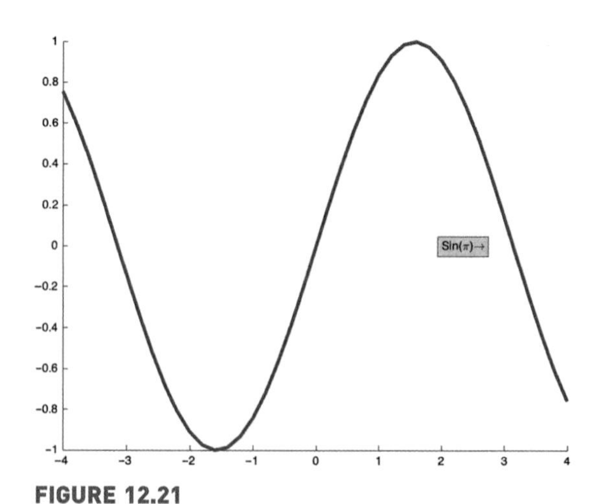

FIGURE 12.21
Text box with a modified edge color and background color.

The **gtext** function allows you to move your mouse to a particular location in a Figure Window, indicating where a string should be displayed. As the mouse is moved into the Figure Window, cross hairs indicate a location; clicking on the mouse will display the text in a box with the lower left corner at that location. The **gtext** function uses the **text** function in conjunction with **ginput**, which allows you to click the mouse at various locations within the Figure Window and store the *x* and *y* coordinates of these points.

Another core graphics object is **rectangle**, which can have a curvature added to it (!!). Just calling the function **rectangle** without any arguments brings up a Figure Window (shown in Fig. 12.22), which, at first glance, doesn't seem to have anything in it:

```
>> recthand = rectangle
recthand =
  Rectangle with properties:

     FaceColor: 'none'
     EdgeColor: [0 0 0]
     LineWidth: 0.5000
     LineStyle: '-'
     Curvature: [0 0]
      Position: [0 0 1 1]

Show all properties
```

The Position of a rectangle is [x y w h], where x and y are the coordinates of the lower left point, w is the width, and h is the height. The default rectangle has a Position of [0 0 1 1]. The default Curvature is [0 0], which means no curvature.

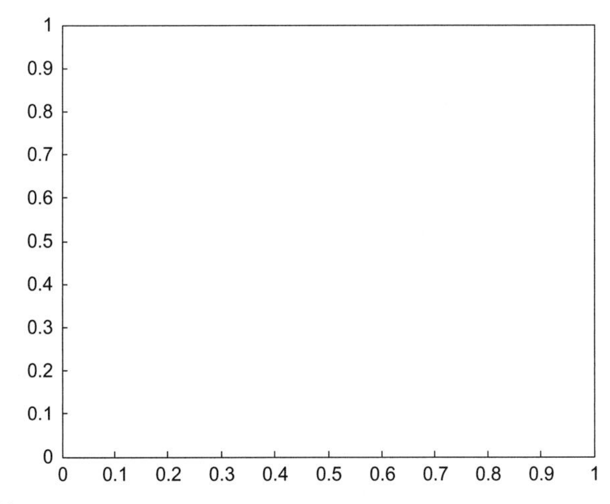

FIGURE 12.22
A **rectangle** object.

The values range from [0 0] (no curvature) to [1 1] (ellipse). A more interesting rectangle object is seen in Fig. 12.23.

Note that properties can be set when calling the **rectangle** function, and also subsequently using the **set** function, as follows:

```
>> rh = rectangle('Position', [0.2, 0.2, 0.5, 0.8],...
    'Curvature',[0.5, 0.5]);
>> axis([0 1.2 0 1.2])
>> set(rh,'LineWidth',3,'LineStyle',':')
```

This creates a curved rectangle and uses dotted lines.

The **patch** function is used to create a patch graphics object, which is made from 2D polygons. A simple patch in 2D space, a triangle, is defined by specifying the coordinates of three points as shown in Fig. 12.24; in this case, the color red is specified for the polygon.

```
>> x = [0 1 0.5];
>> y = [0 0 1];
>> hp = patch(x,y,'r')
hp =
  Patch with properties:

    FaceColor: [1 0 0]
    FaceAlpha: 1
    EdgeColor: [0 0 0]
    LineStyle: '-'
        Faces: [1 2 3]
     Vertices: [3x2 double]

  Show all properties
```

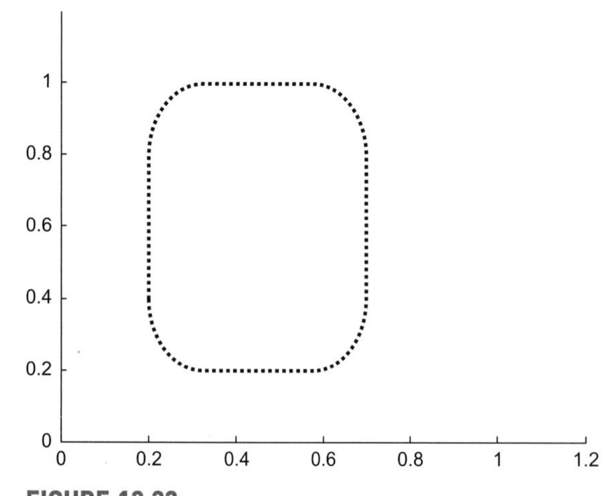

FIGURE 12.23
Rectangle object with curvature.

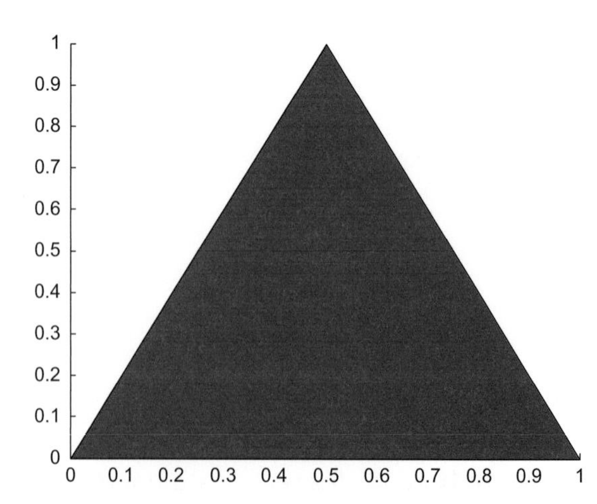

FIGURE 12.24
Simple patch.

The Vertices property stores the three points given by the x and y data vectors.

```
>> hp.Vertices
ans =
         0         0
    1.0000         0
    0.5000    1.0000
```

The Faces property tells how the vertices are connected to create the patch. The vertices are numbered; the first point (0,0) is vertex 1, the point (1,0) is vertex 2, and the last point (0.5, 1) is vertex 3. The Faces property specifies connecting vertex 1 to 2 to 3 (and then by default back to 1).

Another method of creating a patch is to specifically use the Vertices and Faces properties. One way of calling **patch** is `patch(fv)`, where *fv* is a structure variable with fields called *Vertices* and *Faces*. For example, consider a patch object that consists of three connected triangles, and has five vertices given by the coordinates:

```
(1)   (0, 0)
(2)   (2, 0)
(3)   (1, 2)
(4)   (1, -2)
(5)   (3, 1)
```

The order in which the points are given is important, as the faces describe how the vertices are linked. To create these vertices in MATLAB and define faces that connect them, we use a structure variable and then pass it to the **patch** function; the result is shown in Fig. 12.25.

```
mypatch.Vertices = [...
     0 0
     2 0
     1 2
     1 -2
     3 1];
mypatch.Faces = [
     1 2 3
     2 3 5
     1 2 4];

patchhan = patch(mypatch, 'FaceColor', 'r',...
     'EdgeColor','k');
```

The face color is set to red and the edge color to black.

To vary the colors of the faces of the polygons, the FaceColor property is set to 'flat,' which means that every face has a separate color. The *mycolors* variable stores three colors in the rows of the matrix by specifying the red, green, and blue components for each; the first is blue, the second is cyan (a combination

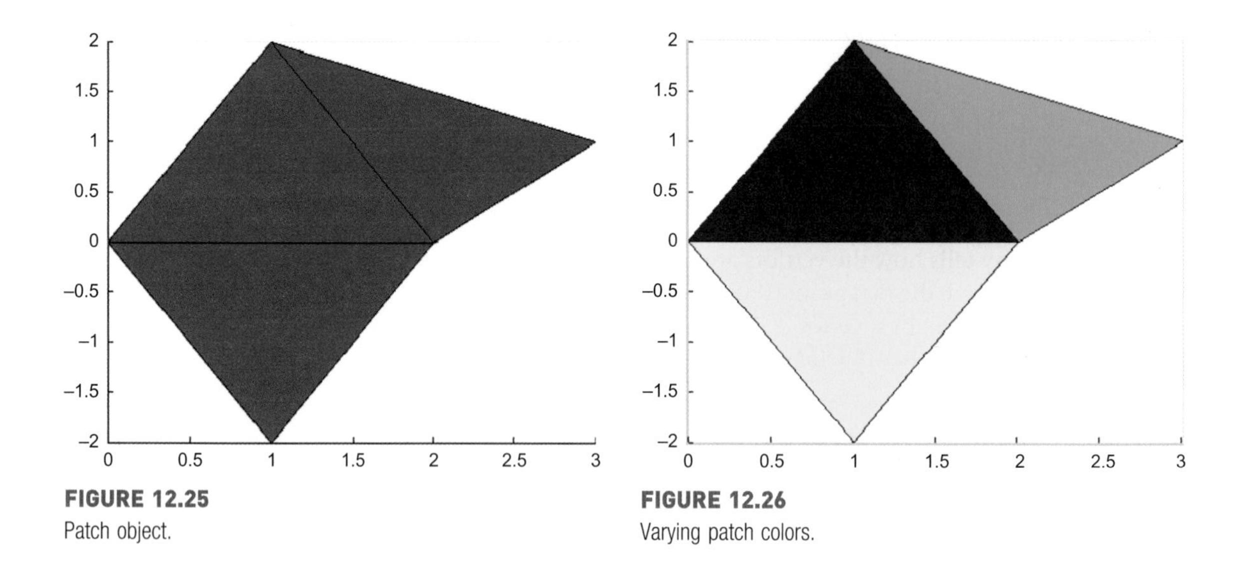

FIGURE 12.25
Patch object.

FIGURE 12.26
Varying patch colors.

of green and blue), and the third is yellow (a combination of red and green). The property FaceVertexCData specifies the color data for the vertices, as seen in Fig. 12.26.

```
>> mycolors = [0 0 1; 0 1 1; 1 1 0];
>> patchhan = patch(mypatch, 'FaceVertexCData', ...
                mycolors, 'FaceColor','flat');
```

Patches can also be defined in 3D space. For example:

```
polyhedron.Vertices = [...
0 0 0
1 0 0
0 1 0
0.5 0.5 1];

polyhedron.Faces = [...
1 2 3
1 2 4
1 3 4
2 3 4];

pobj = patch(polyhedron, ...
'FaceColor',[0.8, 0.8, 0.8],...
'EdgeColor','black');
```

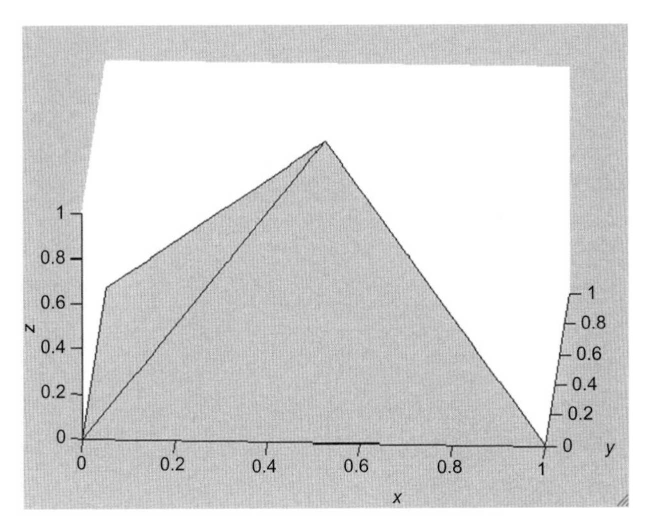

FIGURE 12.27
Rotated **patch** object.

The Figure Window initially shows only two faces. Using the rotate icon, the figure can be rotated so the other edges can be seen as shown in Fig. 12.27.

12.4 PLOT APPLICATIONS

In this section, we will show some examples that integrate plots and many of the other concepts covered to this point in the book. For example, we will have a function that receives an *x* vector, a function handle of a function used to create the *y* vector, and a cell array of plot types as strings and will generate the plots, and we will also show examples of reading data from a file and plotting them.

12.4.1 Plotting From a Function

The following function generates a Figure Window (seen in Fig. 12.28) that shows different types of plots for the same data. The data are passed as input arguments (as an *x* vector and the handle of a function to create the *y* vector) to the function, as is a cell array with the plot type names. The function generates the Figure Window using the cell array with the plot type names. It creates a function handle for each using the **str2func** function.

plotxywithcell.m

```
function plotxywithcell(x, fnhan, rca)
% plotxywithcell receives an x vector, the handle
% of a function (used to create a y vector), and
% a cell array with plot type names; it creates
% a subplot to show all of these plot types
% Format: plotxywithcell(x,fn handle, cell array)

lenrca = length(rca);
y = fnhan(x);
for i = 1:lenrca
    subplot(1,lenrca,i)
    funh = str2func(rca{i});
    funh(x,y)
    title(upper(rca{i}))
    xlabel('x')
    ylabel(func2str(fnhan))
end
end
```

For example, the function could be called as follows:

```
>> anfn = @ (x) x .^3;
>> x = 1:2:9;
>> rca = {'bar', 'area', 'plot'};
>> plotxywithcell(x, anfn, rca)
```

The function is general and works for any number of plot types stored in the cell array.

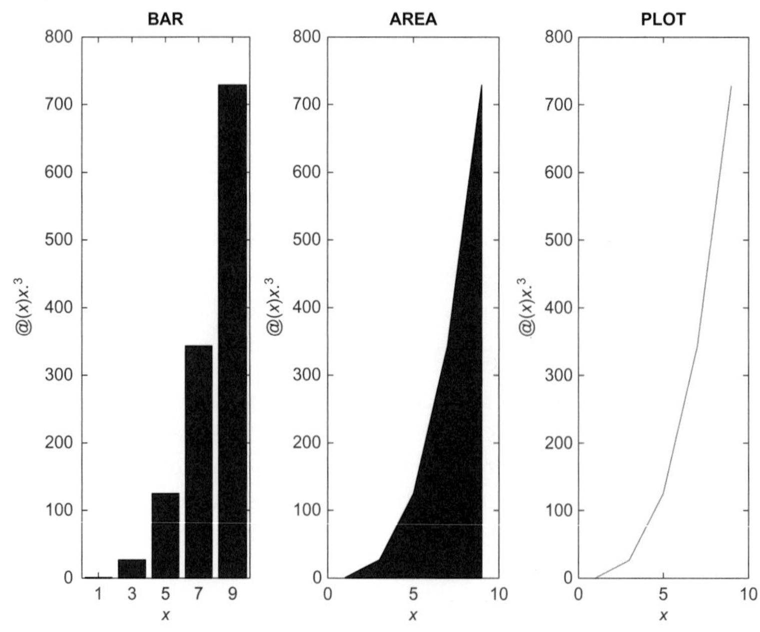

FIGURE 12.28
Subplot showing different file types with their names as titles.

12.4.2 Plotting File Data

It is often necessary to read data from a file and plot them. Normally, this entails knowing the format of the file. For example, let us assume that a company has two divisions, A and B. Assume that the file "ab16.dat" contains four lines, with the sales figures (in millions) for the two divisions for each quarter of the year 2016. For example, the file might look like this (and the format will be exactly like this):

```
A5.2B6.4
A3.2B5.5
A4.4B4.3
A4.5B2.2
```

The following script reads in the data and plots the data as bar charts in one Figure Window. The script prints an error message if the file open is not successful or if the file close was not successful. The **axis** command is used to force the *x*-axis to range from 0 to 3 and the *y*-axis from 0 to 8, which will result in the axes shown here. The numbers 1 and 2 would show on the *x*-axis rather than the division labels A and B by default. The **set** function changes the XTickLabel property to use the strings in the cell array as labels on the tick marks on the *x*-axis; **gca** is used to return the handle to the axes in the current figure.

plotdivab.m

```
% Reads sales figures for 2 divisions of a company one
% line at a time as strings, and plots the data

fid = fopen('ab16.dat');
if fid == -1
    disp('File open not successful')
else
    for i = 1:4
        % Every line is of the form A#B#; this separates
        % the characters and converts the #'s to actual
        % numbers
        aline = fgetl(fid);
        aline = aline(2:length(aline));
        [compa, rest] = strtok(aline,'B');
        compa = str2double(compa);
        compb = rest(2:length(rest));
        compb = str2double(compb);

        % Data from every line is in a separate subplot
        subplot(1,4,i)
        bar([compa,compb])
        set(gca, 'XTickLabel', {'A', 'B'})
        axis([0 3 0 8])
        ylabel('Sales (millions)')
        title(sprintf('Quarter %d',i))
    end
    closeresult = fclose(fid);
    if closeresult ~= 0
        disp('File close not successful')
    end
end
```

Running this produces the subplot shown in Fig. 12.29.

As another example, a data file called "compsales.dat" stores sales figures (in millions) for divisions in a company. Each line in the file stores the sales number, followed by an abbreviation of the division name, in this format:

```
5.2 X
3.3 A
5.8 P
2.9 Q
```

The script that follows uses the **textscan** function to read this information into a cell array, and then uses **subplot** to produce a Figure Window that displays the information in a **bar** chart and in a **pie** chart (shown in Fig. 12.30).

compsalesbarpie.m

```
% Reads sales figures and plots as a bar chart and a pie chart

fid = fopen('compsales.dat');

if fid == -1
    disp('File open not successful')
else
    % Use textscan to read the numbers and division codes
    % into separate elements in a cell array
    filecell = textscan(fid,'%f %s');
    % plot the bar chart with the division codes on the x ticks
    subplot(1,2,1)
    bar(filecell{1})
    xlabel('Division')
    ylabel('Sales (millions)')
    set(gca, 'XTickLabel', filecell{2})
    % plot the pie chart with the division codes as labels
    subplot(1,2,2)
    pie(filecell{1}, filecell{2})
    title('Sales in millions by division')

    closeresult = fclose(fid);
    if closeresult ~= 0
       disp('File close not successful')
    end
end
```

12.5 SAVING AND PRINTING PLOTS

Once any plot has been created in a Figure Window, there are several options for saving it, printing it, and copying and pasting it into a report. When the

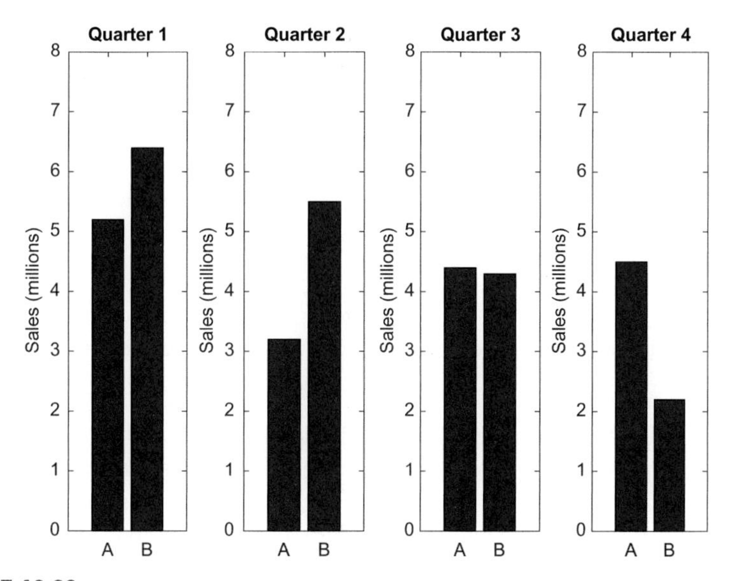

FIGURE 12.29
Subplot with customized *x*-axis tick labels.

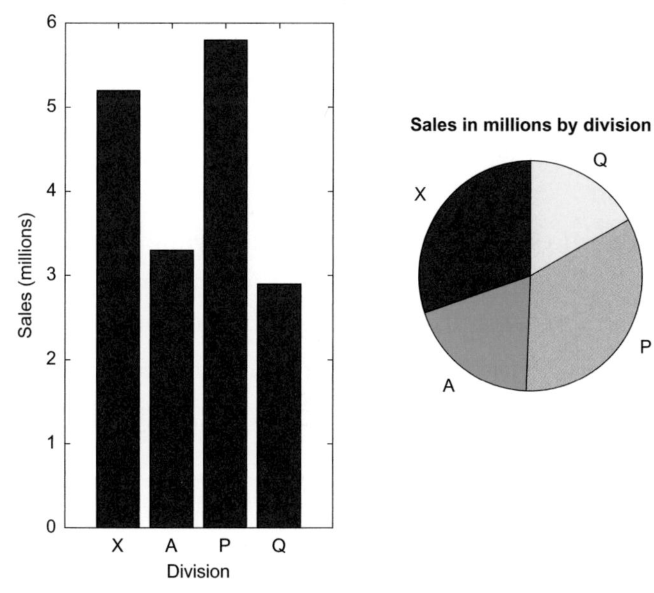

FIGURE 12.30
Bar and pie charts with labels from file data.

Figure Window is open, choosing Edit and then Copy Figure will copy the Figure Window so that it can then be pasted into a word processor. Choosing File and then Save As allows you to save in different formats, including common image types, such as .jpg, .tif, and .png. Another option is to save it as a .fig file, which is a Figure file type used in MATLAB. If the plot was not created programmatically, or the plot properties have been modified using the plot tools icon, choosing File and then Generate Code will generate a script that will re-create the plot.

Choosing File and then Print allows you to print the file on a connected printer. The **print** command can also be used in MATLAB programs. The line

```
print
```

in a script will print the current Figure Window using default formats. Options can also be specified (see the Documentation page on **print** for the options). Also, by specifying a file name, the plot is saved to a file rather than printed. For example, the following would save a plot as a.tif file with 400 dots per inch in a file named 'plot.tif':

```
print -dtiff -r400 plot.tif
```

■ Explore Other Interesting Features

There are many built-in plot functions in MATLAB, and many ways to customize plots. Use the Help facility to find them. Here are some specific suggestions for functions to investigate.

Investigate the **peaks** function, and the use of the resulting matrix as a test for various plot functions.

Investigate how to show confidence intervals for functions using the **errorbar** function.

Find out how to set limits on axes using **xlim**, **ylim**, and **zlim**.

The **plotyy** function allows y axes on both the left and the right of the graph. Find out how to use it, and how to put different labels on the two *y* axes.

Investigate how to use the **gtext** and **ginput** functions.

Investigate the 3D functions **meshc** and **surfc**, which put contour plots under the mesh and/or surface plots.

Investigate using the **datetick** function to use dates to label tick lines. Note that there are many options!

Investigate the use of **pie** charts with categorical arrays. ■

SUMMARY

COMMON PITFALLS

- Closing a Figure Window prematurely—the properties can only be set if the Figure Window is still open!

PROGRAMMING STYLE GUIDELINES

- Always label plots
- Take care to choose the type of plot in order to highlight the most relevant information

MATLAB Functions and Commands		
barh	plot3	cylinder
area	bar3	colorbar
stem	bar3h	line
histogram	pie3	rectangle
pie	comet3	text
loglog	stem3	patch
semilogy	zlabel	image
semilogx	spiral	gtext
comet	mesh	ginput
movie	surf	print
getframe	sphere	

Exercises

1. Create a data file that contains 10 numbers. Write a script that will load the vector from the file, and use **subplot** to do an **area** plot and a **stem** plot with these data in the same Figure Window (Note: a loop is not needed). Prompt the user for a title for each plot.
2. Write a script that will read *x* and *y* data points from a file, and will create an **area** plot with these points. The format of every line in the file is the letter 'x,' a space, the *x* value, space, the letter 'y,' space, and the *y* value. You must assume that the data file is exactly in that format, but you may not assume that the number of lines in the file is known. The number of points will be in the plot title. The script loops until the end of file is reached, using **fgetl** to read each line as a string. For example, *if* the file contains the following lines,

```
x 0 y 1
x 1.3 y 2.2
x 2.2 y 6
x 3.4 y 7.4
```

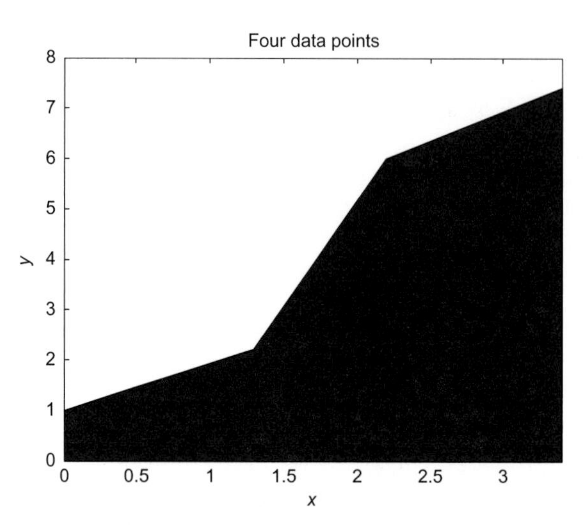

FIGURE 12.31
Area plot produced from *x*, *y* data read as strings from a file.

when running the script, the result will be as shown in Fig. 12.31.

3. Do a quick survey of your friends to find out who prefers cheese pizza, pepperoni, or mushroom (no other possibilities; everyone must pick one of those three choices). Draw a pie chart to show the percentage favoring each. Label the pieces of this pizza pie chart!
4. The number of faculty members in each department at a certain College of Engineering is:

```
ME 22
BM 45
CE 23
EE 33
```

   Experiment with at least 3 different plot types to graphically depict this information. Make sure that you have appropriate titles, labels, and legends on your plots. Which type(s) work best, and why?
5. Experiment with the **comet** function: try the example given when **help comet** is entered and then animate your own function using **comet**.
6. Experiment with the **comet3** function: try the example given when **help comet3** is entered and then animate your own function using **comet3**.
7. Experiment with the **scatter** and **scatter3** functions.
8. Use the **cylinder** function to create *x*, *y*, and *z* matrices and pass them to the **surf** function to get a surface plot. Experiment with different arguments to cylinder.
9. Experiment with **contour** plots.
10. The electricity generated by wind turbines annually in kilowatt-hours per year is given in a file. The amount of electricity is determined by, among other factors,

the diameter of the turbine blade (in feet) and the wind velocity in mph. The file stores on each line the blade diameter, wind velocity, and the approximate electricity generated for the year. For example,

```
5 5 406
5 10 3250
5 15 10970
5 20 26000
10 5 1625
10 10 13000
10 15 43875
10 20 104005
```

Create a file in this format, and determine how to graphically display this data.

11. Create an *x* vector, and then two different vectors (*y* and *z*) based on *x*. Plot them with a legend. Use **help legend** to find out how to position the legend itself on the graph, and experiment with different locations.
12. Create an *x* vector that has 30 linearly spaced points in the range from -2π to 2π, and then *y* as **sin(x)**. Do a **stem** plot of these points, and store the handle in a variable. Use **get** to see the properties of the stem plot, and then **set** to change the face color of the marker. Also do this using the dot operator.
13. When an object with an initial temperature *T* is placed in a substance that has a temperature *S*, according to Newton's law of cooling, in *t* minutes it will reach a temperature T_t using the formula $T_t = S + (T - S)\, e^{(-kt)}$ where *k* is a constant value that depends on properties of the object. For an initial temperature of 100 and $k=0.6$, graphically display the resulting temperatures from 1 to 10 min for two different surrounding temperatures: 50 and 20. Use the **plot** function to plot two different lines for these surrounding temperatures, and store the handle in a variable. Note that two function handles are actually returned and stored in a vector. Change the line width of one of the lines.
14. Write a script that will draw the line $y=x$ between $x=2$ and $x=5$, with a random line width between 1 and 10.
15. Write a script that will plot the data points from *y* and *z* data vectors, and store the handles of the two plots in variables *yhand* and *zhand*. Set the line widths to 3 and 4 respectively. Set the colors and markers to random values (create strings containing possible values and pick a random index).
16. Write a function *plotexvar* that will plot data points represented by *x* and *y* vectors which are passed as input arguments. If a third argument is passed, it is a line width for the plot, and if a fourth argument is also passed, it is a color. The plot title will include the total number of arguments passed to the function. Here is an example of calling the function; the resulting plot is shown in Fig. 12.32.

```
>> x=-pi:pi/50:2*pi;
>> y = sin(x);
>> plotexvar(x,y,12,'r')
```

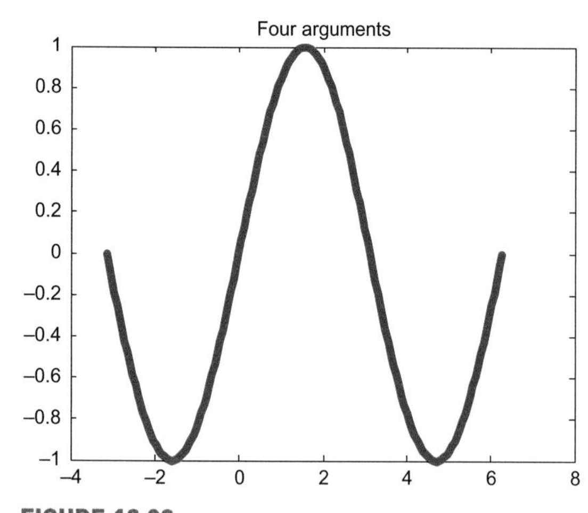

FIGURE 12.32
Sin with color and line width.

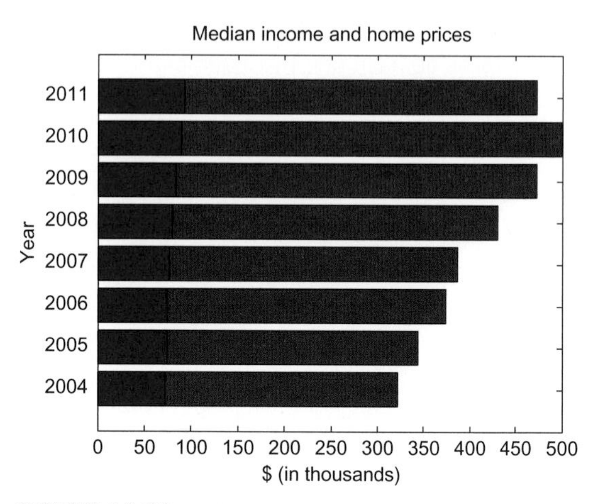

FIGURE 12.33
Horizontal stacked bar chart of median incomes and home prices.

17. A file *houseafford.dat* stores on its three lines years, median incomes, and median home prices for a city. The dollar amounts are in thousands. For example, it might look like this:

```
2004 2005 2006 2007 2008 2009 2010 2011
72 74 74 77 80 83 89 93
250 270 300 310 350 390 410 380
```

    Create a file in this format, and then **load** the information into a matrix. Create a horizontal stacked bar chart to display the information, with an appropriate title. Note: use the 'XData' property to put the years on the axis as shown in Fig. 12.33.
18. Write a function that will plot **cos(x)** for *x* values ranging from −pi to pi in steps of 0.1, using black *'s. It will do this thrice across in one Figure Window, with varying line widths (Note: even if individual points are plotted rather than a solid line, the line width property will change the size of these points.). If no arguments are passed to the function, the line widths will be 1, 2, and 3. If, on the other hand, an argument is passed to the function, it is a multiplier for these values (e.g., if 3 is passed, the line widths will be 3, 6, and 9). The line widths will be printed in the titles on the plots.
19. Create a graph, and then use the **text** function to put some text on it, including some \specchar commands to increase the font size and to print some Greek letters and symbols.
20. Create a **rectangle** object, and use the **axis** function to change the axes so that you can see the rectangle easily. Change the Position, Curvature, EdgeColor, LineStyle, and LineWidth. Experiment with different values for the Curvature.
21. Write a script that will display rectangles with varying curvatures and line widths, as shown in Fig. 12.34. The script will, in a loop, create a 2 by 2 subplot showing rectangles. In all, both the *x* and *y* axes will go from 0 to 1.4. Also, in all,

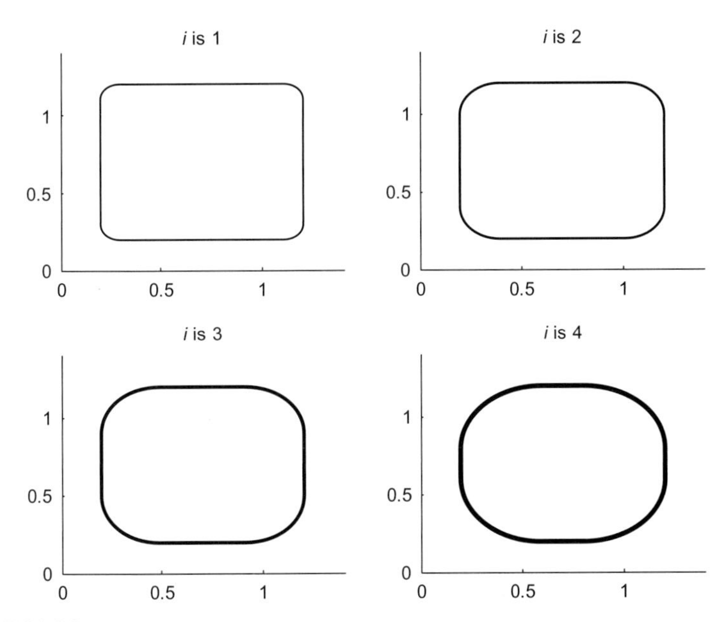

FIGURE 12.34
Varying rectangle curvature.

the lower left corner of the rectangle will be at (0.2, 0.2), and the length and width will both be 1. The line width, *i*, is displayed in the title of each plot. The curvature will be [0.2, 0.2] in the first plot, then [0.4, 0.4], [0.6,0.6], and finally [0.8,0.8].

22. Write a script that will start with a rounded rectangle. Change both the *x* and *y* axes from the default to go from 0 to 3. In a **for** loop, change the position vector by adding 0.1 to all elements 10 times (this will change the location and size of the rectangle each time). Create a movie consisting of the resulting rectangles. The final result should look like the plot shown in Fig. 12.35.
23. A hockey rink looks like a rectangle with curvature. Draw a hockey rink, as in Fig. 12.36.
24. Write a script that will create a 2D **patch** object with just three vertices and one face connecting them. The *x* and *y* coordinates of the three vertices will be random real numbers in the range from 0 to 1. The lines used for the edges should be black with a width of 3, and the face should be gray. The axes (both *x* and *y*) should go from 0 to 1. For example, depending on what the random numbers are, the Figure Window might look like Fig. 12.37.
25. Using the **patch** function, create a black box with unit dimensions (so, there will be eight vertices and six faces). Set the edge color to white so that when you rotate the figure, you can see the edges.
26. Write a function *plotline* that will receive *x* and *y* vectors of data points, and will use the **line** primitive to display a line using these points. If only the *x* and *y* vectors are passed to the function, it will use a line width of 5; otherwise, if a third argument is passed, it is the line width.

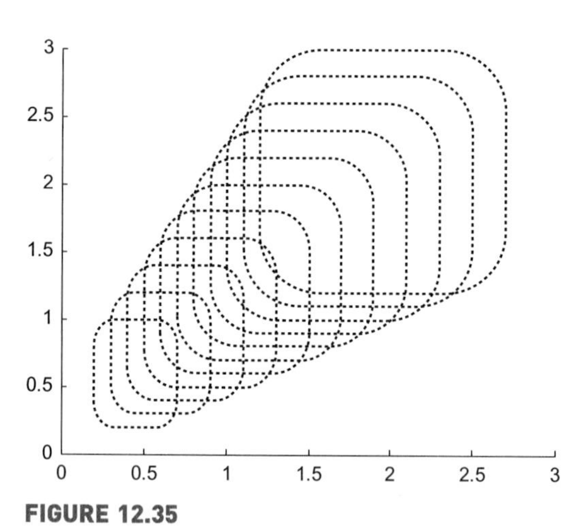

FIGURE 12.35
Curved rectangles produced in a loop.

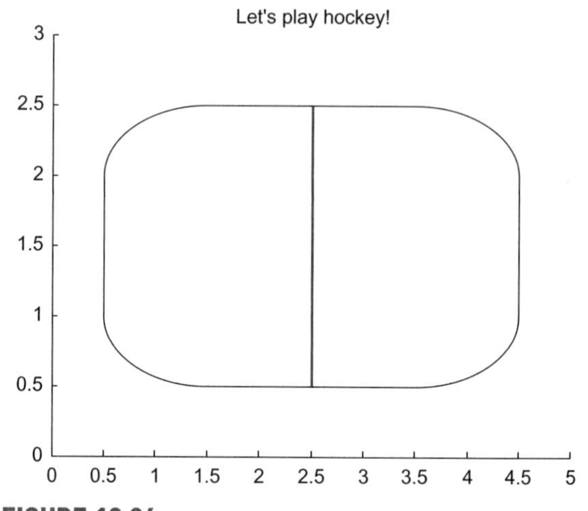

FIGURE 12.36
Hockey rink.

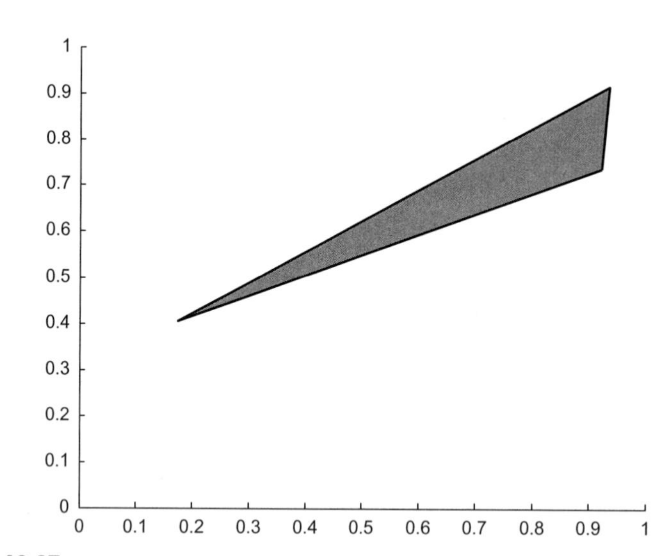

FIGURE 12.37
Patch object with black edge.

CHAPTER 13

Sights and Sounds

KEY TERMS

image processing	colormap	event-driven programming
sound processing	Graphical User Interfaces	components
pixels	event	sound wave
true color	callback function	sampling frequency
RGB		

CONTENTS

The MATLAB® product has functions that manipulate audio or sound files, and also images. MATLAB also has the capability to produce sophisticated graphical user interfaces, or GUIs. GUIs allow users to graphically interact with programs using graphical objects such as push buttons and sliders.

This chapter will start with the introduction of ***image processing*** functions, and the two basic methods for representing color in images.

Next, we will introduce the topic of GUIs from a programming standpoint. Since the focus of this book is programming constructs, the programming aspects of creating GUIs are the main part of this section. However, as of R2014b, the code produced by the graphical user interface development environment, or GUIDE, is easier to understand and modify than previous versions. Also, the exciting new App Designer was introduced in R2016a. This is similar to GUIDE but has much greater functionality and ease of use. Both GUIDE and the App Designer will be introduced.

The chapter ends with a brief introduction to some of the ***sound processing*** functions.

13.1 IMAGE PROCESSING

Color images are represented as grids, or matrices, of picture elements called ***pixels***. In MATLAB, an image is represented by a matrix in which each element

MATLAB®. http://dx.doi.org/10.1016/B978-0-12-804525-1.00013-1

corresponds to a pixel in the image. Each element that represents a particular pixel stores the color for that pixel. There are two basic ways in which the color can be represented:

- ***true color*** or ***RGB***, in which the three color components are stored (red, green, and blue, in that order) in layers in a three-dimensional matrix
- index into a ***colormap***, in which the value stored for each pixel is an integer that refers to a row in another matrix called a colormap; the colormap stores the red, green, and blue components in three separate columns

Thus, for an image that has $m \times n$ pixels, there are two methods for representing color.

In the true color or RGB method, all of the information is stored in one 3D matrix with a size $m \times n \times 3$. The first two indices represent the coordinates of the pixel. The third index is the color component, so `(:,:,1)` is the red, `(:,:,2)` is the green, and `(:,:,3)` is the blue component.

The colormap method instead uses two separate matrices: the image matrix is a 2D matrix with a size of $m \times n$. Every element in this matrix of integers is an index into a colormap matrix, which is of the size $p \times 3$ (where p is the number of colors available in that particular colormap). Each row in the colormap has three numbers representing one color: first the red, then the green, and then the blue component.

Typically colormaps use **double** values in the range from 0 to 1, and RGB matrices use the type **uint8**. These conventions are used in the next two sections that introduce these methods. In Section 13.1.4, alternate number types are explained.

13.1.1 Colormaps

When an image is represented using a colormap, there are two matrices:

- The colormap matrix, which has dimensions $p \times 3$, where p is the number of available colors; every row stores a separate color by storing three real numbers in the range from 0 to 1, representing the red, green, and blue components of the color.
- The image matrix, with dimensions $m \times n$; every element is an index into a particular row in the colormap, which means that it is an integer in the range 1 to p.

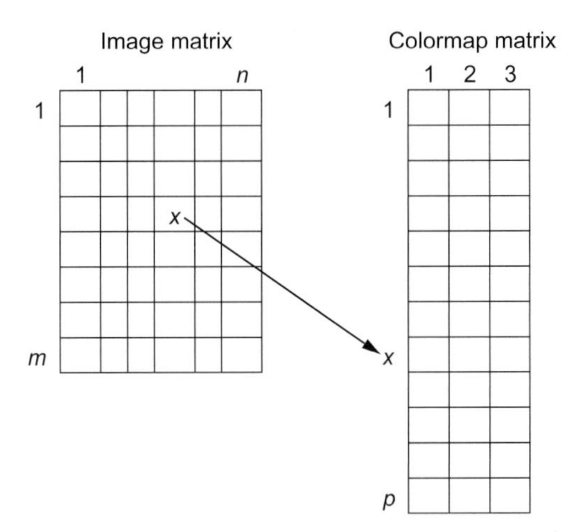

MATLAB has several built-in colormaps that are named; these can be seen and can be set using the built-in function **colormap**. The reference page on **colormap** displays them. Calling the function **colormap** without passing any arguments will return the current colormap. The default colormap is a map named **parula**, which has 64 colors. In versions of MATLAB before R2014b the default was the colormap named **jet**, which is still an available colormap.

The following stores the current colormap in a variable *map*, gets the size of the matrix (which will be the number of rows in this matrix or, in other words, the number of colors, by three columns), and displays the first five rows in this colormap. If the current colormap is the default **parula**, the following will be the result:

```
>> map = colormap;
>> [r, c] = size(map)
r =
      64
c =
       3
>> map(60:64, :)
ans =
      0.9626    0.8705    0.1309
      0.9589    0.8949    0.1132
      0.9598    0.9218    0.0948
      0.9661    0.9514    0.0755
      0.9763    0.9831    0.0538
```

This shows that there are 64 rows or, in other words, 64 colors, in this particular colormap. It also shows that the last five colors are shades of yellow (the combination of almost full red and green but very little blue). Note that **parula** is actually a function that returns a colormap matrix. Passing no arguments results in the 64×3 matrix shown here, although the number of desired colors can be passed as an argument to the **parula** function.

The format of calling the **image** function is

```
image(mat)
```

where the matrix *mat* represents the colors in an $m \times n$ image ($m \times n$ pixels in the image). If the matrix has the size $m \times n$, then each element is an index into the current colormap.

One way to display the colors in the default **parula** colormap (which has 64 colors) is to create a matrix that stores the values 1 through 64, and pass that to the **image** function; the result is shown in Fig. 13.1. When the matrix is passed to the **image** function, the value in each element in the matrix is used as an index into the colormap.

For example, the value in imagemat(1,2) is 9 so the color displayed in location (1,2) in the image will be the color represented by the ninth row in the colormap. By using the numbers 1 through 64, we can see all colors in this colormap.

```
>> imagemat = reshape(1:64, 8,8)
imagemat =

     1     9    17    25    33    41    49    57
     2    10    18    26    34    42    50    58
     3    11    19    27    35    43    51    59
     4    12    20    28    36    44    52    60
     5    13    21    29    37    45    53    61
     6    14    22    30    38    46    54    62
     7    15    23    31    39    47    55    63
     8    16    24    32    40    48    56    64
>> image(imagemat)
```

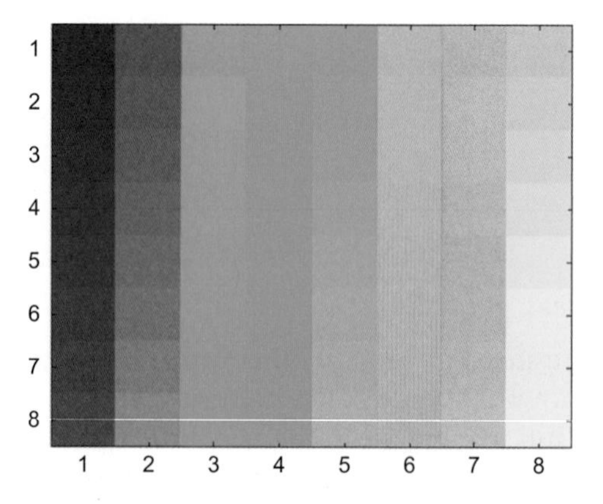

FIGURE 13.1
Columnwise display of the 64 colors in the **parula** colormap.

As with plots, the result of calling the **image** function can be stored in a variable; this allows the properties of the image to be inspected and/or modified using either the dot notation or **get** and **set**.

```
>> im = image(imagemat);
```

In order to see the colors in the colormap **jet** instead, we could make **jet** the current colormap and re-display the image matrix, as seen in Fig. 13.2.

```
>> colormap(jet)
>> image(imagemat)
```

Another example creates a 5×5 matrix of random integers in the range from 1 to the number of colors (stored in a variable r); the resulting image appears in Fig. 13.3 (assuming that the colormap is still **jet**).

```
>> mat = randi(r,5)
    54    33    13    45    32
     2    46    44    25    58
    44    28    20    56    53
    25    20    35    55    42
    54    13    10    38    53
>> image(mat)
```

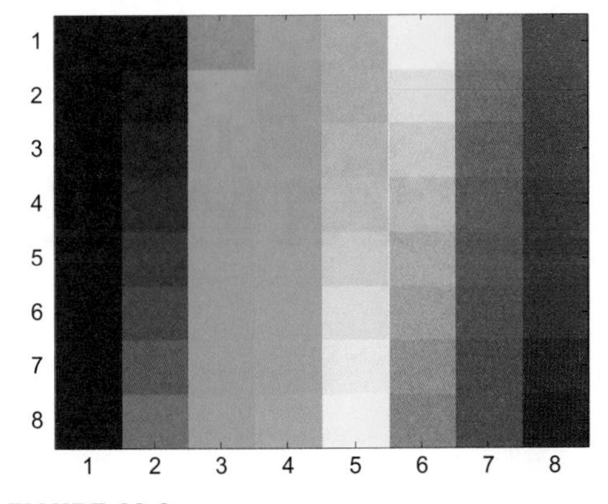

FIGURE 13.2
Columnwise display of the 64 colors in the **jet** colormap.

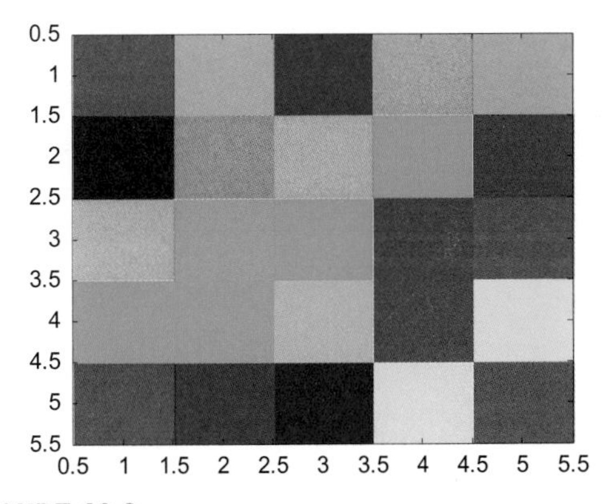

FIGURE 13.3
A 5×5 display of random colors from the **jet** colormap.

Of course, these "images" are rather crude; the elements representing the pixel colors are quite large blocks. A larger matrix would result in something more closely resembling an image, as shown in Fig. 13.4.

```
>> mat = randi(r,500);
>> image(mat)
```

Although MATLAB has built-in colormaps, it is also possible to create others using any color combinations. For example, the following creates a customized colormap with just three colors: black, white, and red. This is then set to be the current colormap by passing the colormap matrix to the **colormap** function. Then, a 40×40 matrix of random integers, each in the range from 1 to 3 (as there are just three colors), is created, and that is passed to the **image** function; the results are shown in Fig. 13.5.

```
>> mycolormap = [0 0 0; 1 1 1; 1 0 0]
mycolormap =
     0     0     0
     1     1     1
     1     0     0
>> colormap(mycolormap)
>> mat = randi(3,40);
>> image(mat)
```

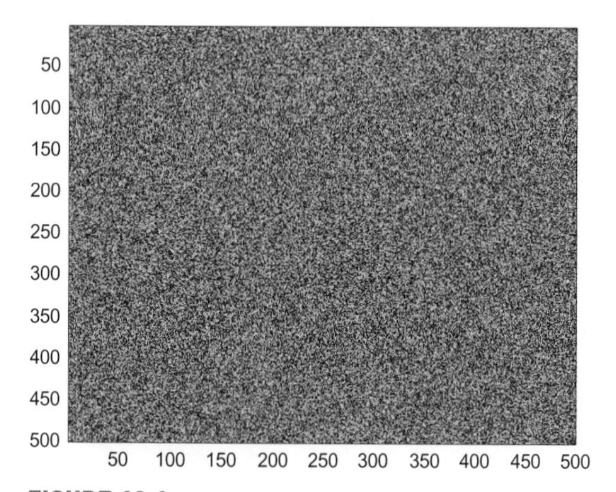

FIGURE 13.4
A 500×500 display of random colors.

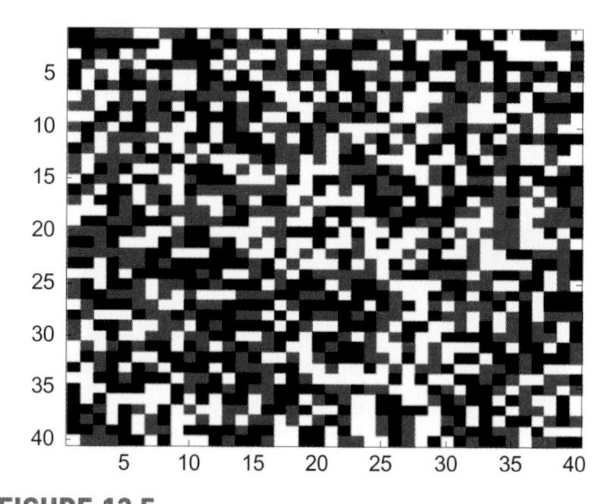

FIGURE 13.5
Random colors from a custom colormap.

The numbers in the colormap do not have to be integers; real numbers represent different shades as seen with the default colormap **parula**. For example, the following colormap gives us a way to visualize different shades of red as shown in Fig. 13.6.

```
>> colors = [0 0 0; 0.2 0 0; 0.4 0 0; ...
             0.6 0 0; 0.8 0 0; 1 0 0];
>> colormap(colors)
>> vec = 1:length(colors);
>> image(vec)
```

PRACTICE 13.1

Given the following colormap, "draw" the scene shown in Fig. 13.7. (Hint: Preallocate the image matrix. The fact that the first color in the colormap is white, makes this easier.)

```
>> mycolors = [1 1 1; 0 1 0; 0 0.5 0; ...
               0 0 1; 0 0 0.5; 0.3 0 0];
```

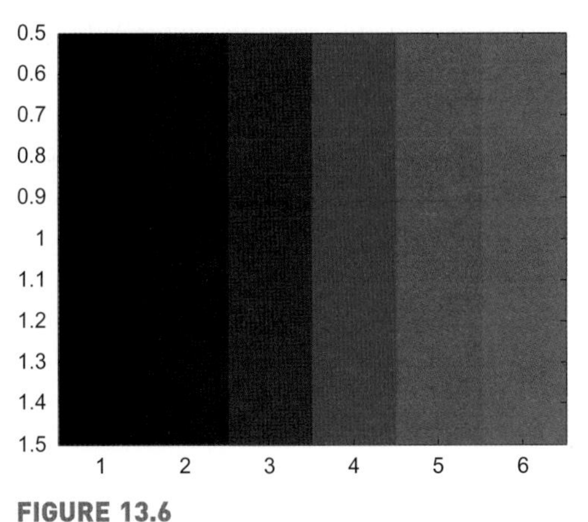

FIGURE 13.6
Shades of red.

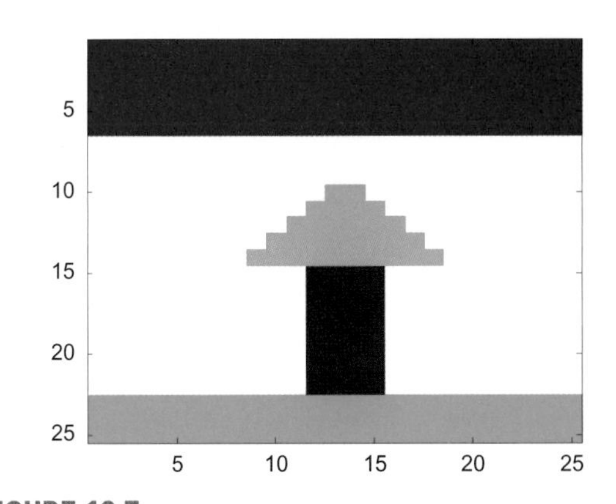

FIGURE 13.7
Draw this tree with grass and sky.

Colormaps are used with many plot functions. Generally, the plots shown assume the default colormap **parula**, but the colormap can be modified. For example, plotting a 3D object using **surf** or **sphere**, and displaying a **colorbar** would normally display the parula colors. The following is an example of modifying this to use the colormap **pink**, as shown in Fig. 13.8.

```
>> [x,y,z] = sphere(20);
>> colormap(pink)
>> surf(x,y,z)
>> title('Pink sphere')
>> colorbar
```

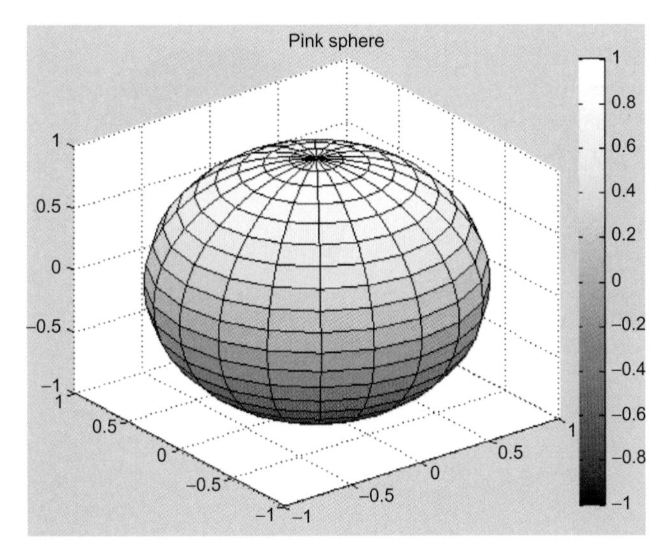

FIGURE 13.8
Pink colormap for sphere function.

As of R2014b, multiple colormaps can be displayed in a single Figure Window by passing the axes handle to the **colormap** function as shown in Fig. 13.9.

```
[x,y,z] = sphere(20);
ax1 = subplot(1,2,1)
colormap(ax1,pink)
surf(x,y,z);
title('Pink sphere')
ax2 = subplot(1,2,2)
colormap(ax2,jet)
surf(x,y,z);
title('Jet sphere')
```

13.1.2 True Color Matrices

True color matrices, or ***RGB matrices***, are another way to represent images. True color matrices are 3D matrices. The first two coordinates are the coordinates of the pixel. The third index is the color component; `(:,:,1)` is the red, `(:,:,2)` is the green, and `(:,:,3)` is the blue component.

In an 8-bit RGB image, each element in the matrix is of the type **uint8**, which is an unsigned integer type storing values in the range from 0 to 255. The minimum value, 0, represents the darkest hue available, so all 0s results in a black pixel. The maximum value 255, represents the brightest hue. For example, if the values for a given pixel coordinates *px* and *py* are: (px,py,1) is 255, (px,py,2) is 0,

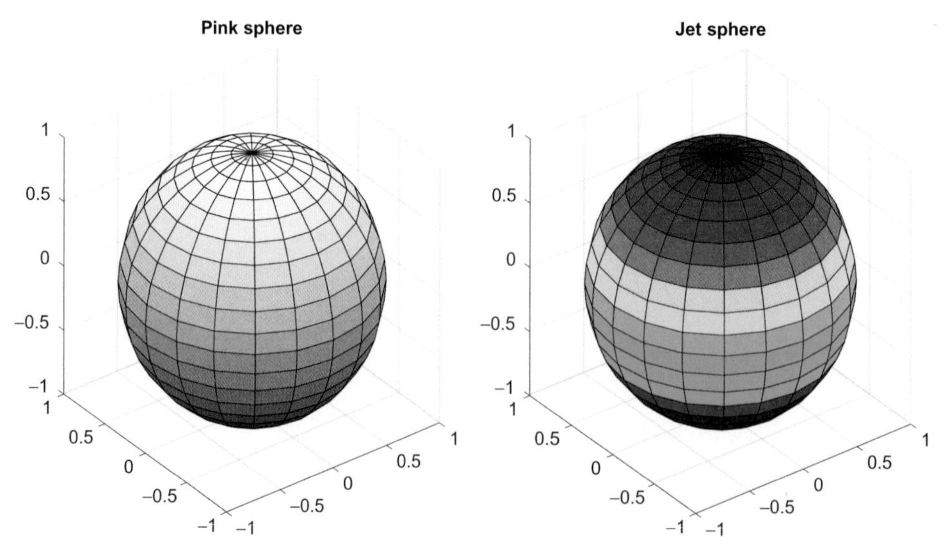

FIGURE 13.9
Subplot with multiple colormaps.

and (px,py,3) is 0, then that pixel will be bright red. All 255s results in a white pixel.

The **image** function displays the information in the 3D matrix as an image. (Note: if a 2D image matrix is passed to **image**, it uses the current colormap to display the image, but if a 3D image matrix is passed to **image**, it ignores the colormap and instead uses the information from the 3D matrix to determine the color of each pixel.)

For example, the following creates a 2×2 image as shown in Fig. 13.10. The matrix is $2 \times 2 \times 3$ where the third dimension is the color. The pixel in location (1,1) is red, the pixel in location (1,2) is blue, the pixel in location (2,1) is green, and the pixel in location (2,2) is black. To make the pixel in location (1,1) red, we could execute the following three statements, which would make the red layer the brightest possible, with no green or blue:

```
>> mat(1,1,1) = 255;
>> mat(1,1,2) = 0;
>> mat(1,1,3) = 0;
```

Similarly we could execute three statements to set the color for each of the other pixels. However, it is much simpler to start by preallocating the entire image matrix to all zeros, and then changing individual values to 255 as necessary. In versions prior to R2016a, it is necessary to cast the matrix to the type **uint8** before passing it to the **image** function.

```
>> mat = zeros(2,2,3);
>> mat(1,1,1) = 255;
>> mat(1,2,3) = 255;
>> mat(2,1,2) = 255;
>> mat = uint8(mat);
>> image(mat)
```

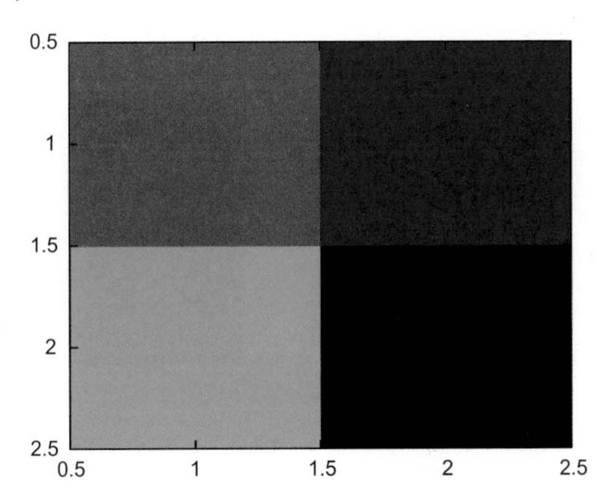

FIGURE 13.10
Image from a true color matrix.

The following shows how to separate the red, green, and blue components from an image matrix. In this case we are using the "image" matrix *mat,* and then use **subplot** to display the original matrix and the red, green, and blue component matrices, as shown in Fig. 13.11.

```
matred = uint8(zeros(2,2,3));
matred(:,:,1) = mat(:,:,1);
matgreen = uint8(zeros(2,2,3));
matgreen(:,:,2) = mat(:,:,2);
matblue = uint8(zeros(2,2,3));
matblue(:,:,3) = mat(:,:,3);
subplot(2,2,1)
image(mat)
subplot(2,2,2)
image(matred)
subplot(2,2,3)
image(matgreen)
subplot(2,2,4)
image(matblue)
```

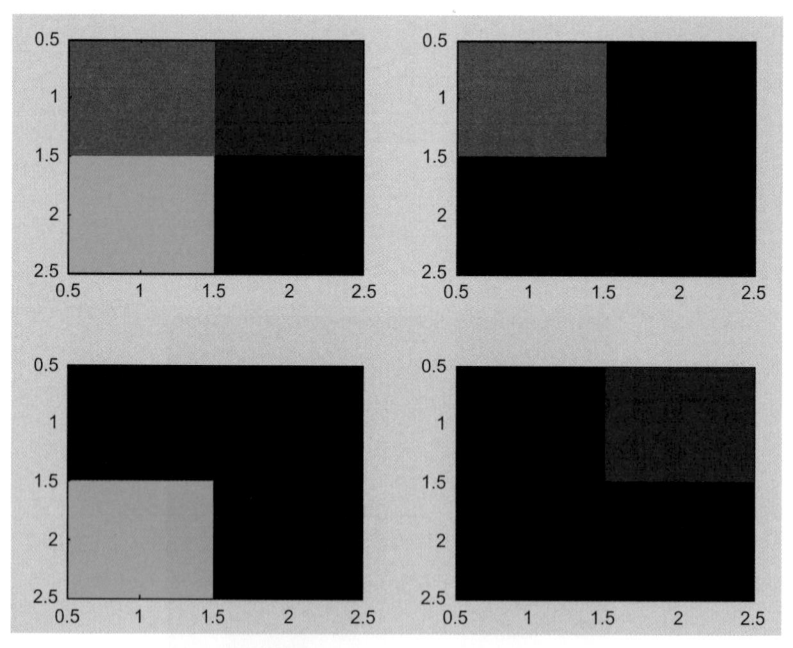

FIGURE 13.11
Separating red, green, and blue components.

Superimposing the images from the three matrices *matred*, *matgreen*, and *matblue* would be achieved by simply adding the three arrays together. The following would result in the image from Fig. 13.10:

```
>> image(matred+matgreen+matblue)
```

PRACTICE 13.2

Create the 3 × 3 (×3) true color matrix shown in Fig. 13.12 (the axes are defaults). Use the type **uint8**.

13.1.3 Image Files

Images that are stored in various formats, such as JPEG, TIFF, PNG, GIF, and BMP, can be manipulated in MATLAB. Built-in functions, such as **imread** and **imwrite**, read from and write to various image file formats. Some images are stored as unsigned 8-bit data (**uint8**), some as unsigned 16-bit (**uint16**), and some are stored as **double**.

For example, the following reads a JPEG image into a 3D matrix; it can be seen from the **size** and **class** functions that this was stored as a **uint8** RGB matrix.

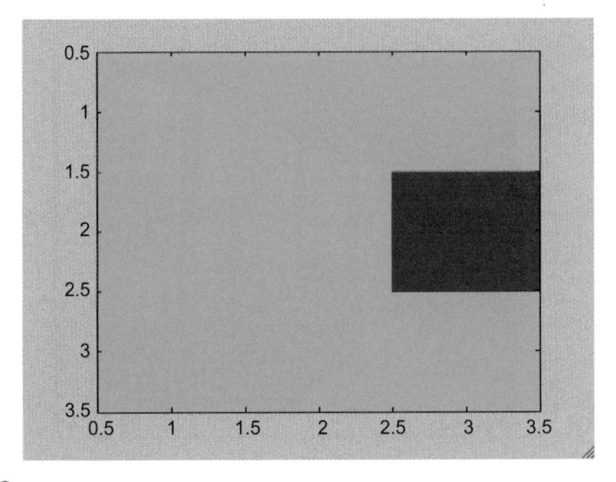

FIGURE 13.12
Create this true color matrix.

```
>> porchimage = imread('snowyporch.JPG');
>> size(porchimage)
ans =
        2848        4272           3
>> class(porchimage)
ans =
uint8
```

The image is stored as a true color matrix and has 2848×4272 pixels. The image function displays the matrix as an image, as shown in Fig. 13.13.

```
>> image(porchimage)
```

The function **imshow** can also be used to display an image. Some image functions, such as **imshow**, **rgb2gray** (converts from RGB to a gray scale), and **im2double** (converts an image matrix to **double**) are now in MATLAB; prior to R2014b they were only in the Image Processing Toolbox.

The image can be manipulated by modifying the numbers in the image matrix. For example, multiplying every number by 0.5 will result in a range of values from 0 to 128 instead of from 0 to 255. As the larger numbers represent brighter hues, this will have the effect of dimming the hues in the pixels, as shown in Fig. 13.14.

```
>> dimmer = 0.5*porchimage;
>> image(dimmer)
```

The function **imwrite** is used to write an image matrix to a file in a specified format:

```
>> imwrite(dimmer, 'dimporch.JPG')
```

FIGURE 13.13
Image from a JPEG file displayed using **image**.

FIGURE 13.14
Image dimmed by manipulating the matrix values.

Images can also be stored as an indexed image rather than RGB. In that case, the colormap is usually stored with the image and will be read in by the **imread** function.

13.1.4 Alternate Number Types for Image Matrices

In the previous two sections, we have described the colormap method using **double** values in the range from 0 to 1 in the colormap matrix and the RGB method using the type **uint8** in the 3D matrix. However, it is not necessary to use these particular types.

Usually, the image matrices used in conjunction with a colormap are the type **double**, as described in Section 13.1.2, and the integers stored in the matrices range from 1 to p, where p is the number of colors in the current colormap. However, it is possible for the image matrix to store the data as the type **uint8** or **uint16**. In that case, the integers stored would range from 0 to $p - 1$, and the **image** function adjusts appropriately (a 0 maps to the first color, 1 maps to the second color, and so forth).

The numbers in an RGB matrix can be of the types **uint8**, **uint16**, or **double**.

The image seen in Fig. 13.10 could also be created as a 16-bit image in which the range of values would be from 0 to 65,535 instead of from 0 to 255.

```
>> clear
>> mat = zeros(2,2,3);
>> mat(1,1,1) = 65535;
>> mat(1,2,3) = 65535;
>> mat(2,1,2) = 65535;
>> mat = uint16(mat);
>> image(mat)
```

In an RGB image matrix in which the numbers range from 0 to 1, the default type is **double** so it is not necessary to typecast the matrix variable. The following would also create the image seen in Fig. 13.9.

```
>> clear
>> mat = zeros(2,2,3);
>> mat(1,1,1) = 1;
>> mat(1,2,3) = 1;
>> mat(2,1,2) = 1;
>> image(mat)
```

The **image** function determines the type of the image matrix and adjusts the colors accordingly when displaying the image.

13.2 INTRODUCTION TO GUIs

GUIs are essentially objects that allow users to have input using graphical interfaces, such as pushbuttons, sliders, radio buttons, toggle buttons, pop-up menus, and so forth. GUIs are an implementation of the handle classes that were introduced in Chapter 11, so there is a hierarchy. For example, the parent may be a Figure Window and its children would be graphics objects, such as pushbuttons and text boxes.

The parent user interface object can be a **figure**, **uipanel**, or **uibuttongroup**. A **figure** is a Figure Window created by the **figure** function. A **uipanel** is a means of grouping together user interface objects (the "ui" stands for user interface). A **uibuttongroup** is a means of grouping together buttons (both radio buttons and toggle buttons).

In MATLAB there are two basic methods for creating GUIs: writing the GUI program from scratch or using the built-in GUIDE. GUIDE allows the user to graphically lay out the GUI and MATLAB generates the code for it automatically. However, to be able to understand and modify this code, it is important to understand the underlying programming concepts. Therefore, this section will concentrate on the programming methodology.

13.2.1 GUI Basics

A Figure Window is the parent of any GUI. Just calling the **figure** function will bring up a blank Figure Window; storing the handle in a variable allows us to manipulate its properties, as we have seen in Chapter 11.

```
>> f = figure
f =
  Figure (1) with properties:

     Number: 1
       Name: ''
      Color: [0.9400 0.9400 0.9400]
   Position: [440 378 560 420]
      Units: 'pixels'

  Show all properties
```

The position vector specifies `[left bottom width height]`. The first two numbers, the left and bottom, are the distance that the lower left corner of the figure box is from the lower left of the monitor screen (first from the left and then from the bottom). The last two are the width and height of the figure box itself. All of these are in the default units of pixels.

```
>> f.Visible
ans =
on
```

The 'Visible' property "on" means that the Figure Window can be seen. When creating a GUI, however, the normal procedure is to create the parent Figure Window, but make it invisible. Then, all user interface objects are added to it and properties are set. When everything has been completed the GUI is made visible.

If the figure is the only Figure Window that has been opened, then it is the current figure. Using **gcf** would be equivalent to *f* in that case. Recall that the parent of a Figure Window is the screen; this could be obtained with either *f.Parent* or **groot**.

Most user interface objects are created using the **uicontrol** function. Pairs of arguments are passed to the **uicontrol** function, consisting of the name of a property as a string and then its value. The default is that the object is created in the current figure; otherwise, a parent can be specified as in **uicontrol(parent,…)**. The 'Style' property defines the type of object, as a string. For example, 'text' is the Style of a static text box, which is normally used as a label for other objects in the GUI or for instructions.

The following example creates a GUI that just consists of a static text box in a Figure Window. The figure is first created, but made invisible. The color is white and it is given a position. Storing the handle of this figure in a variable allows the function to refer to it later on, to set properties, for example.

The **uicontrol** function is used to create a text box, position it (the vector specifies the [left bottom width height] within the Figure Window itself), and put a string in it. Note that the position is within the Figure Window, not within the screen.

A name is put on the top of the figure. The **movegui** function moves the GUI (the figure) to the center of the screen. Finally, when everything has been completed, the GUI is made visible.

simpleGui.m

```
function simpleGui
% simpleGui creates a simple GUI with just a static text box
% Format: simpleGui or simpleGui()

% Create the GUI but make it invisible for now while
%  it is being initialized
f = figure('Visible', 'off','color','white','Position',...
   [300, 400, 450,250]);
htext = uicontrol('Style','text','Position', ...
    [200,50, 100, 25], 'String','My First GUI string');

% Put a name on it and move to the center of the screen
% Note in older versions: set(f,'Name','Simple GUI')
f.Name = 'Simple GUI';
movegui(f,'center')

% Now the GUI is made visible
f.Visible = 'on';
end
```

The Figure Window shown in Fig. 13.15 will appear in the middle of the screen. The static text box requires no interaction with the user, so although this example shows some of the basics, it does not allow any graphical interface with the user.

When the 'Units' property of objects is set to 'Normalized,' this means that rather than specifying in pixels the position, it is done as a percentage. This allows the Figure Window to be resized without changing the way that the GUI appears. It also means that GUIs will look the same on different screen sizes. Normalized units for the figure means as a percentage of the screen, whereas Normalized units for a uicontrol object means as a percentage of the Figure Window. For example, the function *simpleGuiNormalized* is a version of the first GUI example that uses Normalized units:

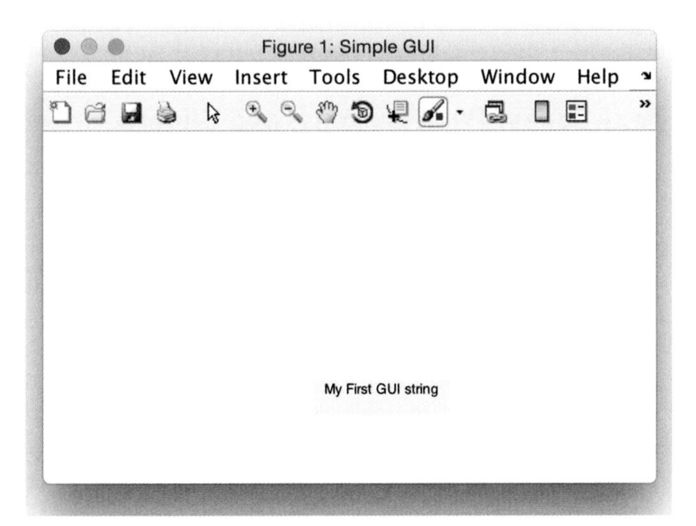

FIGURE 13.15
Simple GUI with a static text box.

simpleGuiNormalized.m

```
function simpleGuiNormalized
% simpleGuiNormalized creates a GUI with just a static text box
% Format: simpleGuiNormalized or simpleGuiNormalized()

% Create the GUI but make it invisible for now while
%  it is being initialized
f = figure('Visible', 'off','color','white','Units',...
  'Normalized', 'Position', [.25, .5, .35, .3]);
htext = uicontrol('Style','text','Units', 'Normalized', ...
   'Position', [.45, .2, .2, .1], ...
   'String','My First GUI string');
% Put a name on it and move to the center of the screen
f.Name = 'Simple GUI Normalized';
movegui(f,'center')

% Now the GUI is made visible
f.Visible = 'on';
end
```

For example, the first two numbers in the position vector for the figure [.25 .5 .35 .3] specifies that the lower left corner of the Figure Window is one quarter of the way from the left and half-way from the bottom of the screen. Of course, the figure is then moved to the middle of the screen, but the position vector is still useful in that it also specifies the width and height of the figure.

Since Normalized units are more general and more intuitive, for the most part they will be used for the remaining examples rather than pixels.

13.2.2 Text Boxes, Push Buttons, and Sliders

Now that we have seen the basic algorithm for a GUI, we will add user interaction.

In the next example, we will allow the user to enter a string in an editable text box and then the GUI will print the user's string in red. In this example there will be user interaction. First, the user must type in a string and, once this happens, the user's entry in the editable text box will no longer be shown, but, instead, the string that the user typed will be displayed in a larger red font in a static text box. When the user's action (which is called an ***event***) causes a response, what happens is that a ***callback function*** is called or invoked. The callback function is the part of the code in which the string is read in and then printed in a larger red font. This is sometimes called ***event-driven programming***: the event triggers an action.

The callback function must be in the path; one way to do this is to make it a nested function within the GUI function. The algorithm for this example is as follows.

- Create the Figure Window, but make it invisible.
- Make the color of the figure white, put a title on it, and move it to the center.
- Create a static text box with an instruction to enter a string.
- Create an editable text box.
  - The Style of this is 'edit.'
  - The callback function must be specified as the user's entry of a string necessitates a response (the function handle of the nested function is used for this).
- Make the GUI visible so that the user can see the instruction and type in a string.
- When the string is entered, the callback function *callbackfn* is called. Note that in the function header, there are two input arguments, *hObject* and *eventdata*. The input argument *hObject* refers to the handle of the **uicontrol** object that called it; *eventdata* stores in a structure information about the event that triggered the call to the callback function (this varies depending on the type of object and is not always used).
- The algorithm for the nested function *callbackfn* is
  - make the previous GUI objects invisible
  - get the string that the user typed (note that either *hObject* or the object handle name *huitext* can be used to refer to the object in which the string was entered)

- create a static text box to print the string in red with a larger font
- make this new object visible

guiWithEditbox.m

```
function guiWithEditbox
% guiWithEditbox has an editable text box
%   and a callback function that prints the user's
%   string in red
% Format: guiWithEditbox or guiWithEditbox()

% Create the GUI but make it invisible for now
f = figure('Visible', 'off','color','white','Units',...
   'Normalized','Position', [.25 .5 .4 .2]);
% Put a name on it and move it to the center of the screen
f.Name = 'GUI with editable text';
movegui(f,'center')
% Create two objects: a box where the user can type and
%   edit a string and also a text title for the edit box
hsttext = uicontrol('Style','text',...
     'BackgroundColor','white','Units','Normalized',...
     'Position',[.2 .6 .6 .2],...
     'String','Enter your string here');
huitext = uicontrol('Style','edit', 'Units',...
     'Normalized','Position',[.3 .3 .4 .2], ...
     'Callback',@callbackfn);
% Now the GUI is made visible
f.Visible = 'on';

     % Call back function
     function callbackfn(hObject,eventdata)
          % callbackfn is called by the 'Callback' property
          % in the editable text box
          set([hsttext huitext],'Visible','off');
          % Get the string that the user entered and print
          %   it in big red letters
          printstr = huitext.String;
          hstr = uicontrol('Style','text',...
               'BackgroundColor','white','Units',...
               'Normalized','Position',[.1 .3 .8 .4],...
               'String',printstr,...
               'ForegroundColor','Red','FontSize',30);
          hstr.Visible = 'on';
     end
end
```

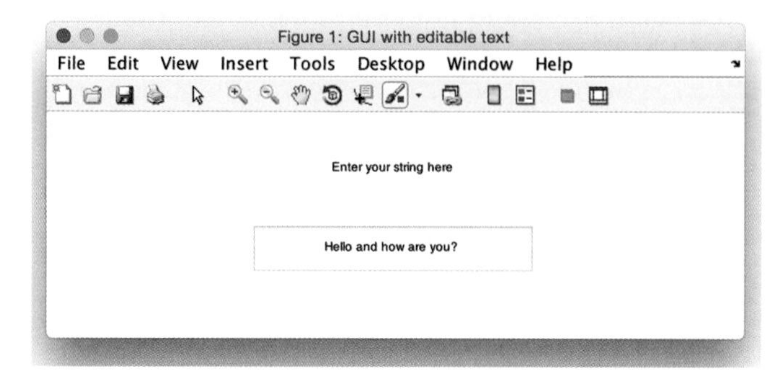

FIGURE 13.16
String entered by user in editable text box.

When the Figure Window is first made visible, the static text and the editable text box are shown. In this case, the user entered 'Hello and how are you?' Note that to enter the string, the user must first click the mouse in the editable text box. The string that was entered by the user is shown in Fig. 13.16.

After the user enters this string and hits the Enter key, the callback function is executed; the results are shown in Fig. 13.17. The callback function sets the Visible property to off for both of the original objects by referring to their handles. As the callback function is a nested function, the handle variables can be used. It then gets the string and writes it in a new static text box in red.

Now we will add a pushbutton to the GUI. This time, the user will enter a string, but the callback function will be invoked when the pushbutton is pushed.

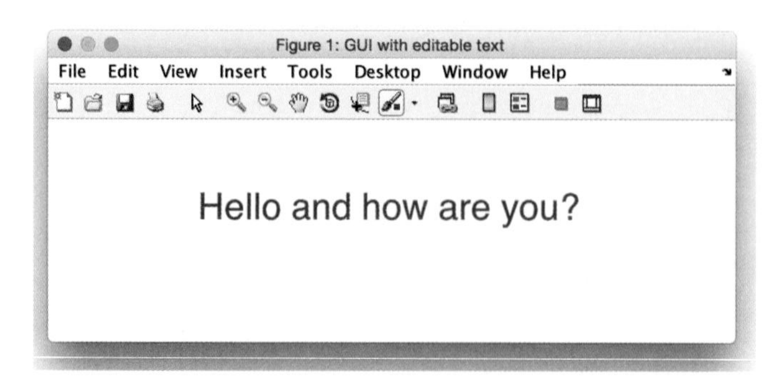

FIGURE 13.17
The result from the callback function.

guiWithPushbutton.m

```
function guiWithPushbutton
% guiWithPushbutton has an editable text box and a pushbutton
% Format: guiWithPushbutton or guiWithPushbutton()

% Create the GUI but make it invisible for now while
%  it is being initialized
f = figure('Visible', 'off','color','white','Units',...
   'Normalized', 'Position',[.25 .5 .5 .3]);
hsttext = ...
uicontrol('Style','text','BackgroundColor','white',...
    'Units','Normalized','Position',[.2 .7 .6 .2],...
    'String','Enter a string here, then push the button');
huitext = uicontrol('Style','edit','Units','Normalized',...
     'Position',[.3 .5 .4 .2]);
f.Name = 'GUI with pushbutton';
movegui(f,'center')

% Create a pushbutton that says "Push me!!"
hbutton = uicontrol('Style','pushbutton','String',...
     'Push me!!', 'Units','Normalized','Position',...
     [.6 .1 .3 .2], 'Callback',@callbackfn);

% Now the GUI is made visible
f.Visible = 'on';

% Call back function
    function callbackfn(hObject,eventdata)
        % callbackfn is called by the 'Callback' property
        % in the pushbutton
        set([hsttext huitext hbutton],'Visible','off');
        printstr = huitext.String;
        if isempty(printstr)
            printstr = 'Enter something next time!';
        end
        hstr = uicontrol('Style','text','BackgroundColor',...
            'white', 'Units', 'Normalized','Position',...
            [.1 .3 .8 .4], 'String', printstr, ...
            'ForegroundColor','Red','FontSize',30);
        hstr.Visible = 'on';
    end
end
```

In this case, the user types the string into the edit box. Hitting Enter, however, does not cause the callback function to be called; instead, the user must push the button with the mouse. The callback function is associated with the pushbutton object. So, pushing the button will bring up the string in a larger red font. The initial configuration with the pushbutton is shown in Fig. 13.18.

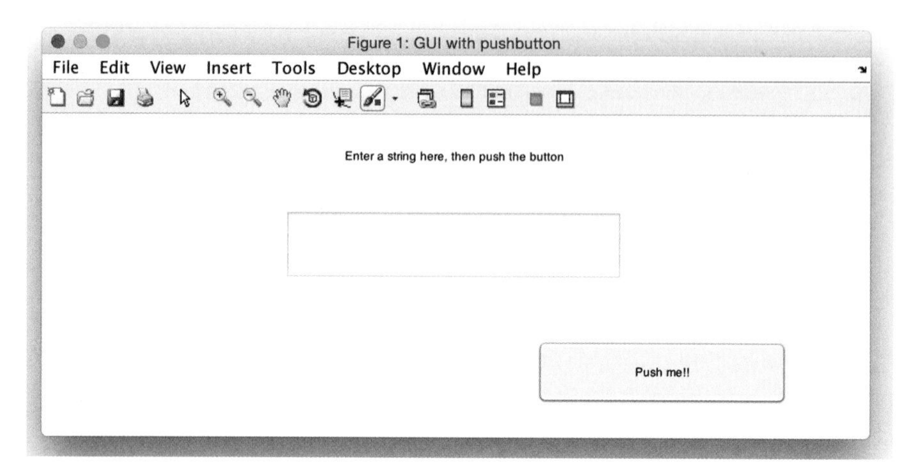

FIGURE 13.18
GUI with a push button.

PRACTICE 13.3

Create a GUI that will convert a length from inches to centimeters. The GUI should have an editable text box in which the user enters a length in inches, and a pushbutton that says "Convert me!" Pushing the button causes the GUI to calculate the length in centimeters and display that. The callback function that accomplishes this should leave all objects visible. That means the user can continue converting lengths until the Figure Window is closed. The GUI should display a default length to begin with (e.g., 1 in.). For example, calling the function might bring up the Figure Window shown in Fig. 13.19.

Then, when the user enters a length (e.g., 5.2 in.) and pushes the button, the Figure Window will show the new calculated length in centimeters (as seen in Fig. 13.20).

Another GUI object that can be created is a slider. The slider object has a numerical value, and can be controlled by either clicking on the arrows to move the value up or down, or by sliding the bar with the mouse. By default, the numerical value ranges from 0 to 1, but these values can be modified using the 'Min' and 'Max' properties.

The function *guiSlider* creates in a Figure Window a slider that has a minimum value of 0 and a maximum value of 5. It uses text boxes to show the minimum and maximum values, and also the current value of the slider.

guiSlider.m

```
function guiSlider
% guiSlider is a GUI with a slider
% Format: guiSlider or guiSlider()

f = figure('Visible', 'off','color','white','Units',...
    'Normalized', 'Position', [.25 .5 .4 .2]);

% Minimum and maximum values for slider
minval = 0;
maxval = 5;

% Create the slider object
slhan = uicontrol('Style','slider', 'Units', 'Normalized', ...
    'Position',[.3 .5 .4 .2], 'Min', minval, 'Max', maxval,...
    'SliderStep', [0.5 0.5], 'Callback', @callbackfn);
% Text boxes to show the minimum and maximum values
hmintext = uicontrol('Style','text','BackgroundColor',...
    'white', 'Units', 'Normalized',...
    'Position', [.1 .5 .1 .1], 'String', num2str(minval));
hmaxtext = uicontrol('Style', 'text','BackgroundColor',...
    'white', 'Units', 'Normalized',...
    'Position', [.8 .5 .1 .1], 'String', num2str(maxval));
% Text box to show the current value (off for now)
hsttext = uicontrol('Style','text','BackgroundColor','white',...
    'Units', 'Normalized',...
    'Position',[.4 .3 .2 .1],'Visible', 'off');

movegui(f,'center')
f.Name = 'Slider Example';
f.Visible = 'on';

% Call back function displays the current slider value
   function callbackfn(hObject,eventdata)
       % callbackfn is called by the 'Callback' property
       % in the slider
       num = slhan.Value;
       set(hsttext,'Visible','on','String',num2str(num))
    end
end
```

Fig. 13.21 shows the configuration once the user has moved the slider.

PRACTICE 13.4

Use the Help browser to find the property that controls the increment value on the slider, and modify the *guiSlider* function to move in increments of 0.5 when the arrow is used.

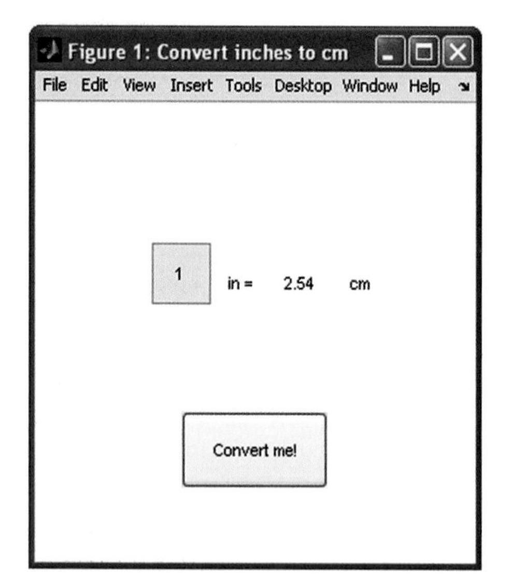

FIGURE 13.19
Length conversion GUI with push button.

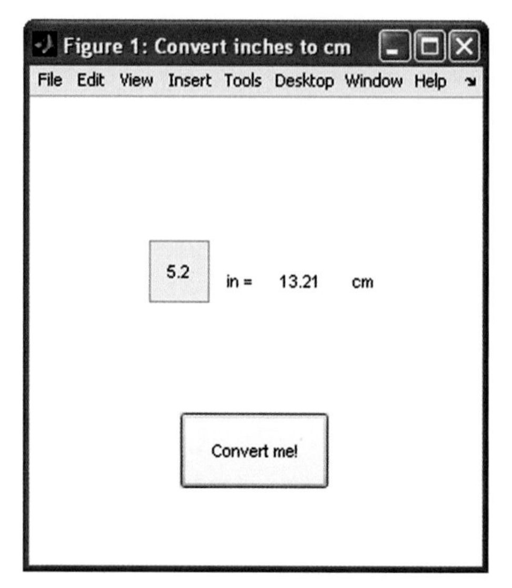

FIGURE 13.20
Result from conversion GUI.

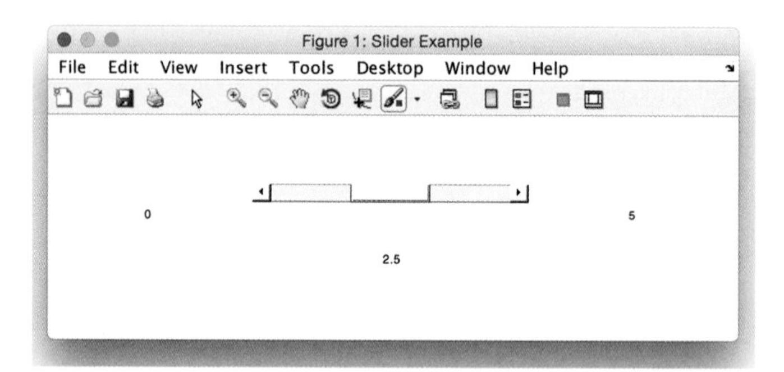

FIGURE 13.21
GUI with slider result shown.

GUI functions can have multiple callback functions. In the example *guiWithTwoPushbuttons*, there are two buttons that could be pushed (see Fig. 13.22). Each of them has a unique callback function associated with it. If the top button is pushed, its callback function prints red exclamation points (as shown in Fig. 13.23). If the bottom button is instead pushed, its callback function prints blue asterisks.

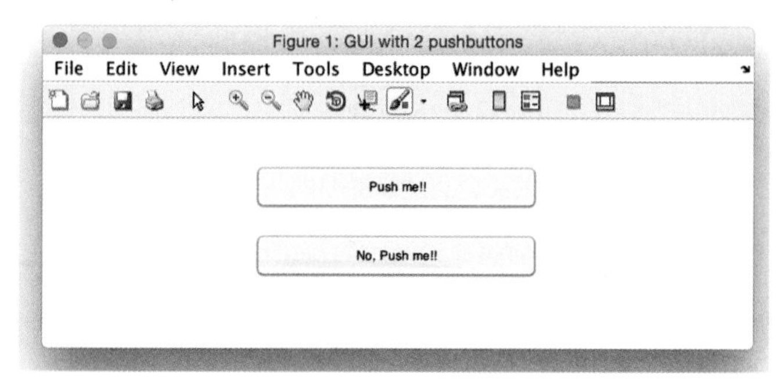

FIGURE 13.22
GUI with two pushbuttons and two callback functions.

FIGURE 13.23
The result from the first callback function.

guiWithTwoPushbuttons.m

```
function guiWithTwoPushbuttons
% guiWithTwoPushbuttons has two pushbuttons, each
%    of which has a separate callback function
% Format: guiWithTwoPushbuttons

% Create the GUI but make it invisible for now while
%  it is being initialized
f = figure('Visible', 'off','color','white','Units',...
   'Normalized', 'Position', [.25 .5 .4 .2]);
f.Name = 'GUI with 2 pushbuttons';
movegui(f,'center')

% Create a pushbutton that says "Push me!!"
hbutton1 = uicontrol('Style','pushbutton','String',...
    'Push me!!', 'Units', 'Normalized',...
    'Position',[.3 .6 .4 .2], ...
    'Callback',@callbackfn1);

% Create a pushbutton that says "No, Push me!!"
hbutton2 = uicontrol('Style','pushbutton','String',...
    'No, Push me!!', 'Units', 'Normalized',...
    'Position',[.3 .3 .4 .2], ...
    'Callback',@callbackfn2);
% Now the GUI is made visible
f.Visible = 'on';

    % Call back function for first button
    function callbackfn1(hObject,eventdata)
        % callbackfn is called by the 'Callback' property
        % in the first pushbutton

         set([hbutton1 hbutton2],'Visible','off');
         hstr = uicontrol('Style','text',...
             'BackgroundColor', 'white', 'Units',...
             'Normalized','Position',[.4 .5 .2 .2],...
             'String','!!!!!', ...
             'ForegroundColor','Red','FontSize',30);
         hstr.Visible = 'on';
    end

 % Call back function for second button
 function callbackfn2(hObject,eventdata)
   % callbackfn is called by the 'Callback' property
   % in the second pushbutton

    set([hbutton1 hbutton2],'Visible','off');
    hstr = uicontrol('Style','text',...
        'BackgroundColor','white', 'Units', ...
        'Normalized', 'Position',[.4 .5 .2 .2],...
        'String','*****', ...
        'ForegroundColor','Blue','FontSize',30);
    hstr.Visible = 'on';
 end

end
```

If the first button is pushed, the first callback function is called, which would produce the result in Fig. 13.23.

It is also possible to have one callback function invoked, or called, by multiple objects. In the example with two pushbuttons, instead of having one callback function associated with each pushbutton, we could have just one callback function. In that case, the value of the input argument *hObject* would be checked to determine which object called it. A modified GUI to accomplish this is shown next.

guiWithTwoPushbuttonsii.m

```
function guiWithTwoPushbuttonsii
% guiWithTwoPushbuttonsii has two pushbuttons
%    but just one callback function
% Format: guiWithTwoPushbuttonsii

% Create the GUI but make it invisible for now while
%  it is being initialized
f = figure('Visible', 'off','color','white','Units',...
    'Normalized', 'Position', [.25 .5 .4 .2]);
f.Name = 'GUI with 2 pushbuttons';
movegui(f,'center')

% Create a pushbutton that says "Push me!!"
hbutton1 = uicontrol('Style','pushbutton','String',...
    'Push me!!', 'Units', 'Normalized',...
    'Position',[.3 .6 .4 .2], ...
    'Callback',@callbackfn);

% Create a pushbutton that says "No, Push me!!"
hbutton2 = uicontrol('Style','pushbutton','String',...
    'No, Push me!!', 'Units', 'Normalized',...
    'Position',[.3 .3 .4 .2], ...
    'Callback',@callbackfn);
% Now the GUI is made visible
f.Visible = 'on';

    % Call back function for both buttons
    function callbackfn(hObject,eventdata)
        % callbackfn is called by the 'Callback' property

        set([hbutton1 hbutton2],'Visible','off');
        hstr = uicontrol('Style','text',...
            'BackgroundColor', 'white', 'Units',...
            'Normalized','Position',[.4 .5 .2 .2],...
            'FontSize',30);
        if hObject == hbutton1
            hstr.String = '!!!!!';
            hstr.ForegroundColor = 'Red';
        else
            hstr.String = '*****';
            hstr.ForegroundColor = 'Blue';
        end
        hstr.Visible = 'on';
    end

end
```

In this example, the value of the input argument *hObject* was used by the callback function. The second input argument, *eventdata*, was not used, however. Whenever any callback function is called, both of these arguments are automatically passed to the callback function. If the callback function is not going to use one or both of them, the tilde can be put in the function header instead of the name of the input argument. In the previous example, the header could have been:

```
function callbackfn(hObject,~)
```

13.2.3 Plots and Images in GUIs

Plots and images can be imbedded in a GUI. The next example *guiSliderPlot* shows a plot of **sin(x)** from 0 to the value of a slider bar. The axes are positioned within the Figure Window using the **axes** function, and then, when the slider is moved the callback function plots. Note that the default Units property for an **axes** object is Normalized.

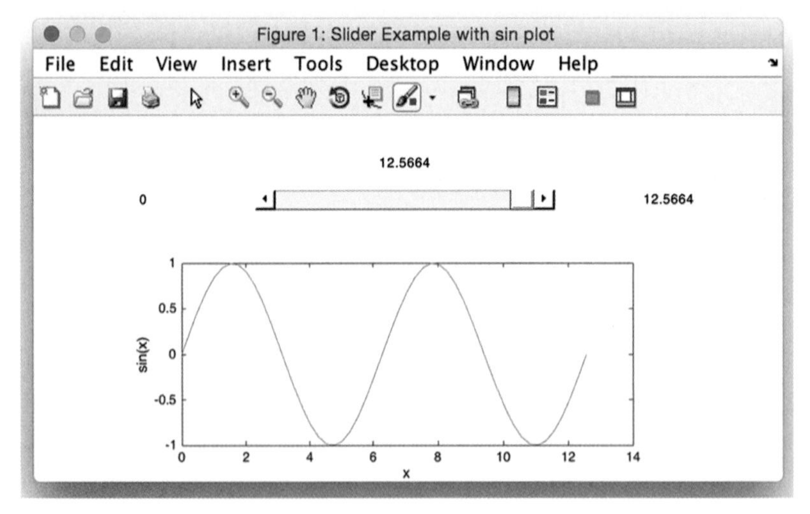

FIGURE 13.24
Plot shown in a GUI Figure Window.

guiSliderPlot.m

```
function guiSliderPlot
% guiSliderPlot has a slider
% It plots sin(x) from 0 to the value of the slider
% Format: guisliderPlot

f = figure('Visible', 'off', 'Units', 'Normalized',...
    'Position', [.2 .5 .4 .3], 'Color', 'white');

% Minimum and maximum values for slider
minval = 0;
maxval = 4*pi;

% Create the slider object
slhan = uicontrol('Style','slider','Units','Normalized',...
    'Position',[.3 .7 .4 .1], ...
    'Min', minval, 'Max', maxval,'Callback', @callbackfn);
% Text boxes to show the min and max values and slider value
hmintext = uicontrol('Style','text','BackgroundColor', 'white', ...
    'Units','Normalized','Position', [.1 .7 .1 .1],...
    'String', num2str(minval));
hmaxtext = uicontrol('Style','text', 'BackgroundColor', 'white',...
    'Units', 'Normalized','Position', [.8 .7 .1 .1], ...
    'String', num2str(maxval));
hsttext = uicontrol('Style','text','BackgroundColor', 'white',...
    'Units','Normalized', 'Position', [.45 .8 .1 .1],...
    'Visible','off');
% Create axes handle for plot
axhan = axes('Position', [.2 .1 .6 .5]);

f.Name = 'Slider Example with sin plot';
movegui(f,'center')
f.Visible = 'on';

% Call back function displays the current slider value & plots sin
   function callbackfn(~,~)
       % callbackfn is called by the 'Callback' property
       % in the slider
       num = slhan.Value;
       set(hsttext,'Visible','on','String',num2str(num))
       x = 0:num/50:num;
       y = sin(x);
       plot(x,y)
       xlabel('x')
       ylabel('sin(x)')
   end
end
```

Fig. 13.24 shows the configuration of the window, with the slider bar, static text boxes to the left and right showing the minimum and maximum values, and the axes positioned underneath. After the slider bar is moved, the callback function plots **sin(x)** from 0 to the position of the slider bar.

Images can also be placed in GUIs, again using **axes** to locate the image. In a variation of the previous example, the next example displays an image and uses a slider to vary the brightness of the image. The result is shown in Fig. 13.25.

guiSliderImage.m

```
function guiSliderImage
% guiSliderImage has a slider
% Displays an image; slider dims it
% Format: guisliderImage

f = figure('Visible', 'off', 'Units', 'Normalized',...
    'Position', [.2 .5 .4 .3], 'Color', 'white');
% Minimum and maximum values for slider
minval = 0;
maxval = 1;
% Create the slider object
slhan = uicontrol('Style','slider','Units','Normalized',...
    'Position',[.3 .7 .4 .1], ...
    'Min', minval, 'Max', maxval,'Callback', @callbackfn);
% Text boxes to show the min and max values and slider value
hmintext = uicontrol('Style','text','BackgroundColor', 'white', ...
    'Units','Normalized','Position', [.1 .7 .1 .1],...
    'String', num2str(minval));
hmaxtext = uicontrol('Style','text', 'BackgroundColor', 'white',...
    'Units', 'Normalized','Position', [.8 .7 .1 .1], ...
    'String', num2str(maxval));
hsttext = uicontrol('Style','text','BackgroundColor', 'white',...
    'Units','Normalized', 'Position', [.45 .8 .1 .1],...
    'Visible','off');
% Create axes handle for plot
axhan = axes('Position', [.2 .1 .6 .5]);

f.Name = 'Slider Example with an image';
movegui(f,'center')
f.Visible = 'on';

% Call back function displays the current slider value & image
   function callbackfn(~,~)
        % callbackfn is called by the 'Callback' property
        % in the slider
        num = slhan.Value;
        set(hsttext,'Visible','on','String',num2str(num))

        myimage1 = imread('snowyporch.JPG');
        dimmer = num * myimage1;
        image(dimmer)
    end
end
```

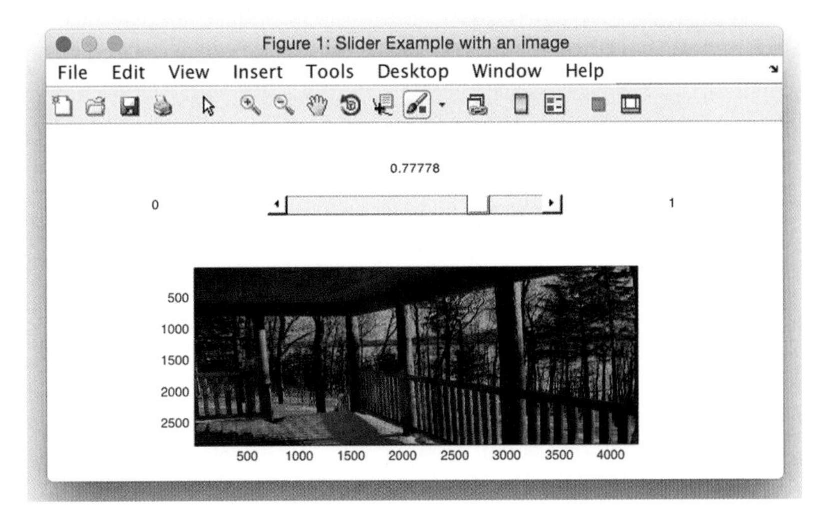

FIGURE 13.25
GUI with an image and slider for brightness.

13.2.4 Button Groups

This section illustrates radio buttons, and grouping objects together; in this case, in a button group.

The next GUI presents the user with a choice of colors using two radio buttons, only one of which can be chosen at any given time. The GUI prints a string to the right of the radio buttons, in the chosen color.

The function **uibuttongroup** creates a mechanism for grouping together the buttons. As only one button can be chosen at a time, there is a type of callback function called 'SelectionChangedFcn' that is called when a button is chosen (note: prior to R2014b this was named 'SelectionChangeFcn').

This function gets from the button group which button is chosen with the 'SelectedObject' property. It then chooses the color based on this. This property is set initially to the empty vector, so that neither button is selected; the default is that the first button would be selected.

guiWithButtongroup.m

```
function guiWithButtongroup
% guiWithButtongroup has a button group with 2 radio buttons
% Format: guiWithButtongroup

% Create the GUI but make it invisible for now while
%  it is being initialized
f = figure('Visible', 'off','color','white','Units',...
    'Normalized', 'Position',[.2 .5 .4 .3]);

%   Create a button group
group = uibuttongroup('Parent',f,'Units','Normalized',...
    'Position',[.2 .5 .4 .4], 'Title','Choose Color',...
    'SelectionChangedFcn',@whattodo);

%   Put two radio buttons in the group
toph = uicontrol(grouph,'Style','radiobutton',...
    'String','Blue','Units','Normalized',...
    'Position', [.2 .7 .4 .2]);

both = uicontrol(grouph, 'Style','radiobutton',...
    'String','Green','Units','Normalized',...
    'Position',[.2 .4 .4 .2]);

%   Put a static text box to the right
texth = uicontrol('Style','text','Units','Normalized',...
    'Position',[.6 .5 .3 .3],'String','Hello',...
    'Visible','off','BackgroundColor','white');

group.SelectedObject = []; % No button selected yet
f.Name = 'GUI with button group';
movegui(f,'center')

% Now the GUI is made visible
f.Visible = 'on';

    function whattodo(~, ~)
    % whattodo is called by the 'SelectionChangedFcn' property
    % in the button group

    which = get(grouph,'SelectedObject');

    if which == toph
        texth.ForegroundColor = 'blue';
    else
        texth.ForegroundColor = 'green';
    end

    texth.Visible = 'on';
    end

end
```

Fig. 13.26 shows the initial configuration of the GUI: the button group is in place, as are the buttons (but neither is chosen).

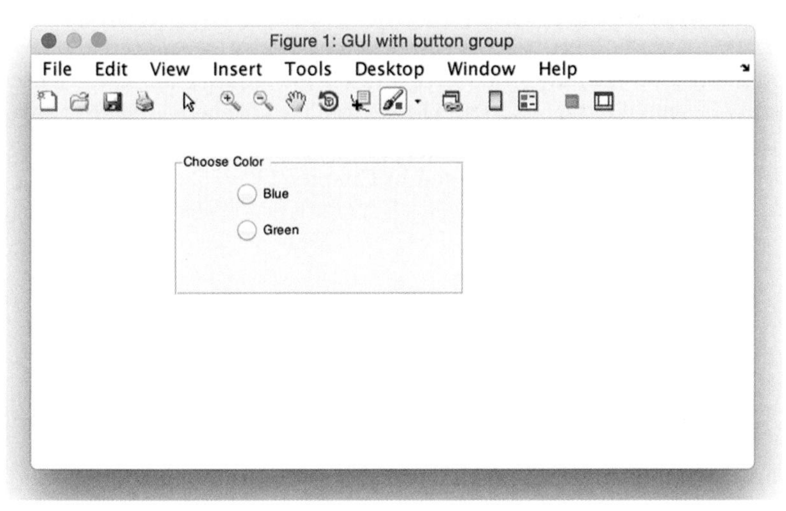

FIGURE 13.26
Button group with radio buttons.

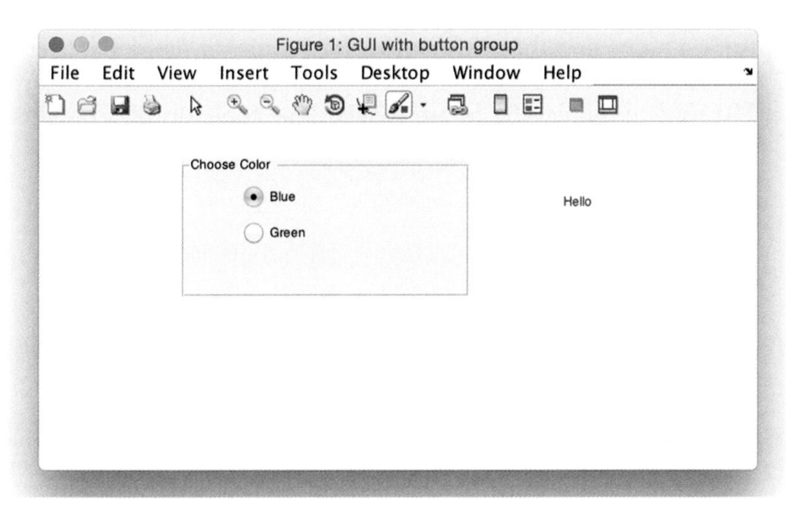

FIGURE 13.27
Button group: choice of color for string.

Once a radio button has been chosen, the *whattodo* function chooses the color for the string, which is printed in a static text box on the right, as shown in Fig. 13.27.

13.3 GUIDE AND APP DESIGNER

Current versions of MATLAB, as of R2016a, have two development environments in which GUIs can be graphically created and the code will be automatically generated. GUIDE creates code for GUIs that is procedural and similar to the examples shown in the previous section. The App Designer uses similar functions, but is all object-based. GUIDE will be briefly introduced to show that the programming techniques covered in the previous section can be used to modify code generated by GUIDE. More complete coverage is given in this section to App Designer, which creates app classes derived from a superclass, and reinforces the OOP concepts covered in Chapter 11.

13.3.1 GUIDE

In order to get into GUIDE, the command **guide** can be entered in the Command Window.

```
>> guide
```

This brings up a dialog box in which you should choose "Blank GUI (Default)" and then click the "OK" button. This brings up a GUIDE layout window with icons for objects on the left that can be dragged into the grid, to graphically lay out a GUI. However, without creating anything, we will see the basic code that GUIDE creates for a blank GUI. Clicking on the green arrow "Run" button prompts for a name and then saves two separate files: a figure file (.fig) and a function (.m) file. For example, if we enter the name "guideGUI1," the files guideGUI1.fig and guideGUI1.m are created. The function file contains the following:

```
function varargout = guideGUI1(varargin)
% GUIDEGUI1 MATLAB code for guideGUI1.fig
%      GUIDEGUI1, by itself, creates a new GUIDEGUI1 or raises the existing
%      singleton*.
%
%      H = GUIDEGUI1 returns the handle to a new GUIDEGUI1 or the handle to
%      the existing singleton*.
%
%      GUIDEGUI1('CALLBACK',hObject,eventData,handles,...) calls the
%      local function named CALLBACK in GUIDEGUI1.M with the given input
%      arguments.
%
%      GUIDEGUI1('Property','Value',...) creates a new GUIDEGUI1 or
%      raises the existing singleton*. Starting from the left, property
%      value pairs are applied to the GUI before guideGUI1_OpeningFcn gets
%      called. An unrecognized property name or invalid value makes property
%      application stop. All inputs are passed to guideGUI1_OpeningFcn
%      via varargin.
%
%      *See GUI Options on GUIDE's Tools menu. Choose "GUI allows only one
%      instance to run (singleton)".
%
% See also: GUIDE, GUIDATA, GUIHANDLES
```

```
% Edit the above text to modify the response to help guideGUI1

% Last Modified by GUIDE v2.5 29-Jan-2016 08:55:11

% Begin initialization code - DO NOT EDIT
gui_Singleton = 1;
gui_State = struct('gui_Name',       mfilename, ...
                   'gui_Singleton',  gui_Singleton, ...
                   'gui_OpeningFcn', @guideGUI1_OpeningFcn, ...
                   'gui_OutputFcn',  @guideGUI1_OutputFcn, ...
                   'gui_LayoutFcn',  [] , ...
                   'gui_Callback',   []);
if nargin && ischar(varargin{1})
    gui_State.gui_Callback = str2func(varargin{1});
end

if nargout
    [varargout{1:nargout}] = gui_mainfcn(gui_State, varargin{:});
else
    gui_mainfcn(gui_State, varargin{:});
end
% End initialization code - DO NOT EDIT

% --- Executes just before guideGUI1 is made visible.
function guideGUI1_OpeningFcn(hObject, eventdata, handles, varargin)
% This function has no output args, see OutputFcn.
% hObject    handle to figure
% eventdata  reserved - to be defined in a future version of MATLAB
% handles    structure with handles and user data (see GUIDATA)
% varargin   command line arguments to guideGUI1 (see VARARGIN)

% Choose default command line output for guideGUI1
handles.output = hObject;

% Update handles structure
guidata(hObject, handles);

% UIWAIT makes guideGUI1 wait for user response (see UIRESUME)
% uiwait(handles.figure1);

% --- Outputs from this function are returned to the command line.
function varargout = guideGUI1_OutputFcn(hObject, eventdata, handles)
% varargout  cell array for returning output args (see VARARGOUT);
% hObject    handle to figure
% eventdata  reserved - to be defined in a future version of MATLAB
% handles    structure with handles and user data (see GUIDATA)

% Get default command line output from handles structure
varargout{1} = handles.output;
```

The *guideGUI1* function creates a structure that stores some properties, as well as functions *guideGUI1_OpeningFcn* and *guideGUI1_OutputFcn* which execute when the GUI is made visible and when the *guideGUI1* function finishes executing, respectively.

If, instead of creating a blank GUI, we dragged a static text box and an editable text box into the GUIDE layout, the code is similar to the previous example. However, it also creates a callback function stub and a create object function stub for the editable text box. Fig. 13.28 shows the object icons on the left, and part of the grid layout into which we have dragged a static text box and an editable text box, without resizing or changing the labels.

```
function edit1_Callback(hObject, eventdata, handles)
% hObject    handle to edit1 (see GCBO)
% eventdata  reserved - to be defined in a future version of MATLAB
% handles    structure with handles and user data (see GUIDATA)

% Hints: get(hObject,'String') returns contents of edit1 as text
%        str2double(get(hObject,'String')) returns contents of edit1 as a
double


% --- Executes during object creation, after setting all properties.
function edit1_CreateFcn(hObject, eventdata, handles)
% hObject    handle to edit1 (see GCBO)
% eventdata  reserved - to be defined in a future version of MATLAB
% handles    empty - handles not created until after all CreateFcns called

% Hint: edit controls usually have a white background on Windows.
%       See ISPC and COMPUTER.
if ispc && isequal(get(hObject,'BackgroundColor'),
get(0,'defaultUicontrolBackgroundColor'))
    set(hObject,'BackgroundColor','white');
end
```

The callback function *edit1_Callback* does not do anything by default, but it does contain a hint that since this is an editable text box, a reasonable thing to do would be to get the string from the box and then do something with it. At this point, it is possible to edit the code directly using the programming techniques from the previous section, or make modifications using GUIDE. The GUIDE documentation explains its features.

13.3.2 App Designer

The App Designer is similar to GUIDE in that it allows the user to graphically layout objects, which are called ***components***. However, there are more types of components than GUIDE has, and the code that is generated is all object-based. Also, App Designer creates UI Figure windows, whereas GUIDE uses Figure Windows.

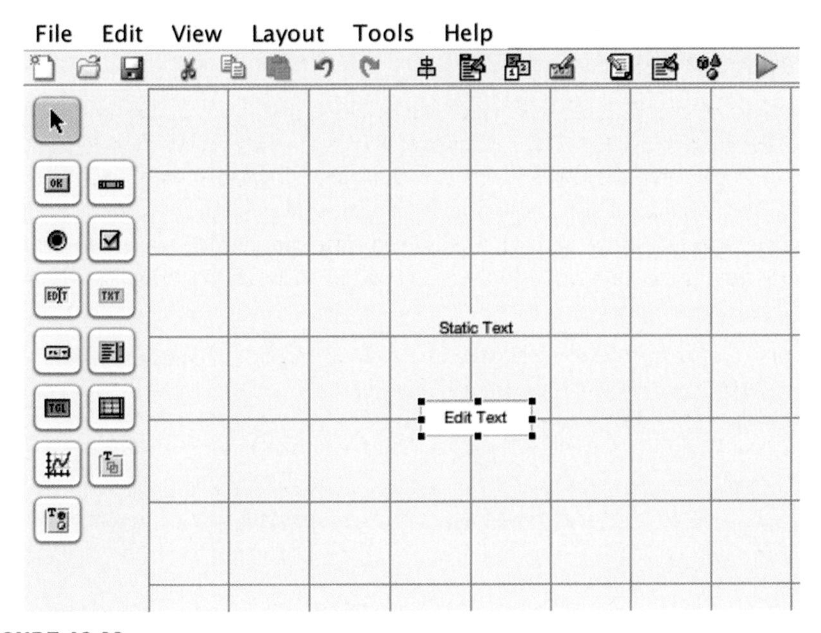

FIGURE 13.28
GUIDE layout of text boxes.

13.3.2.1 Intro to App Designer

Typing the command **appdesigner** in the Command Window will bring up the App Designer. Fig. 13.29 shows the App Designer; in the middle is the blank layout under "Design View." The Component Library on the left shows the icons of the components that can be created. On the right are the Component Browser and Component Properties Windows. In the Component Browser, the default name for the blank figure, *app.UIFigure*, can be seen.

```
>> appdesigner
```

Fig. 13.30 shows the remainder of the Component Library, including the Container and Instrumentation components.

Clicking on "Code View" instead of "Design View" shows the code that has been created for the blank app. App Designer creates a class named *App1* that is derived from a MATLAB apps superclass called **matlab.apps.AppBase**. The properties will consist of all of the components; for now, there is just one property, which is the UI Figure Window. There are also two private methods blocks and one public methods block.

FIGURE 13.29
App Designer Layout.

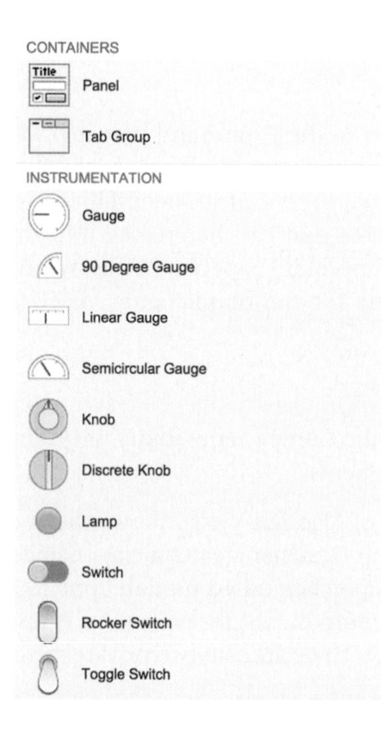

FIGURE 13.30
Bottom of Component Library.

```
classdef App1 < matlab.apps.AppBase
    % Properties that correspond to app components
    properties (Access = public)
        UIFigure matlab.ui.Figure % UI Figure
    end

    methods (Access = private)

        % Code that executes after component creation
        function startupFcn(app)

        end

    end

    % App initialization and construction
    methods (Access = private)

        % Create UIFigure and components
        function createComponents(app)

            % Create UIFigure
            app.UIFigure = uifigure;
            app.UIFigure.Position = [100 100 640 480];
            app.UIFigure.Name = 'UI Figure';
            setAutoResize(app, app.UIFigure, true)
        end
    end

    methods (Access = public)

        % Construct app
        function app = App1()

            % Create and configure components
            createComponents(app)

            % Register the app with App Designer
            registerApp(app, app.UIFigure)

            % Execute the startup function
            runStartupFcn(app, @startupFcn)

            if nargout == 0
                clear app
            end
        end

        % Code that executes before app deletion
        function delete(app)
            % Delete UIFigure when app is deleted
            delete(app.UIFigure)
        end
    end
end
```

The methods are used for all of the app functions, including eventually callback functions. For now, with just a blank UI Figure Window, the public methods include the constructor function *App1*, and a function *delete* that deletes the figure when the app is deleted. The constructor calls a private function *createComponents*, which creates a UI Figure using the **uifigure** function. It then uses the dot notation to set properties for the *app.UIFigure*, including its *Name* and *Position*. Note that the dot notation is used twice here, as in *app.UIFigure.Name*.

The function *startupFcn* executes when components are created. It is called by a function in the constructor. Within the Code View, most of the window is grey, which means that the code cannot be modified. However, within the *startupFcn* function there is a white box, which means that code can be inserted in that function. Fig. 13.31 shows the *startupFcn* function in Code View.

We have seen the code that is generated with just a blank canvas. In the Design View of the App Designer environment, components can be dragged from the Component Library onto the blank layout. When this is done, the code will be updated to create a new property for each component, and to initialize some of its properties (for example, the Position property will be based on the location to which the component was dragged). Properties can then be modified in the Component Properties Window to the right of the layout.

For example, dragging a Label (generically named 'Label'!) into the design area causes the Design View to look like Fig. 13.32.

Clicking on the label itself in the Design View will change "Component Properties" to "Label Properties." Properties such as the text on the label, its justification, font name, font size, and so forth can then be modified from the Label Properties window. Each of these will modify the code that is created.

Assuming that no properties have been modified, and there is just a generic label in the layout as shown in Fig. 13.32, most of the code remains the same as with just a blank UI Figure Window. Choosing Code View will show that the class *App1* now has a new property Label, and the *createComponents* method creates a Label using the **uilabel** function and gives it a Position. The added code is shown in bold here (not in App Designer itself).

```
8       methods (Access = private)
9
10          % Code that executes after component creation
11          function startupFcn(app)
12
13          end
14      end
15
```

FIGURE 13.31
Editable portion of default app code.

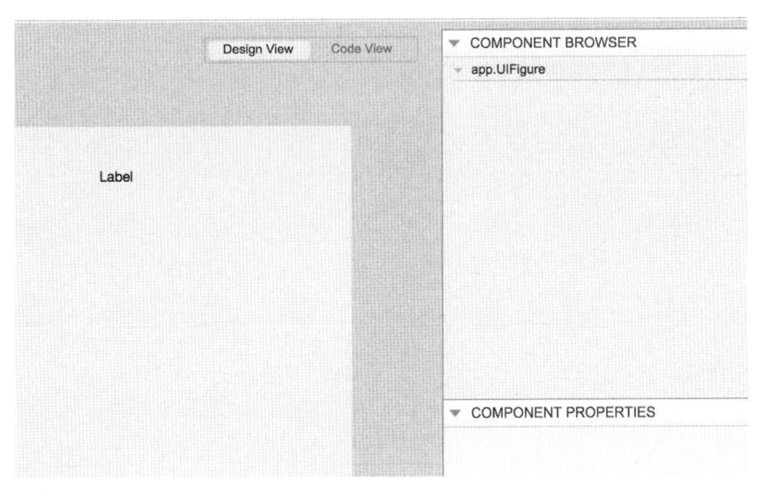

FIGURE 13.32
Simple Label in app.

```
classdef App1 < matlab.apps.AppBase
% Properties that correspond to app components
properties (Access = public)
    UIFigure matlab.ui.Figure % UI Figure
    Label   matlab.ui.control.Label % Label
end

% Create UIFigure and components
    function createComponents(app)
        % Create UIFigure
        app.UIFigure = uifigure;
        app.UIFigure.Position = [100 100 640 480];
        app.UIFigure.Name = 'UI Figure';
        setAutoResize(app, app.UIFigure, true)
        % Create Label
        app.Label = uilabel(app.UIFigure);
        app.Label.Position = [421 431 30 15];
    end
end
```

Modifying a property, such as changing the text of the label in the Label Properties window to be "Hello" instead of "Label" will modify the code in *createComponents* to:

```
% Create Label
app.Label = uilabel(app.UIFigure);
app.Label.Position = [421 431 30 15];
app.Label.Text = 'Hello';
```

It is also possible programmatically to modify properties such as *app.Label.Text* in the *startupFcn* function, but it is preferable and easier to do this from the Design View.

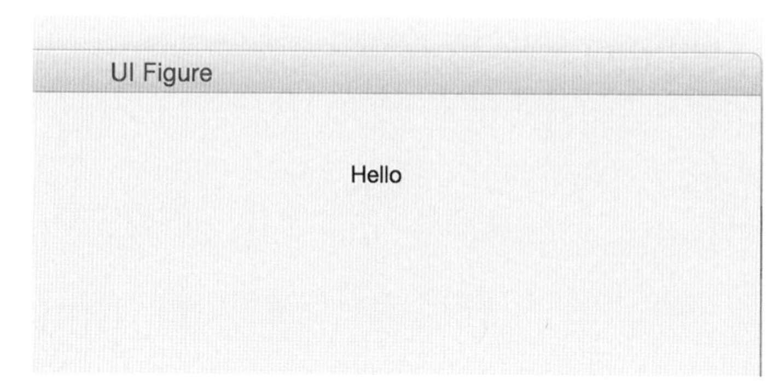

FIGURE 13.33
Text of label in UI Figure Window.

To run the app, click on the green Run arrow. This brings up a dialog box that asks for the name of the file that will be created; the default name that shows is "App1.mlapp." Changing the name to "HelloLabel.mlapp" will create a file with this name and bring up a UI Figure Window. The upper right portion of this is shown in Fig. 13.33.

Note that the App Designer creates one file with the extension '.mlapp,' unlike GUIDE, which creates two files: a .m file and a .fig file.

While this example shows the basics of App Designer and the object-based code that it creates, it did not involve any callbacks.

13.3.2.2 UI Figure Functions

As we have seen in Section 13.2, traditional GUIs create different types of objects by specifying the Style property in the **uicontrol** function. For example, a slider is created with **uicontrol('Style', 'slider')**. By contrast, App Designer uses separate functions to create the different component types; for example, a slider is created with the **uislider** function. In the previous section, we saw two of these functions: **uifigure**, which creates the UI Figure Window, and **uilabel**, which creates a static text box.

All of these functions can be called directly from the Command Window or from any script or function. For example, the following code will bring up the UI Figure Window shown in Fig. 13.34.

```
>> uif = uifigure;
>> uif.Name = 'Simple UI Fig Slider';
>> uislid = uislider(uif);
```

By creating a UI Figure Window first, and then passing its handle to the **uislider** function, the slider is put inside that UI Figure Window (otherwise, a new UI

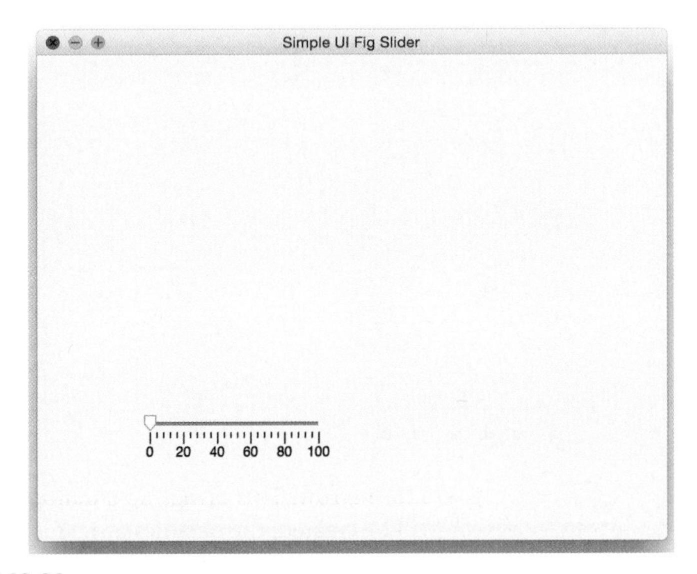

FIGURE 13.34
Simple slider in UI Figure Window.

Figure Window would be created for the slider). The Value property of the slider can be inspected, e.g.,

```
>> uislid.Value
ans =
     0
```

Of course, to do anything with the value of the slider would require more complicated code with callback functions. For sliders, there are two:

'ValueChangedFcn': executes when the slider value has been changed
'ValueChangingFcn': executes as the slider is being moved

For example, the function *uisliderуilabel* creates a UI Figure Window with a slider and a label. When the slider has been moved, the 'ValueChangedFcn' callback is called, which puts the value of the slider in the label as seen in Fig. 13.35. Many of the concepts here are similar to those used in GUI callbacks. The function *uisliderуilabel* is a function that has a nested callback function, *whatslid*. Since the *whatslid* function is nested, the variable scope is such that the variables *uilab* and *uislid* can be used within the callback function without passing them. The callback function has two input arguments representing the source of the callback (in this case *uislid*) and the eventdata, just like GUI callback functions.

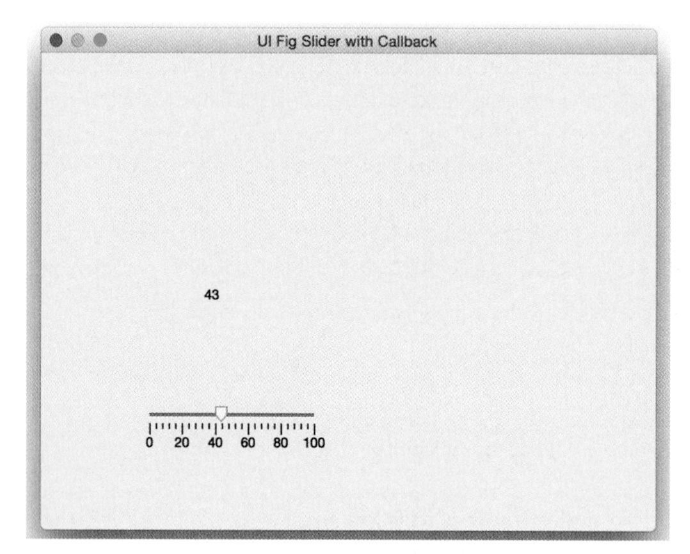

FIGURE 13.35
Slider with label example.

uislideruilabel.m

```
function uislideruilabel
  uif = uifigure;
  uif.Name = 'UI Fig Slider with Callback';
  uislid = uislider(uif);
  uilab = uilabel(uif,'Position',[150 200 30 15]);
  uilab.Text = '0';

  uislid.ValueChangedFcn = @whatslid;

  function whatslid(source,event)
  uilab.Text = num2str(round(uislid.Value));
  end
  end
```

The list of UI functions that create object components can be found in the App Designer documentation. The documentation for each function describes the properties that can be used for each, including all of the possible callback functions.

13.3.2.3 App Designer Example

An advantage of App Designer is that it will create much of the code automatically, and since it is based on objects, the private properties and methods protect the apps that are created.

As an example of an app that uses a callback, an app will be created that is similar to the previous example in which there is a slider and a label that shows the value of the slider. Since it makes sense to label many components, such as sliders, App Designer automatically creates a label when the component is dragged into the design area (unless the Ctrl key is pressed while dragging the component). By default the label is to the left of the slider. In Design View, since there are two components, the slider and its label, the properties window is initially labeled "Multiple Components Properties."

The Code View shows that three properties have been created:

```
properties (Access = public)
    UIFigure    matlab.ui.Figure          % UI Figure
    LabelSlider matlab.ui.control.Label   % Slider
    Slider      matlab.ui.control.Slider  % [0 100]
end
```

The constructor and other methods are similar to what was seen in a previous example. Unlike GUIDE, App Designer does not automatically create callback function stubs. Instead, to create a callback, it is necessary to right-click on the component in Design View. Selecting 'Callbacks' brings up two possible callback functions. For example, enter the name "SliderValueChanged" for the ValueChangedFcn. This creates a new function, seen in Code View:

```
% Slider value changed function
function SliderValueChanged(app)
    value = app.Slider.Value;

end
```

In the function, an example is given that shows how to reference the Value property of the slider, but the function is editable. Modifying the function body as follows will put the value of the slider in the label.

```
function SliderValueChanged(app)
    app.LabelSlider.Text = ...
            num2str(round(app.Slider.Value));
end
```

In Design View, clicking on either the slider or its label will change the Properties window. For example, clicking on the label allows one to change the initial label of "Slider" to "0," which is the initial value of the slider.

To change properties of any of the objects, one can click on the object name in the Component Browser. For example, clicking on "app.UIFigure" in the Component Browser then shows "UI Figure Properties." That allows us to change the title of the UI Figure Window, for example to "App with Slider."

When the app is executed, the UI Figure window seen in Fig. 13.36 is created.

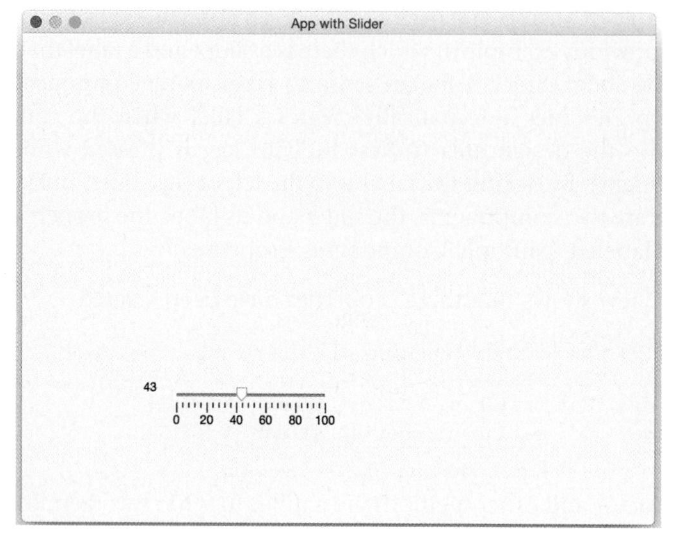

FIGURE 13.36
App with slider and label.

Putting plots into apps uses a method that is very similar to GUIs: it is necessary to first drag axes into the design layout, and then any plot will be placed within the current axes.

13.4 SOUND FILES

A ***sound wave*** is an example of a continuous signal that can be sampled to result in a discrete signal. In this case, sound waves traveling through the air are recorded as a set of measurements that can then be used to reconstruct the original sound signal, as closely as possible. The sampling rate, or ***sampling frequency***, is the number of samples taken per time unit, for example per second. Sound signals are usually measured in Hertz (Hz).

In MATLAB, the discrete sound signal is represented by a vector and the frequency is measured in Hertz. MATLAB has several MAT-files that store for various sounds the signal vector in a variable *y* and the frequency in a variable *Fs*. These MAT-files include **chirp**, **gong**, **laughter**, **splat**, **train**, and **handel**. There is a built-in function, **sound**, that will send a sound signal to an output device such as speakers.

The function call:

```
>> sound(y,Fs)
```

will play the sound represented by the vector *y* at the frequency *Fs*. For example, to hear a gong, load the variables from the MAT-file and then play the sound using the **sound** function:

```
>> load gong
>> sound(y,Fs)
```

Sound is a wave; the amplitudes are what are stored in the sound signal variable *y*. These are supposed to be in the range from −1 to 1. The **plot** function can be used to display the data. For example, the following script creates a **subplot** that displays the signals from **chirp** and from **train**, as shown in Fig. 13.37.

chirptrain.m

```
% Display the sound signals from chirp and train
subplot(2,1,1)
load chirp
plot(y)
ylabel('Amplitude')
title('Chirp')

subplot(2,1,2)
load train
plot(y)
ylabel('Amplitude')
title('Train')
```

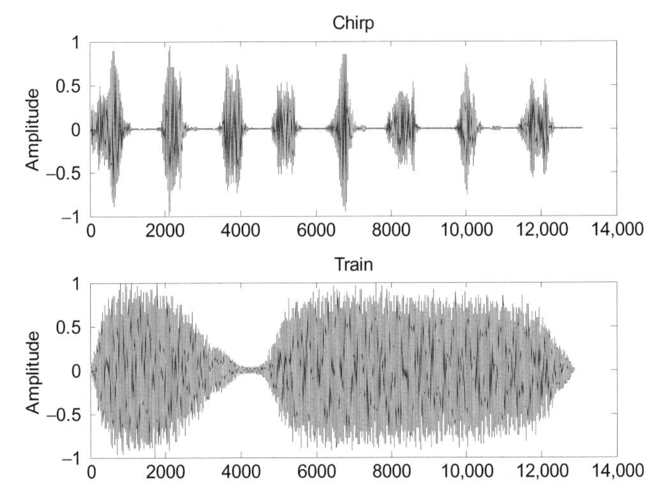

FIGURE 13.37
Amplitudes from **chirp** and **train**.

The first argument to the **sound** function can be an $n \times 2$ matrix for stereo sound. Also, the second argument can be omitted when calling the **sound** function, in which case the default sample frequency of 8192 Hz is used. This is the frequency stored in the built-in sound MAT-files.

```
>> load train
Fs
Fs =
        8192
```

PRACTICE 13.5

If you have speakers try loading one of the sound MAT-files and use the **sound** function to play the sound. Then, change the frequency; for instance, multiply the variable *Fs* by 2 and by 0.5, and play these sounds again.

```
>> load train
>> sound(y, Fs)
>> sound(y, Fs*2)
>> sound(y, Fs*.5)
```

■ Explore Other Interesting Features

Several ***audio file formats*** are used in industry on different computer platforms. Audio files with the extension ".au" were developed by Sun Microsystems; typically, they are used with Java and Unix, whereas Windows PCs typically use ".wav" files that were developed by Microsoft. Investigate the MATLAB functions **audioread**, **audioinfo**, and **audiowrite**.

Investigate the **colorcube** function, which returns a colormap with regularly spaced R, G, and B colors.

Investigate the **imfinfo** function, which will return information about an image file in a structure variable.

Investigate how colormaps work with image matrices of types **uint8** and **uint16**.

In addition to true color images and indexed images into a colormap, a third type of image is an intensity image, which is used frequently for grayscale images. Investigate how to use the image scale function **imagesc**.

The **uibuttongroup** function is used specifically to group together buttons; other objects can be grouped together similarly using the **uipanel** function. Investigate how this works.

When a GUI has a lot of objects, creating a structure to store the handles can be useful. Investigate the **guihandles** function which accomplishes this.

Investigate the **uitable** function. Use it to create a GUI that demonstrates a matrix operation.

Beginning with MATLAB Version R2012b, GUIs can be packaged as apps! Under the Search Documentation under GUI Building, read how to do this in the category "Packaging GUIs as Apps." Apps can be shared with other users. There is also a lot of information on apps (creating them, downloading them, modifying them, and so forth) under "Desktop Environment." ■

SUMMARY

COMMON PITFALLS

- Confusing true color and colormap images.
- Forgetting that **uicontrol** object positions are within the Figure Window, not within the screen.

PROGRAMMING STYLE GUIDELINES

- Make a GUI invisible while it is being created, so that everything becomes visible at once.

MATLAB Functions and Commands		
colormap	uipanel	laughter
parula	uibuttongroup	splat
jet	root	train
image	uicontrol	handel
pink	movegui	sound
imread	chirp	
imwrite	gong	

Exercises

1. Create a custom colormap for a sphere that consists of the first 25 colors in the colormap **jet**. Display **sphere(25)** with a colorbar as seen in Fig. 13.38.
2. Write a script that will create the image seen in Fig. 13.39 using a colormap.
3. Write a function *numimage* that will receive two input arguments: a colormap matrix, and an integer "*n*"; the function will create an image that shows *n*

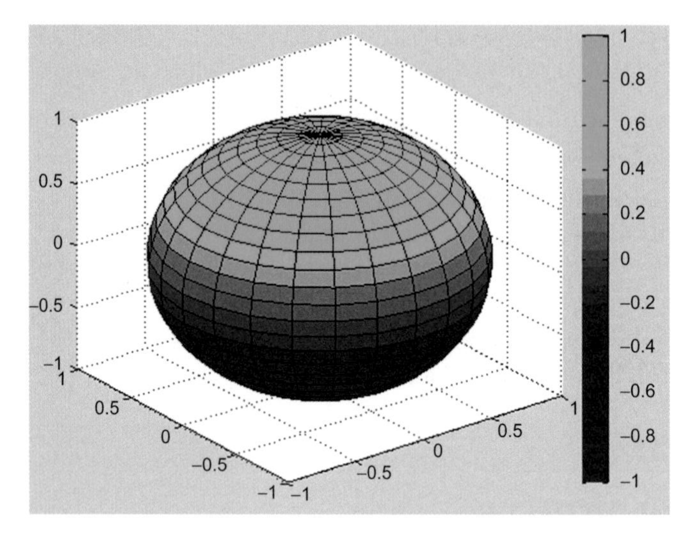

FIGURE 13.38
Custom blue map for a sphere.

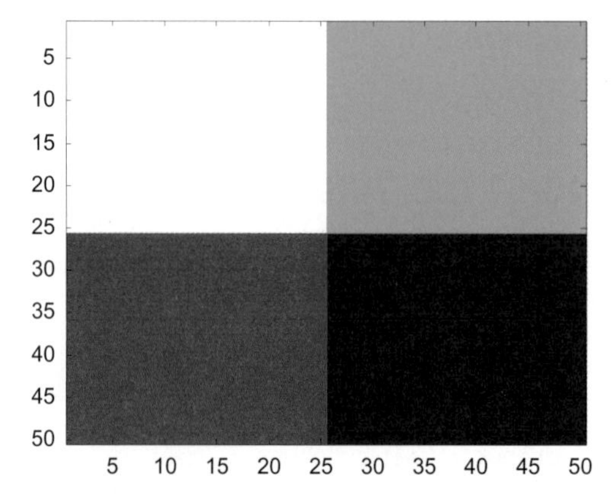

FIGURE 13.39
Image displaying four colors using a custom colormap.

"rings" of color, using the first *n* colors from the colormap. For example, if the function is called as follows:

```
>> cm = [0 0 0; 1 0 0; 0 1 0; 0 0 1; ...
1 1 0; 1 0 1; 0 1 1];
>> numimage(cm,5)
```

the image as seen in Fig. 13.40 will be created. Each "ring" has the thickness of one pixel. In this case, since *n* was 5, the image shows the first 5 colors from the colormap: the outermost ring is the first color, the next ring is the second color, and the innermost pixel is the fifth color. Note that since *n* was 5, the image matrix is 9×9.

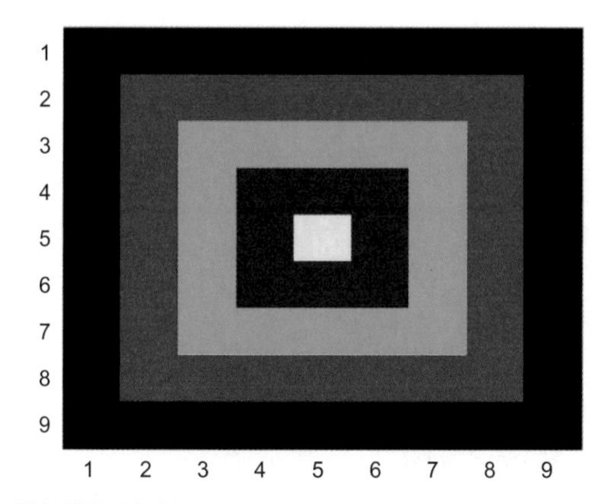

FIGURE 13.40
Color rings.

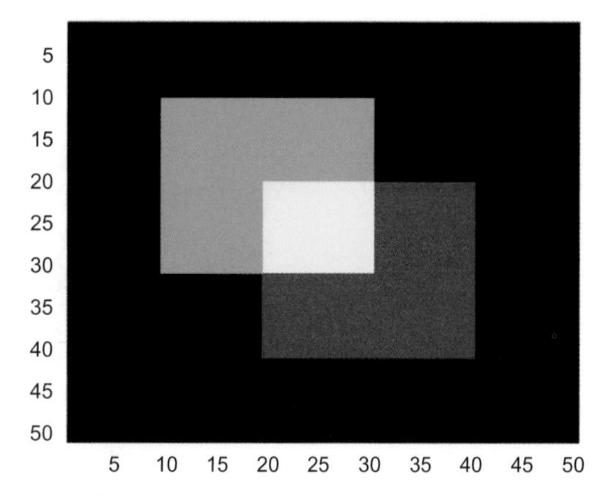

FIGURE 13.41
Green, yellow, and red blocks.

4. Write a script that would produce the 50 × 50 "image" seen in Fig. 13.41, using the RGB or true color method (NOT the colormap method).
5. A script *rancolors* displays random colors in the Figure Window as shown in Fig. 13.42. It starts with a variable *nColors* which is the number of random colors to display (e.g., below this is 10). It then creates a colormap variable *mycolormap*, which has that many random colors, meaning that all three of the color components (red, green, and blue) are random real numbers in the range from 0 to 1. The script then displays these colors in an image in the Figure Window.

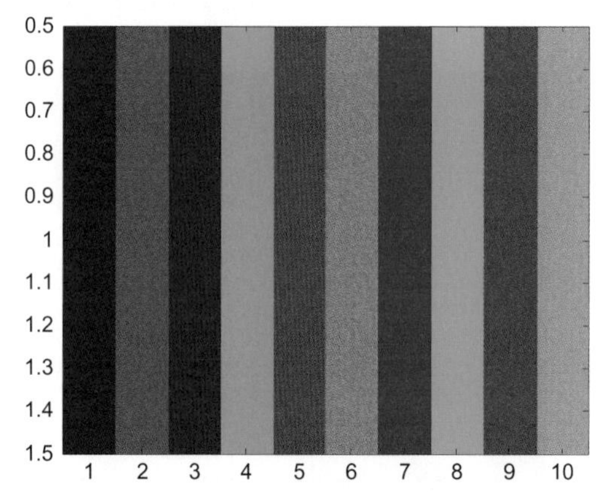

FIGURE 13.42
Rainbow of random colors.

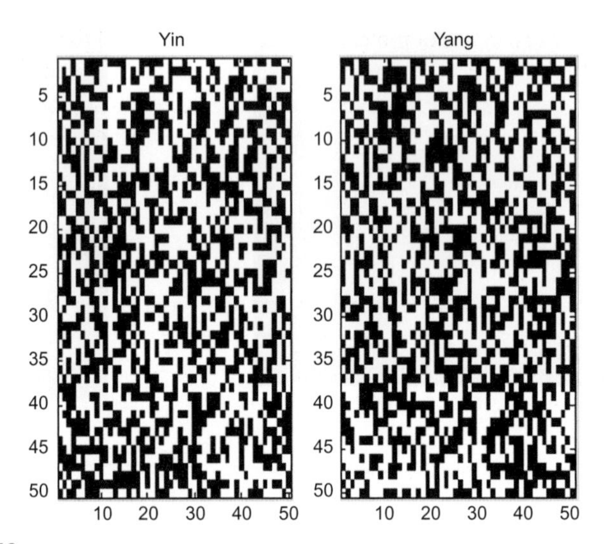

FIGURE 13.43
Reverse black and white pixels.

6. Write a script that will create a colormap that has just two colors: white and black. The script will then create a 50 × 50 image matrix in which each element is randomly either white or black. In one Figure Window, display this image on the left. On the right, display another image matrix in which the colors have been reversed (all white pixels become black and vice versa). For example, the images might look like Fig. 13.43 (the axes are defaults; note the titles):
   Do not use any loops or **if** statements. For the image matrix that you created, what would you expect the overall mean of the matrix elements to be?
7. Write a script that will show shades of green and blue as seen in Fig. 13.44. First, create a colormap that has 30 colors (10 blue, 10 aqua, and then 10 green).

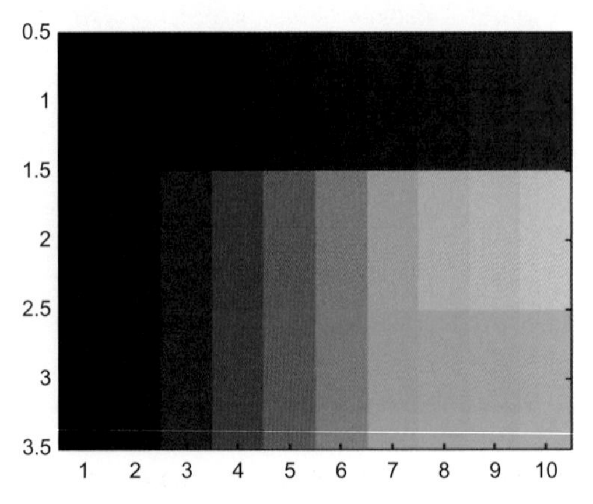

FIGURE 13.44
Shades of blue, aqua, and green.

There is no red in any of the colors. The first 10 rows of the colormap have no green, and the blue component iterates from 0.1 to 1 in steps of 0.1. In the second 10 rows, both the green and blue components iterate from 0.1 to 1 in steps of 0.1. In the last 10 rows, there is no blue, but the green component iterates from 0.1 to 1 in steps of 0.1. Then, display all of the colors from this colormap in a 3×10 image matrix in which the blues are in the first row, aquas in the second, and greens in the third, as follows (the axes are the defaults). Do not use loops.

8. A part of an image is represented by an $n \times n$ matrix. After performing data compression and then data reconstruction techniques, the resulting matrix has values that are close to but not exactly equal to the original matrix. For example, the following 4×4 matrix variable *orig_im* represents a small part of a true color image, and *fin_im* represents the matrix after it has undergone data compression and then reconstruction.

```
orig_im =
   156    44   129    87
    18   158   118   102
    80    62   138    78
   155   150   241   105

fin_im =
   153    43   130    92
    16   152   118   102
    73    66   143    75
   152   155   247   114
```

   Write a script that will simulate this by creating a square matrix of random integers, each in the range from 0 to 255. It will then modify this to create the new matrix by randomly adding or subtracting a random number (in a relatively small range, say 0–10) from every element in the original matrix. Then, calculate the average difference between the two matrices.

9. It is sometimes difficult for the human eye to perceive the brightness of an object correctly. For example, in Fig. 13.45, the middle of both images is of the same color, and yet, because of the surrounding colors, the one on the left looks lighter than the one on the right. Write a script to generate a Figure Window similar to this one. Two 3×3 matrices were created. Use **subplot** to display both images side-by-side (the axes shown here are the defaults). Use the RGB method.

10. Put a JPEG file in your Current Folder and use **imread** to load it into a matrix. Calculate and print the **mean** separately of the red, green, and blue components in the matrix.

11. Some image acquisition systems are not very accurate, and the result is ***noisy*** images. To see this effect, put a JPEG file in your Current Folder and use **imread** to load it. Then, create a new image matrix by randomly adding or subtracting a value *n* to every element in this matrix. Experiment with

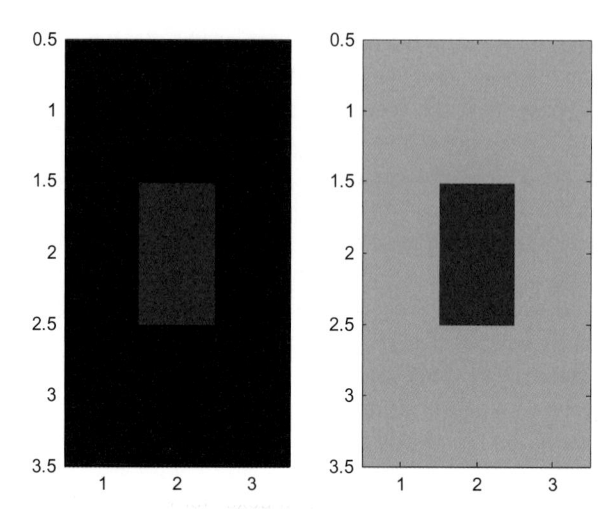

FIGURE 13.45
Depiction of brightness perception.

different values of *n*. Create a script that will use **subplot** to display both images side-by-side.

12. Put a JPEG file into your Current Folder. Type in the following script, using your own JPEG file name.

```
I1 = imread('xxx.jpg');
[r c h] = size(I1);
Inew(:,:,:) = I1(:,c:-1:1,:);
figure(1)
subplot(2,1,1)
image(I1);
subplot(2,1,2)
image(Inew);
```

Determine what the script does. Put comments into the script to explain it step-by-step.

13. Write a function that will create a simple GUI with one static text box near the middle of the Figure Window. Put your name in the string and make the background color of the text box white.
14. Write a function that will create a GUI with one editable text box near the middle of the Figure Window. Put your name in the string. The GUI should have a callback function that prints the user's string twice, one under the other.
15. Fill in the callback function so that it gets the value of the slider, prints that value in the text box, and uses it to set the LineWidth of the plot (so, e.g., if the slider value is its maximum, the line width of the plot would be 5).

```
function sliderlinewidth
f =  figure('Visible', 'off','Position',[20,20,500,400]);
```

```
slhan = uicontrol('Style','slider','Units','Normalized',...
     'Position',[.3 .3 .4 .1], ...
     'Min', 1, 'Max', 5,'Value',3,'Callback', @callbackfn);
slval = uicontrol('Style','text',...
     'Units','Normalized','Position', [.4 .1 .2 .1]);
axhan = axes('Units', 'Normalized','Position', [.3 .5 .4 .3]);
x = -2*pi:0.1:2*pi;
y = cos(x);
phan = plot(x,y)
set(f,'Visible','on');
function callbackfn(source,eventdata)

end
end
```

16. Write a function that creates a GUI to calculate the area of a rectangle. It should have edit text boxes for the length and width, and a push button that causes the area to be calculated and printed in a static text box.
17. Write a function that creates a simple calculator with a GUI. The GUI should have two editable text boxes in which the user enters numbers. There should be four pushbuttons to show the four operations (+, −, *, /). When one of the four pushbuttons is pressed, the type of operation should be shown in a static text box between the two editable text boxes and the result of the operation should be displayed in a static text box. If the user tries to divide by zero, display an error message in a static text box.
18. Modify any example GUI to use the 'Horizontal Alignment' property to left-justify text within an edit text box.
19. The wind chill factor (WCF) measures how cold it feels with a given air temperature T (in °F) and wind speed (V, in miles/h). The formula is approximately

    $$WCF = 35.7 + 0.6\,T - 35.7\,(V^{0.16}) + 0.43\,T\,(V^{0.16})$$

    Write a GUI function that will display sliders for the temperature and wind speed. The GUI will calculate the WCF for the given values, and display the result in a text box. Choose appropriate minimum and maximum values for the two sliders.
20. Write a GUI function that will graphically demonstrate the difference between a **for** loop and a **while** loop. The function will have two push buttons: one that says 'for,' and the other that says 'while.' There are two separate callback functions, one associated with each of the pushbuttons. The callback function associated with the 'for' button prints the integers 1 through 5, using **pause(1)** to pause for 1 s between each, and then prints 'Done.' The callback function associated with the 'while' button prints integers beginning with 1 and also pauses between each. This function, however, also has another pushbutton that says 'mystery' on it. This function continues printing integers until the 'mystery' button is pushed, and then it prints 'Finally!'

21. Write a function that will create a GUI in which there is a plot of cos(*x*). There should be two editable text boxes in which the user can enter the range for *x*.
22. Write a function that will create a GUI in which there is a plot. Use a button group to allow the user to choose among several functions to plot.
23. Write a GUI function that will create a **rectangle** object. The GUI has a slider on top that ranges from 2 to 10. The value of the slider determines the width of the **rectangle**. You will need to create axes for the rectangle. In the callback function, use **cla** to clear the children from the current axes so that a thinner rectangle can be viewed.
24. Write a GUI that displays an image in which all of the elements are of the same color. Put 3 sliders in that allow the user to specify the amount of red, green, and blue in the image. Use the RGB method.
25. Put two different JPEG files into your Current Folder. Read both into matrix variables. To superimpose the images, if the matrices are of the same size, the elements can simply be added element-by-element. However, if they are not of the same size, one method of handling this is to crop the larger matrix to be of the same size as that of the smaller, and then add them. Write a script to do this.

In a random walk, every time a "step" is taken, a direction is randomly chosen. Watching a random walk as it evolves, by viewing it as an image, can be very entertaining. However, there are actually very practical applications of random walks; they can be used to simulate diverse events such as the spread of a forest fire or the growth of a dendritic crystal.

26. The following function simulates a "random walk," using a matrix to store the random walk as it progresses. To begin with all elements are initialized to 1. Then, the "middle" element is chosen to be the starting point for the random walk; a 2 is placed in that element. (Note: these numbers will eventually represent colors.) Then, from this starting point another element next to the current one is chosen randomly and the *color* stored in that element is incremented; this repeats until one of the edges of the matrix is reached. Every time an element is chosen for the next element, it is done randomly by either adding or subtracting one to/from each coordinate (*x* and *y*), or leaving it alone. The resulting matrix that is returned is an *n* by *n* matrix.

ranwalk.m

```
function walkmat = ranwalk(n)
walkmat = ones(n);
x = floor(n/2);
y = floor(n/2);
color = 2;
walkmat(x,y) = color;
while x ~= 1 && x ~= n && y ~= 1 && y ~= n
    x = x + randi([-1 1]);
    y = y + randi([-1 1]);
    color = color + 1;
    walkmat(x,y) = mod(color,65);
end
```

You are to write a script that will call this function twice (once passing 8 and once passing 100) and display the resulting matrices as images side-by-side. Your script must create a custom colormap that has 65 colors; the first is white and the rest are from the colormap **jet**. For example, the result may look like Fig. 13.46. (Note that with the 8×8 matrix the colors are not likely to get out of the blue range, but with 100×00 it cycles through all colors multiple times until an edge is reached.):

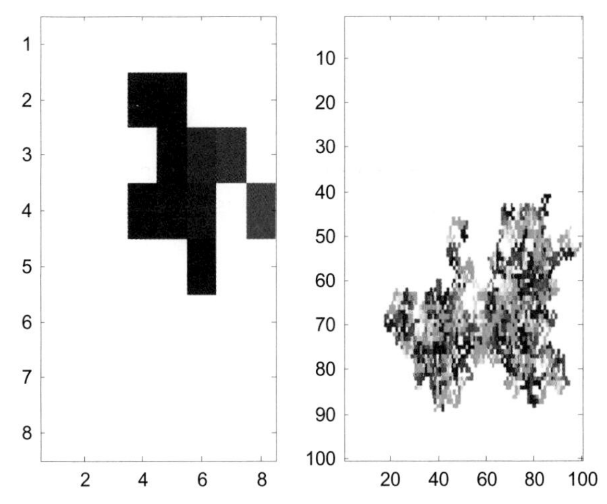

FIGURE 13.46
Random walk.

27. Use App Designer to create a text editor. Create an app that has a large text box as seen in Fig. 13.47. Under it there will be a slider that controls the font size of the text, and buttons to make the text bold and/or italic. Start by dragging a text box into the design area. Click on the box, and then in Design View look at the Edit Field (Text) Properties browser. By changing the properties such as the Style, Name, and Size, and then inspecting the code in Code View, you can see what to change in the callback functions.
28. Create a stoplight app as seen in Fig. 13.48. There are two pushbuttons labeled 'Stop' and 'Go,' and three lamps. When the 'Go' button is pushed, the green lamp is lit. When the 'Stop' button is pushed, the yellow lamp is lit briefly, and then the red lamp is lit.
29. Load two of the built-in MAT-file sound files (e.g., **gong** and **chirp**). Store the sound vectors in two separate variables. Determine how to concatenate these so that the **sound** function will play one immediately followed by the other; fill in the blank here:

```
sound(   , 8192)
```

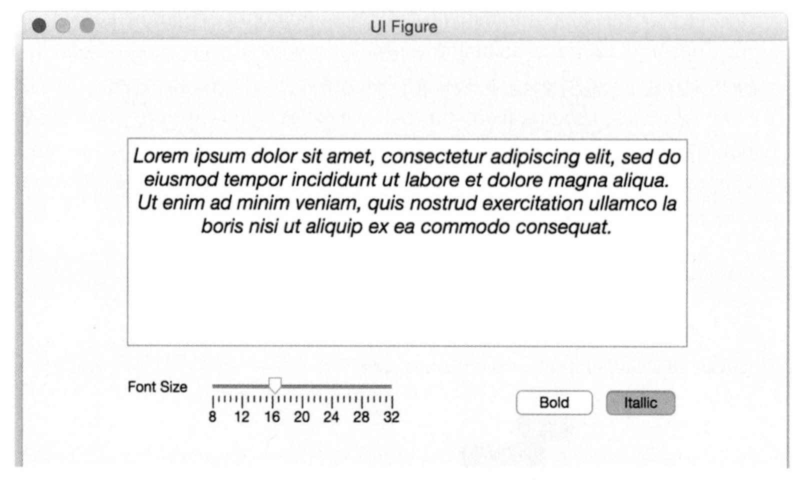

FIGURE 13.47
Text editor app.

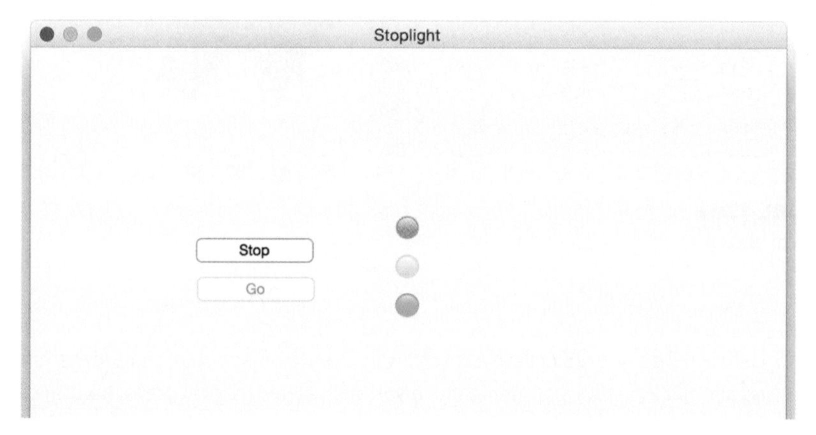

FIGURE 13.48
Stoplight app.

30. The following function *playsound* below plays one of the built-in sounds. The function has a cell array that stores the names. When the function is called, an integer is passed, which is an index into this cell array indicating the sound to be played. The default is 'train,' so if the user passes an invalid index, the default is used. The appropriate MAT-file is loaded. If the user passes a second argument, it is the frequency at which the sound should be played (otherwise, the default frequency is used). The function prints what sound is

about to be played and at which frequency, and then actually plays this sound. You are to fill in the rest of the following function. Here are examples of calling it (you can't hear it here, but the sound will be played!)

```
>> playsound(-4)
You are about to hear train at frequency 8192.0
>> playsound(2)
You are about to hear gong at frequency 8192.0
>> playsound(3,8000)
You are about to hear laughter at frequency 8000.0
```

playsound.m

```
function playsound(caind, varargin)
% This function plays a sound from a cell array
% of mat-file names
% Format playsound(index into cell array) or
%    playsound(index into cell array, frequency)
% Does not return any values

soundarray = {'chirp','gong','laughter','splat','train'};
if caind < 1 || caind > length(soundarray)
    caind = length(soundarray);
end
mysound = soundarray{caind};
eval(['load ' mysound])

% Fill in the rest
```

CHAPTER 14

Advanced Mathematics

KEY TERMS

curve fitting
best fit
symbolic mathematics
mean
sorting
index vectors
searching
arithmetic mean
average
outlier
harmonic mean
geometric mean
standard deviation
variance
mode
median
set operations
polynomials
degree
order
discrete
continuous
data sampling
interpolation
extrapolation
complex number
real part
imaginary part
purely imaginary
complex conjugate
magnitude
complex plane
linear algebraic equation
square matrix
main diagonal
diagonal matrix
trace
identity matrix
banded matrix
tridiagonal matrix
lower triangular matrix
upper triangular matrix
symmetric matrix
matrix inverse
matrix augmentation
coefficients
unknowns
determinant
Gauss-Jordan method
reduced row echelon form
integration
differentiation

CONTENTS

In this chapter, selected advanced mathematical concepts and related built-in functions in the MATLAB® software are introduced. This chapter will cover some simple statistics, as well as set operations that can be performed on data sets.

In many applications, data are sampled, which results in discrete data points. Fitting a curve to the data is often desired. ***Curve fitting*** is finding the curve that ***best fits*** the data. The first section in this chapter first explores fitting simple polynomial curves to data.

MATLAB®. http://dx.doi.org/10.1016/B978-0-12-804525-1.00014-3

Other topics include complex numbers and a brief introduction to differentiation and integration in calculus. ***Symbolic mathematics*** means doing mathematics on symbols. Some of the symbolic math functions, all of which are in Symbolic Math Toolbox™ in MATLAB, are also introduced. (Note that this is a Toolbox and, as a result, may not be available universally.)

Solutions to sets of linear algebraic equations are important in many applications. To solve systems of equations using MATLAB, there are basically two methods, both of which will be covered in this chapter: using a matrix representation and using the **solve** function (which is part of Symbolic Math Toolbox™).

14.1 STATISTICAL FUNCTIONS

There are a lot of statistical analyses that can be performed on data sets. In MATLAB, the statistical functions are in the data analysis help topic called **datafun**.

Statistics can be used to characterize properties of a data set. For example, consider a set of exam grades {33, 75, 77, 82, 83, 85, 85, 91, 100}. What is a "normal," "expected," or "average" exam grade? There are several ways that this could be interpreted. Perhaps the most common is the ***mean*** grade, which is found by summing the grades and dividing by the number of them (the result of that would be 79). Another way of interpreting that would be the grade found the most often, which would be 85. Also, the value in the middle of the sorted list, 83, could be used. Another property that is useful to know is how spread out the data values are within the data set.

MATLAB has built-in functions for many statistics; the simplest of which we have already seen (e.g., **min** and **max** to find the minimum or maximum value in a data set).

Both of these functions also return the index of the smallest or largest value; if there is more than one occurrence, it returns the first. For example, in the following data set 10 is the largest value; it is found in three elements in the vector, but the index returned is the first element in which it is found (which is 2):

```
>> x = [9  10  10  9  8  7  3  10  9  8  5  10];
>> [maxval, maxind] = max(x)
maxval =
    10
maxind =
     2
```

For matrices, the **min** and **max** functions operate columnwise by default:

```
>> mat = [9  10  17  5;  19  9  11  14]
mat =
     9    10    17     5
    19     9    11    14
```

```
>> [minval, minind] = min(mat)
minval =
     9     9    11     5

minind =
     1     2     2     1
```

These functions can also compare vectors or matrices (with the same dimensions) and return the minimum (or maximum) values from corresponding elements. For example, the following iterates through all elements in the two vectors, comparing corresponding elements, and returning the minimum for each:

```
>> x = [3 5 8 2 11];
>> y = [2 6 4 5 10];
>> min(x,y)
ans =
     2     5     4     2    10
```

Some of the other functions in the **datafun** help topic that have been described already include **sum**, **prod**, **cumsum**, **cumprod**, and **histogram**. Other statistical operations, and the functions that perform them in MATLAB, will be described in the rest of this section.

14.1.1 Mean

The ***arithmetic mean*** of a data set is what is usually called the ***average*** of the values or, in other words, the sum of the values divided by the number of values in the data set. Mathematically, we would write this as $\frac{\sum_{i=1}^{n} x_i}{n}$.

THE PROGRAMMING CONCEPT

Calculating a mean, or average, would normally be accomplished by looping through the elements of a vector, adding them together, and then dividing by the number of elements:

mymean.m

```
function outv = mymean(vec)
% mymean returns the mean of a vector
% Format: mymean(vector)

mysum = 0;
for i=1:length(vec)
    mysum = mysum + vec(i);
end
outv = mysum/length(vec);
end
```

```
>> x = [9  10  10  9  8  7  3  10  9  8  5  10];
>> mymean(x)
ans =
    8.1667
```

THE EFFICIENT METHOD

There is a built-in function **mean**, in MATLAB to accomplish this:

```
>> mean(x)
ans =
    8.1667
```

For a matrix, the **mean** function operates columnwise. To find the mean of each row, the dimension of 2 is passed as the second argument to the function, as is the case with the functions **sum**, **prod**, **cumsum**, and **cumprod** (the [] as a middle argument is not necessary for these functions like it is for **min** and **max**).

```
>> mat = [8  9  3; 10  2  3; 6  10  9]
mat =
     8     9     3
    10     2     3
     6    10     9
>> mean(mat)
ans =
     8     7     5
>> mean(mat,2)
ans =
    6.6667
    5.0000
    8.3333
```

Sometimes a value that is much larger or smaller than the rest of the data (called an ***outlier***) can throw off the mean. For example, in the following all of the numbers in the data set are in the range from 3 to 10, with the exception of the 100 in the middle. Because of this outlier, the mean of the values in this vector is actually larger than any of the other values in the vector.

```
>> xwithbig = [9  10  10  9  8  100  7  3  10  9  8  5  10];
>> mean(xwithbig)
ans =
   15.2308
```

Typically, an outlier like this represents an error of some kind, perhaps in the data collection. In order to handle this, sometimes the minimum and maximum values from a data set are discarded before the mean is computed. In this example, a **logical** vector indicating which elements are neither the largest nor smallest value is used to index into the original data set, resulting in removing the minimum and the maximum.

```
>> xwithbig = [9  10  10  9  8  100  7  3  10  9  8  5  10];
>> newx = xwithbig(xwithbig ~= min(xwithbig)  &  ...
                  xwithbig ~= max(xwithbig))
newx =
 9  10  10  9  8  7  10  9  8  5  10
```

Instead of just removing the minimum and maximum values, sometimes the largest and smallest 1% or 2% of values are removed, especially if the data set is very large.

There are several other means that can be computed. The ***harmonic mean*** of the n values in a vector or data set x is defined as

$$\frac{n}{\frac{1}{x_1}+\frac{1}{x_2}+\frac{1}{x_3}+\ldots+\frac{1}{x_n}}$$

The ***geometric mean*** of the n values in a vector x is defined as the nth root of the product of the data set values.

$$\sqrt[n]{x_1 * x_2 * x_3 \ldots * x_n}$$

Both of these could be implemented as anonymous functions:

```
>> x = [9  10  10  9  8  7  3  10  9  8  5  10];
>> harmhand = @ (x) length(x) / sum(1 ./ x);
>> harmhand(x)
ans =
    7.2310
>> geomhand = @ (x) nthroot(prod(x), length(x));
>> geomhand(x)
ans =
    7.7775
```

Note that Statistics Toolbox™ has functions for these means, called **harmmean** and **geomean**, as well as a function **trimmean** which trims the highest and lowest n% of data values, where the percentage n is specified as an argument.

14.1.2 Variance and Standard deviation

The ***standard deviation*** and ***variance*** are ways of determining the spread of the data. The variance is usually defined in terms of the arithmetic mean as:

$$\text{var} = \frac{\sum_{i=1}^{n} (x_i - mean)^2}{n-1}$$

Sometimes, the denominator is defined as n rather than $n-1$. The default definition in MATLAB uses $n-1$ for the denominator, so we will use that definition here.

For example, for the vector [8 7 5 4 6], there are $n=5$ values so $n-1$ is 4. Also, the mean of this data set is 6. The variance would be

$$\text{var} = \frac{(8-6)^2 + (7-6)^2 + (5-6)^2 + (4-6)^2 + (6-6)^2}{4}$$

$$= \frac{4+1+1+4+0}{4} = 2.5$$

The built-in function to calculate the variance is called **var**:

```
>> xvals = [8  7  5  4  6];
>> myvar = var(xvals)
yvar =
    2.5000
```

The standard deviation is the square root of the variance:

$$\text{sd} = \sqrt{\text{var}}$$

The built-in function in MATLAB for the standard deviation is called **std**; the standard deviation can be found either as the **sqrt** of the variance or using **std**:

```
>> shortx = [2 5 1 4];
>> myvar = var(shortx)
myvar =
    3.3333
>> sqrt(myvar)
ans =
   1.8257
>> std(shortx)
ans =
   1.8257
```

The less spread out the numbers are, the smaller the standard deviation will be, as it is a way of determining the spread of the data. Likewise, the more spread out the numbers are, the larger the standard deviation will be. For example, the following are two data sets that have the same number of values and also the same mean, but the standard deviations are quite different:

```
>> x1 = [9  10  9.4  9.6];
>> mean(x1)
ans =
    9.5000
>> std(x1)
ans =
   0.4163

>> x2 = [2  17  -1.5  20.5];
>> mean(x2)
ans =
   9.5000
>> std(x2)
ans =
  10.8704
```

14.1.3 Mode

The ***mode*** of a data set is the value that appears most frequently. The built-in function in MATLAB for this is called **mode**.

```
>> x = [9  10  10  9  8  7  3  10  9  8  5  10];
>> mode(x)
ans =
    10
```

If there is more than one value with the same (highest) frequency, the smaller value is the mode. In the following case, as 3 and 8 appear twice in the vector, the smaller value (3) is the mode:

```
>> x = [3  8  5  3  4  1  8];
>> mode(x)
ans =
     3
```

Therefore, if no value appears more frequently than any other, the mode of the vector will be the same as the minimum.

14.1.4 Median

The ***median*** is defined only for a data set that has been ***sorted*** first, meaning that the values are in order. The median of a sorted set of n data values is defined as the value in the middle, if n is odd, or the average of the two values in the middle if n is even. For example, for the vector [1 4 5 9 12], the middle value is 5. The function in MATLAB is called **median**:

```
>> median([1  4  5  9  12 ])
ans =
     5
```

For the vector [1 4 5 9 12 33], the median is the average of the 5 and 9 in the middle:

```
>> median([1  4  5  9  12  33])
ans =
     7
```

If the vector is not in sorted order to begin with, the **median** function will still return the correct result (it will sort the vector automatically). For example, scrambling the order of the values in the first example will still result in a median value of 5.

```
>> median([9  4  1  5  12])
ans =
     5
```

PRACTICE 14.1

For the vector [2 4 8 3 8], find the following

- minimum
- maximum
- arithmetic mean
- variance
- mode
- median

In MATLAB, find the harmonic mean and the geometric mean for this vector (either using **harmmean** and **geomean** if you have Statistics Toolbox, or by creating anonymous functions if you do not).

PRACTICE 14.2

For matrices, the statistical functions will operate on each column. Create a 5 × 4 matrix of random integers, each in the range from 1 to 30. Write an expression that will find the mode of all numbers in the matrix (not column-by-column).

14.2 SET OPERATIONS

MATLAB has several built-in functions that perform ***set operations*** on vectors. These include **union**, **intersect**, **unique**, **setdiff**, and **setxor**. All of these functions can be useful when working with data sets. By default, in earlier versions of MATLAB, all returned vectors were sorted from lowest to highest (ascending order). Beginning with MATLAB Version R2013a, however, these set functions provide the option of having the results in sorted order or in the original order. Additionally, there are two "is" functions that work on sets: **ismember** and **issorted**.

For example, given the following vectors:

```
>> v1 = 6:-1:2
     6     5     4     3     2
>> v2 = 1:2:7
v2 =
     1     3     5     7
```

the **union** function returns a vector that contains all of the values from the two input argument vectors, without repeating any.

```
>> union(v1,v2)
ans =
     1     2     3     4     5     6     7
```

By default, the result is in sorted order, so passing the arguments in the reverse order would not affect the result. This is the same as calling the function as:

```
>> union(v1,v2, 'sorted')
```

If, instead, the string 'stable' is passed to the function, the result would be in the original order; this means that the order of the arguments would affect the result.

```
>> union(v1,v2,'stable')
ans =
     6     5     4     3     2     1     7
>> union(v2,v1,'stable')
ans =
     1     3     5     7     6     4     2
```

The **intersect** function instead returns all of the values that can be found in both the two input argument vectors.

```
>> intersect(v1,v2)
ans =
     3     5
```

The **setdiff** function receives two vectors as input arguments, and returns a vector consisting of all of the values that are contained in the first vector argument but not the second. Therefore, the result that is returned (not just the order) will depend on the order of the two input arguments.

```
>> setdiff(v1,v2)
ans =
     2     4     6
>> setdiff(v2,v1)
ans =
     1     7
```

The function **setxor** receives two vectors as input arguments, and returns a vector consisting of all of the values from the two vectors that are not in the intersection of these two vectors. In other words, it is the union of the two vectors obtained using **setdiff** when passing the vectors in different orders, as seen before.

```
>> setxor(v1,v2)
ans =
     1     2     4     6     7
>> union(setdiff(v1,v2), setdiff(v2,v1))
ans =
     1     2     4     6     7
```

The set function **unique** returns all of the unique values from a set argument:

```
>> v3 = [1:5  3:6]
v3 =
     1     2     3     4     5     3     4     5     6
>> unique(v3)
ans =
     1     2     3     4     5     6
```

All of these functions—**union**, **intersect**, **unique**, **setdiff**, and **setxor**—can be called with 'stable' to have the result returned in the order given by the original vector(s).

Many of the set functions return vectors that can be used to index into the original vectors as optional output arguments. However, be careful with this: the resulting index vectors will be changed in a future version of MATLAB (one change is that they will be returned as column vectors).

For example, the two vectors *v1* and *v2* were defined previously as:

```
>> v1
v1 =
      6      5      4      3      2
>> v2
v2 =
      1      3      5      7
```

The **intersect** function returns, in addition to the vector containing the values in the intersection of *v1* and *v2*, an index vector into *v1*, and an index vector into *v2* such that *outvec* is the same as *v1(index1)* and also *v2(index2)*.

```
>> [outvec, index1, index2] = intersect(v1,v2)
outvec =
       3      5

index1 =
       4      2

index2 =
       2      3
```

Using these vectors to index into *v1* and *v2* will return the values from the intersection. For example, this expression returns the second and fourth elements of *v1* (it puts them in ascending order):

```
>> v1(index1)
ans =
      3      5
```

This returns the second and third elements of *v2*:

```
>> v2(index2)
ans =
      3      5
```

The function **ismember** receives two vectors as input arguments and returns a **logical** vector that is of the same length as the first argument, containing **logical** 1 for **true** if the element in the first vector is also in the second, or **logical** 0 for **false** if not. The order of the arguments matters for this function.

```
>> v1
v1 =
       6      5      4      3      2
```

```
>> v2
v2 =
     1     3     5     7
>> ismember(v1,v2)
ans =
     0     1     0     1     0
>> ismember(v2,v1)
ans =
     0     1     1     0
```

Using the result from the **ismember** function as an index into the first vector argument will return the same values as the **intersect** function (although not necessarily sorted).

```
>> logv = ismember(v1,v2)
logv =
     0     1     0     1     0
>> v1(logv)
ans =
     5     3
>> logv = ismember(v2,v1)
logv =
     0     1     1     0
>> v2(logv)
ans =
     3     5
```

The **issorted** function will return **logical** 1 for **true** if the argument is sorted in ascending order, or **logical** 0 for **false** if not.

```
>> v3 = [1:5  3:6]
v3 =
     1     2     3     4     5     3     4     5     6
>> issorted(v3)
ans =
     0
>> issorted(v2)
ans =
     1
```

PRACTICE 14.3

Create two vector variables *vec1* and *vec2* that contain five random integers, each in the range from 1 to 20. Do each of the following operations by hand first and then check in MATLAB (if you have one of the latest versions, do this with both 'stable' and 'sorted'):

- union
- intersection
- setdiff
- setxor
- unique (for each)

14.3 FITTING CURVES TO DATA

MATLAB has several curve-fitting functions; Curve Fitting Toolbox™ has many more of these functions. Some of the simplest curves are polynomials of different degrees, which are described next.

14.3.1 Polynomials

Simple curves are ***polynomials*** of different ***degrees*** or ***orders***. The degree is the integer of the highest exponent in the expression. For example:

- a straight line is a first-order (or degree 1) polynomial of the form `ax + b`, or, more explicitly, ax^1+b
- a quadratic is a second-order (or degree 2) polynomial of the form ax^2+bx+c
- a cubic (degree 3) is of the form ax^3+bx^2+cx+d

MATLAB represents a polynomial as a row vector of coefficients. For example, the polynomial x^3+2x^2-4x+3 would be represented by the vector `[1 2 -4 3]`. The polynomial $2x^4-x^2+5$ would be represented by `[2 0 -1 0 5]`; note the zero terms for x^3 and x^1.

The **roots** function in MATLAB can be used to find the roots of an equation represented by a polynomial. For example, for the mathematical function:

$$f(x) = 4x^3 - 2x^2 - 8x + 3$$

to solve the equation $f(x)=0$:

```
>> roots([4 -2 -8 3])
ans =
  -1.3660
   1.5000
   0.3660
```

The function **polyval** will evaluate a polynomial *p* at *x*; the form is **polyval(p,x)**. For example, the polynomial $-2x^2+x+4$ is evaluated at `x = 3`, which yields

$-2 * 3^2+3+4$, or -11:

```
>> p = [-2  1  4];
>> polyval(p,3)
ans =
   -11
```

The argument *x* can be a vector:

```
>> polyval(p,1:3)
ans =
     3    -2    -11
```

14.3.2 Curve Fitting

Data that we acquire to analyze can be either ***discrete*** (e.g., a set of object weights) or ***continuous***. In many applications, continuous properties are ***sampled***, such as:

- The temperature recorded every hour
- The speed of a car recorded every one-tenth of a mile
- The mass of a radioactive material recorded every second as it decays
- Audio from a sound wave as it is converted to a digital audio file

Sampling provides data in the form of (x,y) points, which could then be plotted. For example, let us say the temperature was recorded every hour one afternoon from 2:00 pm to 6:00 pm; the vectors might be:

```
>> x = 2:6;
>> y = [65   67   72   71   63];
```

14.3.3 Interpolation and Extrapolation

In many cases, estimating values other than at the sampled data points is desired. For example, we might want to estimate what the temperature was at 2:30 pm, or at 1:00 pm. ***Interpolation*** means estimating the values in between recorded data points. ***Extrapolation*** is estimating outside of the bounds of the recorded data. One way to do this is to fit a curve to the data and use this for the estimations. Curve fitting is finding the curve that "best fits" the data.

Simple curves are polynomials of different degrees, as described previously. Thus, curve fitting involves finding the best polynomials to fit the data; for example, for a quadratic polynomial in the form ax^2+bx+c, it means finding the values of a, b, and c that yield the best fit. Finding the best straight line that goes through data would mean finding the values of a and b in the equation $ax + b$.

MATLAB has a function to do this called **polyfit**. The function **polyfit** finds the coefficients of the polynomial of the specified degree that best fits the data using a least squares algorithm. There are three arguments passed to the function: x and y the vectors that represent the data, and the degree of the desired polynomial. For example, to fit a straight line (degree 1) through the points representing temperatures, the call to the **polyfit** function would be

```
>> polyfit(x,y,1)
ans =
    0.0000    67.6000
```

which says that the best straight line is of the form $0x + 67.6$.

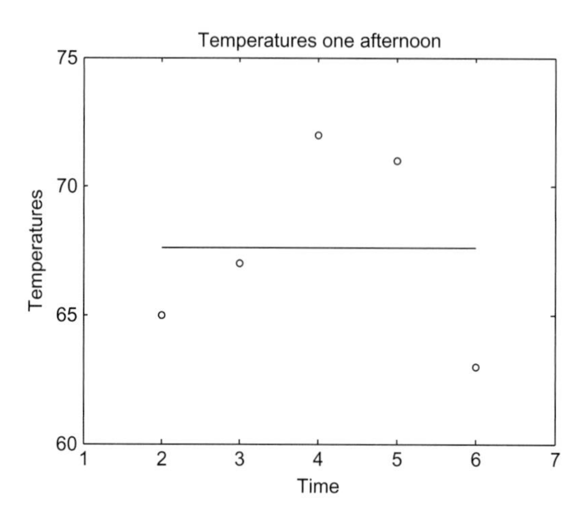

FIGURE 14.1
Sampled temperatures with straight line fit.

However, from the plot (shown in Fig. 14.1), it looks like a quadratic would be a much better fit. The following would create the vectors and then fit a polynomial of degree 2 through the data points, storing the values in a vector called *coefs*.

```
>> x = 2:6;
>> y = [65   67   72   71   63];
>> coefs = polyfit(x,y,2)
coefs =
   -1.8571   14.8571   41.6000
```

This says that the **polyfit** function has determined that the best quadratic that fits these data points is $-1.8571x^2+14.8571x+41.6$. So, the variable *coefs* now stores a coefficient vector that represents this polynomial.

The function **polyval** can then be used to evaluate the polynomial at specified values. For example, we could evaluate at every value in the *x* vector:

```
>> curve = polyval(coefs,x)
curve =
   63.8857   69.4571   71.3153   69.4571   63.8857
```

This results in *y* values for each point in the *x* vector and stores them in a vector called *curve*. Putting all of this together, the following script called *polytemp* creates the *x* and *y* vectors, fits a second-order polynomial through these points, and plots both the points and the curve on the same figure. Running this results in the plot seen in Fig. 14.2. The curve doesn't look very smooth on this plot, but that is because there are only five points in the *x* vector.

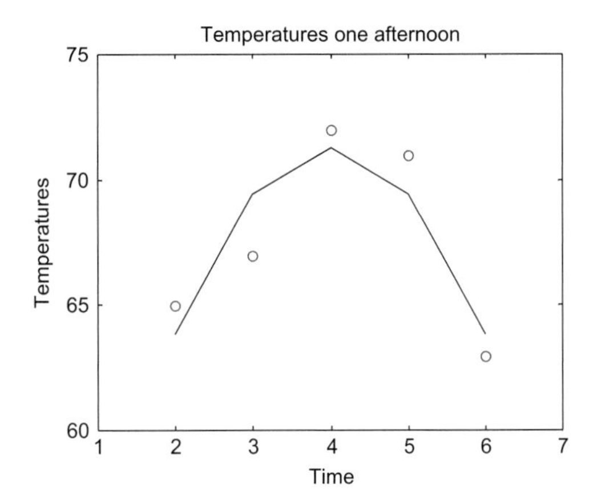

FIGURE 14.2
Sampled temperatures with quadratic curve.

polytemp.m

```
% Fits a quadratic curve to temperature data
x= 2:6;
y =[65   67   72   71   63];
coefs = polyfit(x,y,2);
curve = polyval(coefs,x);
plot(x,y,'ro',x,curve)
xlabel('Time')
ylabel('Temperatures')
title('Temperatures one afternoon')
axis([1 7 60 75])
```

PRACTICE 14.4

To make the curve smoother, modify the script *polytemp* to create a new *x* vector with more points for plotting the curve. Note that the original *x* vector for the data points must remain as is.

To estimate the temperature at different times, **polyval** can be used for discrete *x* points; it does not have to be used with the entire *x* vector. For example, to interpolate between the given data points and estimate what the temperature was at 2:30 pm, 2.5 would be used.

```
>> polyval(coefs,2.5)
ans =
    67.1357
```

Also, **polyval** can be used to extrapolate beyond the given data points. For example, to estimate the temperature at 1:00 pm:

```
>> polyval(coefs,1)
ans =
    54.6000
```

The better the curve fit, the more accurate these interpolated and extrapolated values will be.

Using the **subplot** function, we can loop to show the difference between fitting curves of degrees 1, 2, and 3 to some data. For example, the following script will accomplish this for the temperature data. (Note that the variable *morex* stores 100 points so the graph will be smooth.)

polytempsubplot.m

```
% Fits curves of degrees 1-3 to temperature
%  data and plots in a subplot
x = 2:6;
y = [65  67  72  71  63];
morex = linspace(min(x),max(x));
for pd = 1:3
    coefs = polyfit(x,y,pd);
    curve = polyval(coefs,morex);
    subplot(1,3,pd)
    plot(x,y,'ro',morex,curve)
    xlabel('Time')
    ylabel('Temperatures')
    title(sprintf('Degree %d',pd))
    axis([1  7  60  75])
end
```

Executing the script

```
>> polytempsubplot
```

creates the Figure Window shown in Fig. 14.3.

14.4 COMPLEX NUMBERS

A ***complex number*** is generally written in the form

```
z = a + bi
```

where a is called the ***real part*** of the number z, b is the ***imaginary part*** of z and i is $\sqrt{-1}$. A complex number is ***purely imaginary*** if it is of the form $z=bi$ (in other words if a is 0).

Note

This is the way mathematicians usually write a complex number; in engineering it is often written as $a+bj$, where j is $\sqrt{-1}$.

We have seen that in MATLAB both **i** and **j** are built-in functions that return $\sqrt{-1}$ (so, they can be thought of as built-in constants). Complex numbers

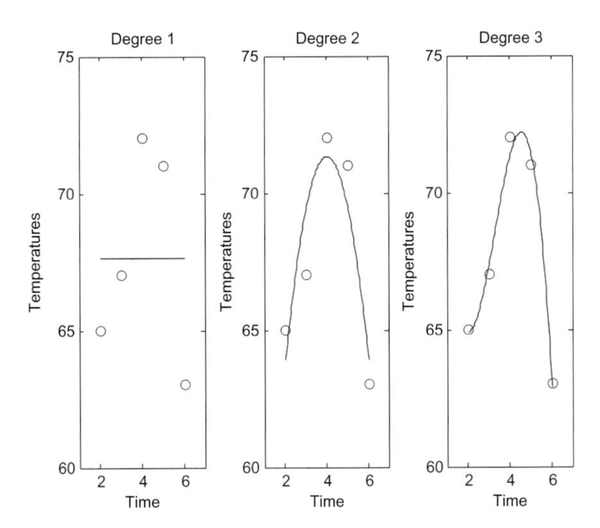

FIGURE 14.3
Subplot to show temperatures with curves of degrees 1, 2, and 3.

can be created using **i** or **j**, such as "5+2i" or "3−4j." The multiplication operator is not required between the value of the imaginary part and the constant **i** or **j**.

QUICK QUESTION!

Is the value of the expression "3*i*" the same as "3**i*"?

Answer: It depends on whether *i* has been used as a variable name or not. If *i* has been used as a variable (e.g., an iterator variable in a **for** loop), then the expression "3**i*" will use the defined value for the variable and the result will not be a complex number. The expression "3*i*" will always be complex. Therefore, it is a good idea when working with complex numbers to use 1**i** or 1**j** rather than just **i** or **j**. The expressions 1**i** and 1**j** always result in a complex number, regardless of whether *i* and *j* have been used as variables. So, use "3*1*i*" or "3*i*."

```
>> i = 5;
>> i
i =
     5
>> 1i
ans =
      0 + 1.0000i
```

MATLAB also has a function **complex** that will return a complex number. It receives two numbers, the real and imaginary parts in that order, or just one number, which is the real part (in which case the imaginary part would be 0). Here are some examples of creating complex numbers in MATLAB:

```
>> z1 = 4 + 2i
z1 =
   4.0000 + 2.0000i

>> z2 = sqrt(-5)
z2 =
   0 + 2.2361i

>> z3 = complex(3,-3)
z3 =
   3.0000 - 3.0000i

>> z4 = 2 + 3j
z4 =
   2.0000 + 3.0000i

>> z5 = (-4) ^ (1/2)
ans =
    0.0000 + 2.0000i

>> myz = input('Enter a complex number: ')
Enter a complex number: 3 + 4i
myz =
    3.0000 + 4.0000i
```

Note that even when **j** is used in an expression, **i** is used in the result. MATLAB shows the type of the variables created here in the Workspace Window (or using **whos**) as **double (complex)**. MATLAB has functions **real** and **imag** that return the real and imaginary parts of complex numbers.

```
>> real(z1)
ans =
    4

>> imag(z3)
ans =
   -3
```

In order to print an imaginary number, the **disp** function will display both parts automatically:

```
>> disp(z1)
   4.0000 + 2.0000i
```

The **fprintf** function will only print the real part unless both parts are printed separately:

```
>> fprintf('%f\n', z1)
4.000000

>> fprintf('%f + %fi\n', real(z1), imag(z1))
4.000000 + 2.000000i
```

The function **isreal** returns **logical** 1 for **true** if there is no imaginary part of the argument or **logical** 0 for **false** if the argument does have an imaginary part (even if it is 0). For example,

```
>> isreal(z1)
ans =
     0

>> z6 = complex(3)
z5 =
    3
>> isreal(z6)
ans =
    0

>> isreal(3.3)
ans =
    1
```

For the preceding variable *z6*, even though it shows the answer as 3, it is really stored as 3 + 0i, and that is how it is displayed in the Workspace Window. Therefore, **isreal** returns **logical false** as it is stored as a complex number.

14.4.1 Equality for Complex Numbers

Two complex numbers are equal if both their real parts and imaginary parts are equal. In MATLAB, the equality operator can be used.

```
>> z1 == z2
ans =
    0

>> complex(0,4) == sqrt(-16)
ans =
    1
```

14.4.2 Adding and Subtracting Complex Numbers

For two complex numbers z1 = a + bi and z2 = c + di,

```
z1 + z2 = (a + c) + (b + d)i

z1 - z2 = (a - c) + (b - d)i
```

As an example, we will write a function in MATLAB to add two complex numbers together and return the resulting complex number.

THE PROGRAMMING CONCEPT

In most cases, to add two complex numbers together you would have to separate the real and imaginary parts, and add them to return your result.

addcomp.m

```
function outc = addcomp(z1, z2)
% addcomp adds two complex numbers z1 and z2 &
%    returns the result
% Adds the real and imaginary parts separately
% Format: addcomp(z1,z2)

realpart = real(z1) + real(z2);
imagpart = imag(z1) + imag(z2);
outc = realpart + imagpart * 1i;
end
```

```
>> addcomp(3+4i, 2-3i)
ans =
    5.0000 + 1.0000i
```

THE EFFICIENT METHOD

MATLAB does this automatically to add two complex numbers together (or subtract).

```
>> z1 = 3 + 4i;
>> z2 = 2 - 3i;
>> z1+z2
ans =
    5.0000 + 1.0000i
```

14.4.3 Multiplying Complex Numbers

For two complex numbers z1 = a + bi and z2 = c + di,

```
z1 * z2 = (a + bi) * (c + di)
        = a*c + a*di + c*bi + bi*di
        = a*c + a*di + c*bi – b*d
        = (a*c – b*d) + (a*d + c*b)i
```

For example, for the complex numbers

```
z1 = 3 + 4i
z2 = 1 - 2i
```

the result of the multiplication would be defined mathematically as

```
z1 * z2 = (3*1 - -8) + (3*-2 + 4*1)i = 11 -2i
```

This is, of course, automatic in MATLAB:

```
>> z1*z2
ans =
   11.0000 - 2.0000i
```

14.4.4 Complex Conjugate and Absolute Value

The ***complex conjugate*** of a complex number $z = a + bi$ is $\bar{z} = a - bi$. The ***magnitude*** or absolute value of a complex number z is $|z| = \sqrt{a^2 + b^2}$. In MATLAB, there is a built-in function **conj** for the complex conjugate, and the **abs** function returns the absolute value.

```
>> z1 = 3 + 4i
z1 =
    3.0000 + 4.0000i

>> conj(z1)
ans =
    3.0000 - 4.0000i

>> abs(z1)
ans =
      5
```

14.4.5 Complex Equations Represented as Polynomials

We have seen that MATLAB represents a polynomial as a row vector of coefficients; this can be used when the expressions or equations involve complex numbers, also. For example, the polynomial $z^2 + z - 3 + 2i$ would be represented by the vector `[1  1  -3+2i]`. The **roots** function in MATLAB can be used to find the roots of an equation represented by a polynomial. For example, to solve the equation $z^2 + z - 3 + 2i = 0$:

```
>> roots([1 1 -3+2i])
ans =
   -2.3796 + 0.5320i
    1.3796 - 0.5320i
```

The **polyval** function can also be used with this polynomial, for example

```
>> cp = [1 1 -3+2i]
cp =
   1.0000          1.0000          -3.0000 + 2.0000i

>> polyval(cp,3)
ans =
    9.0000 + 2.0000i
```

14.4.6 Polar Form

Any complex number z = a + bi can be thought of as a point (a,b) or vector in a ***complex plane*** in which the horizontal axis is the real part of z, and the vertical axis is the imaginary part of z. So, *a* and *b* are the Cartesian or rectangular coordinates. As a vector can be represented by either its rectangular or polar coordinates, a complex number can also be given by its polar coordinates *r* and θ, where *r* is the magnitude of the vector and θ is an angle.

To convert from the polar coordinates to the rectangular coordinates:

$$a = r\cos\theta$$
$$b = r\sin\theta$$

To convert from the rectangular to polar coordinates:

$$r = |z| = \sqrt{a^2 + b^2}$$

$$\theta = \arctan\left(\frac{b}{a}\right)$$

So, a complex number z = a + bi can be written as $r\cos\theta + (r\sin\theta)i$ or

$$z = r(\cos\theta + i\sin\theta)$$

As $e^{i\theta} = \cos\theta + i\sin\theta$, a complex number can also be written as $z = re^{i\theta}$. In MATLAB, *r* can be found using the **abs** function, while there is a built-in function called **angle** to find θ.

```
>> z1 = 3 + 4i;
r = abs(z1)
r =
     5
>> theta = angle(z1)
theta =
     0.9273
>> r*exp(i*theta)
ans =
   3.0000 + 4.0000i
```

14.4.7 Plotting

Several methods are used commonly for plotting complex data:

- plot the real parts versus the imaginary parts using **plot**
- plot only the real parts using **plot**
- plot the real and the imaginary parts in one figure with a legend, using **plot**
- plot the magnitude and angle using **polarplot** (or, prior to R2016a, **polar**)

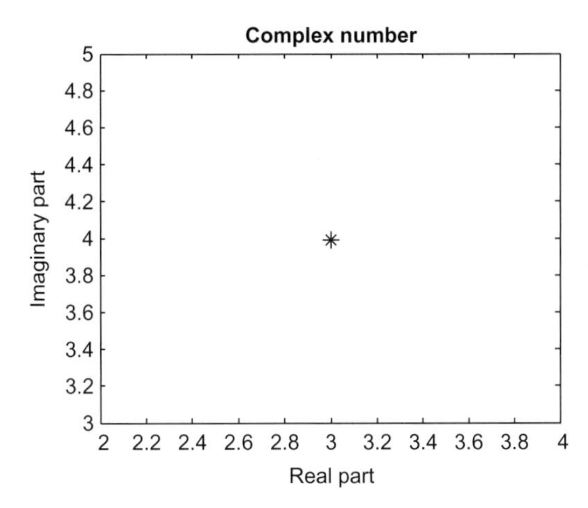

FIGURE 14.4
Plot of complex number.

Using the **plot** function with a single complex number or a vector of complex numbers will result in plotting the real parts versus the imaginary parts, for example **plot(z)** is the same as **plot(real(z), imag(z))**. Thus, for the complex number `z1 = 3 + 4i`, this will plot the point (3,4) (using a large asterisk so we can see it!), as shown in Fig. 14.4.

```
>> z1 = 3 + 4i;
>> plot(z1,'*', 'MarkerSize', 12)
>> xlabel('Real part')
>> ylabel('Imaginary part')
>> title('Complex number')
```

PRACTICE 14.5

Create the following complex variables

```
c1 = complex(0,2);
c2 = 3 + 2i;
c3 = sqrt(-4);
```

Then, carry out the following:

- Get the real and imaginary parts of *c2*.
- Print the value of *c1* using **disp**.
- Print the value of *c2* in the form 'a+bi'.
- Determine whether any of the variables are equal to each other.
- Subtract *c2* from *c1*.
- Multiply *c2* by *c3*.
- Get the complex conjugate and magnitude of *c2*.
- Put *c1* in polar form.
- Plot the real part versus the imaginary part for *c2*.

14.5 MATRIX SOLUTIONS TO SYSTEMS OF LINEAR ALGEBRAIC EQUATIONS

A ***linear algebraic equation*** is an equation of the form

$$a_1x_1 + a_2x_2 + a_3x_3 + \ldots + a_nx_n = b$$

Solutions to sets of equations in this form are important in many applications. In the MATLAB® product, to solve systems of equations, there are basically two methods:

- using a matrix representation
- using the **solve** function (which is part of Symbolic Math Toolbox™)

In this section, we will first investigate some relevant matrix properties and then use these to solve linear algebraic equations. The use of ***symbolic mathematics*** including the **solve** function will be covered in the next section.

14.5.1 Matrix Properties

In Chapter 2 we saw several common operations on matrices. In this section we will examine some properties that will be useful in solving equations using a matrix form.

14.5.1.1 Square Matrices

If a matrix has the same number of rows and columns (e.g., if $m == n$), the matrix is ***square***. The definitions that follow in this section only apply to square matrices.

The ***main diagonal*** of a square matrix (sometimes called just the ***diagonal***) is the set of terms a_{ii} for which the row and column indices are the same, so from the upper left element to the lower right. For example, for the following matrix the diagonal consists of 1, 6, 11, and 16.

$$\begin{bmatrix} 1 & 2 & 3 & 4 \\ 5 & 6 & 7 & 8 \\ 9 & 10 & 11 & 12 \\ 13 & 14 & 15 & 16 \end{bmatrix}$$

A square matrix is a ***diagonal matrix*** if all values that are not on the diagonal are 0. The numbers on the diagonal, however, do not have to be all nonzero, although frequently they are. Mathematically, this is written as $a_{ij} = 0$ for $i \sim= j$. The following is an example of a diagonal matrix:

$$\begin{bmatrix} 4 & 0 & 0 \\ 0 & 9 & 0 \\ 0 & 0 & 5 \end{bmatrix}$$

MATLAB has a function **diag** that will return the diagonal of a matrix as a column vector; transposing will result in a row vector instead.

```
>> mymat = reshape(1:16,4,4)'
mymat =
     1     2     3     4
     5     6     7     8
     9    10    11    12
    13    14    15    16
>> diag(mymat)'
ans =
     1     6    11    16
```

The **diag** function can also be used to take a vector of length n and create an $n \times n$ square diagonal matrix with the values from the vector on the diagonal:

```
>> v=1:4;
>> diag(v)
ans =
     1     0     0     0
     0     2     0     0
     0     0     3     0
     0     0     0     4
```

So, the **diag** function can be used in two ways: (i) pass a matrix and it returns a vector, or (ii) pass a vector and it returns a matrix!

The ***trace*** of a square matrix is the sum of all of the elements on the diagonal. For example, for the diagonal matrix created using v it is $1+2+3+4$, or 10.

QUICK QUESTION!

How could we calculate the trace of a square matrix?

Answer: See the following Programming Concept and Efficient Method.

THE PROGRAMMING CONCEPT

To calculate the trace of a square matrix, only one loop is necessary as the only elements in the matrix we are referring to have subscripts (i, i). So, once the size has been determined, the loop variable can iterate from 1 through the number of rows or from 1 through the number of columns (it doesn't matter which, as they have the same value!). The following function calculates and returns the trace of a square matrix or an empty vector if the matrix argument is not square.

Continued

THE PROGRAMMING CONCEPT—CONT'D

mytrace.m

```
function outsum = mytrace(mymat)
% mytrace calculates the trace of a square matrix
% or an empty vector if the matrix is not square
% Format: mytrace(matrix)

[r, c] = size(mymat);
if r ~= c
    outsum = [];
else
    outsum = 0;
    for i = 1:r
        outsum = outsum + mymat(i,i);
    end
end
end
```

```
>> mymat = reshape(1:16,4,4)'
mymat =
     1     2     3     4
     5     6     7     8
     9    10    11    12
    13    14    15    16
>> mytrace(mymat)
ans =
    34
```

THE EFFICIENT METHOD

In MATLAB, there is a built-in function **trace** to calculate the trace of a square matrix:

```
>> trace(mymat)
ans =
    34
```

A square matrix is an ***identity*** matrix called I if $a_{ij} = 1$ for $i == j$ and $a_{ij} = 0$ for $i \sim= j$. In other words, all of the numbers on the diagonal are 1 and all others are 0. The following is a 3×3 identity matrix:

$$\begin{bmatrix} 1 & 0 & 0 \\ 0 & 1 & 0 \\ 0 & 0 & 1 \end{bmatrix}$$

Note that any identity matrix is a special case of a diagonal matrix.

Identity matrices are very important and useful. MATLAB has a built-in function **eye** that will create an $n \times n$ identity matrix, given the value of n:

```
>> eye(5)
ans =
     1     0     0     0     0
     0     1     0     0     0
     0     0     1     0     0
     0     0     0     1     0
     0     0     0     0     1
```

Note that **i** is built into MATLAB as the square root of -1, so another name is used for the function that creates an identity matrix: **eye**, which sounds like "*i*" (… get it?)

QUICK QUESTION!

What happens if a matrix *M* is multiplied by an identity matrix (of the appropriate size)?

Answer: For the size to be appropriate, the dimensions of the identity matrix would be the same as the number of columns of *M*. The result of the multiplication will always be the original matrix *M* (thus, it is similar to multiplying a scalar by 1).

```
>> M = [1 2 3 1; 4 5 1 2; 0 2 3 0]
M =
     1     2     3     1
     4     5     1     2
     0     2     3     0

>> [r, c] = size(M);
>> M * eye(c)
ans =
     1     2     3     1
     4     5     1     2
     0     2     3     0
```

Several special cases of matrices are related to diagonal matrices.

A ***banded matrix*** is a matrix of all 0s, with the exception of the main diagonal and other diagonals next to (above and below) the main. For example, the following matrix has 0s except for the band of three diagonals; this is a particular kind of banded matrix called a ***tridiagonal matrix***.

$$\begin{bmatrix} 1 & 2 & 0 & 0 \\ 5 & 6 & 7 & 0 \\ 0 & 10 & 11 & 12 \\ 0 & 0 & 15 & 16 \end{bmatrix}$$

A ***lower triangular matrix*** has all 0s above the main diagonal. For example,

$$\begin{bmatrix} 1 & 0 & 0 & 0 \\ 5 & 6 & 0 & 0 \\ 9 & 10 & 11 & 0 \\ 13 & 14 & 15 & 16 \end{bmatrix}$$

An ***upper triangular matrix*** has all 0s below the main diagonal. For example,

$$\begin{bmatrix} 1 & 2 & 3 & 4 \\ 0 & 6 & 7 & 8 \\ 0 & 0 & 11 & 12 \\ 0 & 0 & 0 & 16 \end{bmatrix}$$

It is possible for there to be 0s on the diagonal and in the upper part or lower part and still be a lower or upper triangular matrix, respectively.

MATLAB has functions **triu** and **tril** that will take a matrix and make it into an upper triangular or lower triangular matrix by replacing the appropriate elements with 0s. For example, the results from the **triu** function are shown:

```
>> mymat
mymat =
      1     2     3     4
      5     6     7     8
      9    10    11    12
     13    14    15    16
>> triu(mymat)
ans =
      1     2     3     4
      0     6     7     8
      0     0    11    12
      0     0     0    16
```

A square matrix is ***symmetric*** if $a_{ij} = a_{ji}$ for all i, j. In other words, all of the values opposite the diagonal from each other must be equal. In this example, there are three pairs of values opposite the diagonals, all of which are equal (the 2s, the 9s, and the 4s).

$$\begin{bmatrix} 1 & 2 & 9 \\ 2 & 5 & 4 \\ 9 & 4 & 6 \end{bmatrix}$$

PRACTICE 14.6

For the following matrices:

$$A = \begin{bmatrix} 4 & 3 \\ 3 & 2 \end{bmatrix} \quad B = \begin{bmatrix} 1 & 2 & 3 \\ 4 & 5 & 6 \end{bmatrix} \quad C = \begin{bmatrix} 1 & 0 & 0 \\ 4 & 6 & 0 \\ 3 & 1 & 3 \end{bmatrix}$$

Which are equal?

Which are square?

For all square matrices:
- Calculate the trace.
- Which are symmetric?
- Which are diagonal?
- Which are lower triangular?
- Which are upper triangular?

MATLAB has several "is" functions that determine whether or not matrices have some of the properties explained in this section, for example, **isdiag**, **issymmetric**, **istril**, **istriu**, **isbanded**; these functions were introduced in R2014b.

14.5.1.2 Matrix Operations

There are several common operations on matrices, some of which we have seen already. These include matrix transpose, matrix augmentation, and matrix inverse.

A matrix transpose interchanges the rows and columns of a matrix. For a matrix A, its transpose is written A^T in mathematics. For example, if

$$A = \begin{bmatrix} 1 & 2 & 3 \\ 4 & 5 & 6 \end{bmatrix}$$

then

$$A^T = \begin{bmatrix} 1 & 4 \\ 2 & 5 \\ 3 & 6 \end{bmatrix}$$

In MATLAB, as we have seen, there is a built-in transpose operator, the apostrophe.

If the result of multiplying a matrix A by another matrix is the identity matrix I, then the second matrix is the ***inverse*** of matrix A. The inverse of a matrix A is written as A^{-1}, so

$$AA^{-1} = I$$

How to actually compute the inverse A^{-1} of a matrix by hand is not so easy. MATLAB, however, has a function **inv** to compute a matrix inverse. For example, next a matrix is created, its inverse is found, and then multiplied by the original matrix to verify that the product is in fact the identity matrix:

```
>> a = [1 2; 2 2]
a =
     1    2
     2    2
>> ainv = inv(a)
ainv =
   -1.0000    1.0000
    1.0000   -0.5000
>> a*ainv
ans =
     1    0
     0    1
```

Matrix augmentation means adding column(s) to the original matrix. In MATLAB, matrix augmentation can be accomplished using square brackets to concatenate the two matrices. The square matrix A is concatenated with an identity matrix which has the same size as the matrix A:

```
>> A = [1 3 7; 2 5 4; 9 8 6]
A =
      1    3    7
      2    5    4
      9    8    6

>> [A eye(size(A))]
ans =
      1    3    7    1    0    0
      2    5    4    0    1    0
      9    8    6    0    0    1
```

14.5.2 Linear Algebraic Equations

A ***linear algebraic equation*** is an equation of the form

$$a_1x_1 + a_2x_2 + a_3x_3 + \ldots + a_nx_n = b$$

where the *as* are constant ***coefficients***, the *xs* are the ***unknowns***, and *b* is a constant. A solution is a sequence of numbers that satisfies the equation. For example,

$$4x_1 + 5x_2 - 2x_3 = 16$$

is such an equation in which there are three unknowns: x_1, x_2, and x_3. One solution to this equation is $x_1 = 3$, $x_2 = 4$, and $x_3 = 8$, as 4*3 + 5*4 – 2*8 is equal to 16.

A system of linear algebraic equations is a set of equations of the form:

$$\begin{array}{l}
a_{11}x_1 + a_{12}x_2 + a_{13}x_3 + \ldots. + a_{1n}x_n = b_1 \\
a_{21}x_1 + a_{22}x_2 + a_{23}x_3 + \ldots. + a_{2n}x_n = b_2 \\
a_{31}x_1 + a_{32}x_2 + a_{33}x_3 + \ldots. + a_{3n}x_n = b_3 \\
\quad\vdots \qquad\quad \vdots \qquad\quad \vdots \qquad\qquad\qquad \vdots \qquad \vdots \\
a_{m1}x_1 + a_{m2}x_2 + a_{m3}x_3 + \ldots. + a_{mn}x_n = b_m
\end{array}$$

This is called an $m \times n$ system of equations; there are m equations and n unknowns.

Because of the way that matrix multiplication works, these equations can be represented in matrix form as A x=b where A is a matrix of the coefficients, x is a column vector of the unknowns, and b is a column vector of the constants from the right side of the equations:

$$\overset{A}{\begin{bmatrix} a_{11} & a_{12} & a_{13} & \cdots & a_{1n} \\ a_{21} & a_{22} & a_{23} & \cdots & a_{2n} \\ a_{31} & a_{32} & a_{33} & \cdots & a_{3n} \\ \vdots & \vdots & \vdots & \vdots & \vdots \\ a_{m1} & a_{m2} & a_{m3} & \cdots & a_{mn} \end{bmatrix}} \overset{x}{\begin{bmatrix} x_1 \\ x_2 \\ x_3 \\ \vdots \\ x_n \end{bmatrix}} \overset{=}{=} \overset{b}{\begin{bmatrix} b_1 \\ b_2 \\ b_3 \\ \vdots \\ b_m \end{bmatrix}}$$

Once the system of equations has been written in matrix form, what we want is to solve the equation `Ax = b` for the unknowns x. To do this, we need to isolate x on one side of the equation. If we were working with scalars, we would divide both sides of the equation by A. In fact, with MATLAB we can use the **divided into** operator to do this. However, most languages cannot do this with matrices, so, instead, we multiply both sides of the equation by the inverse of the coefficient matrix A:

$$A^{-1}A\,x = A^{-1}b$$

Then, because multiplying a matrix by its inverse results in the identity matrix I, and because multiplying any matrix by I results in the original matrix, we have:

$$I\,x = A^{-1}b$$

or

$$x = A^{-1}b$$

For example, consider the following three equations with three unknowns—x_1, x_2, and x_3:

$$\begin{aligned} 4x_1 - 2x_2 + 1x_3 &= 7 \\ 1x_1 + 1x_2 + 5x_3 &= 10 \\ -2x_1 + 3x_2 - 1x_3 &= 2 \end{aligned}$$

We write this in the form $Ax=b$, where A is a matrix of the coefficients, x is a column vector of the unknowns x_i, and b is a column vector of the values on the right side of the equations:

$$\overset{A}{\begin{bmatrix} 4 & -2 & 1 \\ 1 & 1 & 5 \\ -2 & 3 & -1 \end{bmatrix}} \overset{x}{\begin{bmatrix} x_1 \\ x_2 \\ x_3 \end{bmatrix}} = \overset{b}{\begin{bmatrix} 7 \\ 10 \\ 2 \end{bmatrix}}$$

The solution is then `x = A^-1 b`. In MATLAB there are two simple ways to solve this. The built-in function **inv** can be used to get the inverse of *A* and then we multiply this by *b*, or we can use the divided into operator.

```
>> A = [4 -2 1; 1 1 5; -2 3 -1];
>> b = [7;10;2];
>> x = inv(A)*b
x =
    3.0244
    2.9512
    0.8049

>> x = A\b
x =
    3.0244
    2.9512
    0.8049
```

14.5.2.1 Solving 2×2 Systems of Equations

Although this may seem easy in MATLAB, in general finding solutions to systems of equations is not. However, 2 × 2 systems are fairly straightforward, and there are several solving methods for these systems for which MATLAB has built-in functions.

Consider the following 2 × 2 system of equations:

$$x_1 + 2x_2 = 2$$
$$2x_1 + 2x_2 = 6$$

This system of equations in matrix form is:

$$\underset{A}{\begin{bmatrix} 1 & 2 \\ 2 & 2 \end{bmatrix}} \underset{x}{\begin{bmatrix} x_1 \\ x_2 \end{bmatrix}} = \underset{b}{\begin{bmatrix} 2 \\ 6 \end{bmatrix}}$$

We have already seen that the solution is $x = A^{-1} b$, so we can solve this if we can find the inverse of *A*. One method of finding the inverse for a 2 × 2 matrix involves calculating the ***determinant*** D.

For a 2 × 2 matrix

$$A = \begin{bmatrix} a_{11} & a_{12} \\ a_{21} & a_{22} \end{bmatrix}$$

the determinant *D* is defined as:

$$D = \begin{vmatrix} a_{11} & a_{12} \\ a_{21} & a_{22} \end{vmatrix} = a_{11}a_{22} - a_{12}a_{21}$$

It is written using vertical lines around the coefficients of the matrix and is defined as the product of the values on the diagonal minus the product of the other two numbers.

For a 2 × 2 matrix, the matrix inverse is defined in terms of D as

$$A^{-1} = \frac{1}{D}\begin{bmatrix} a_{22} & -a_{12} \\ -a_{21} & a_{11} \end{bmatrix}$$

The inverse is therefore the result of multiplying the scalar $1/D$ by every element in the previous matrix. Note that this is not the matrix A, but is determined using the elements from A in the following manner: the values on the diagonal are reversed and the negation operator is used on the other two values.

Notice that if the determinant D is 0, it will not be possible to find the inverse of the matrix A.

For our coefficient matrix $A = \begin{bmatrix} 1 & 2 \\ 2 & 2 \end{bmatrix}$, $D = \begin{vmatrix} 1 & 2 \\ 2 & 2 \end{vmatrix} = 1*2 - 2*2$ or -2 so

$$A^{-1} = \frac{1}{1*2-2*2}\begin{bmatrix} 2 & -2 \\ -2 & 1 \end{bmatrix} = \frac{1}{-2}\begin{bmatrix} 2 & -2 \\ -2 & 1 \end{bmatrix} = \begin{bmatrix} -1 & 1 \\ 1 & -\frac{1}{2} \end{bmatrix}$$

and

$$\begin{bmatrix} x_1 \\ x_2 \end{bmatrix} = \begin{bmatrix} -1 & 1 \\ 1 & -\frac{1}{2} \end{bmatrix}\begin{bmatrix} 2 \\ 6 \end{bmatrix}$$

The unknowns are found by performing this matrix multiplication. Consequently,

$$x_1 = -1 * 2 + 1 * 6 = 4$$
$$x_2 = 1 * 2 + (-1/2) * 6 = -1$$

To do this in MATLAB, we would first create the coefficient matrix variable A and column vector b.

```
>> A = [1 2; 2 2];
>> b = [2;6];
```

THE PROGRAMMING METHOD

For 2 × 2 matrices, the determinant and inverse are found using simple expressions.

```
>> deta = A(1,1)*A(2,2) - A(1,2)*A(2,1)
deta =
    -2

>> inva = (1/deta) * [A(2,2) -A(1,2); -A(2,1) A(1,1)]
inva =
   -1.0000      1.0000
    1.0000     -0.5000
```

THE EFFICIENT METHOD

We have already seen that MATLAB has a built-in function, **inv**, to find a matrix inverse. It also has a built-in function **det** to find a determinant:

```
>> det(A)
ans =
    -2

>> inv(A)
ans =
   -1.0000      1.0000
    1.0000     -0.5000
```

And then, the unknowns x are found:

```
>> x = inv(A) * b
x =
     4
    -1
```

PRACTICE 14.7

For the following 2 × 2 system of equations:

$$x_1 + 2x_2 = 4$$
$$-x_1 \quad = 3$$

Do the following on paper:

- Write the equations in matrix form $Ax=b$.
- Solve by finding the inverse A^{-1} and then $x=A^{-1}\,b$.

Next, get into MATLAB and check your answers.

14.5.2.2 Reduced Row Echelon Form

For 2 × 2 systems of equations, there are solving methods that are well-defined and simple. However, for larger systems of equations, finding solutions is frequently not as straightforward.

Several methods of solving are based on the observation that systems of equations are equivalent if they have the same solution set. Performing some simple operations on rows of the matrix form of a set of equations results in equivalent systems. The ***Gauss-Jordan method*** starts by augmenting the coefficient matrix A with the column vector b, and performing operations until the square part of the matrix becomes diagonal.

Reduced Row Echelon Form takes this one step further to result in all 1s on the diagonal, or in other words until the square part is the identity matrix. In this case the column of b's is the solution. For example, for a 3 × 3 matrix.

$$\begin{bmatrix} a_{11} & a_{12} & a_{13} & b_1 \\ a_{21} & a_{22} & a_{23} & b_2 \\ a_{31} & a_{32} & a_{33} & b_3 \end{bmatrix} \rightarrow \begin{bmatrix} 1 & 0 & 0 & b'_1 \\ 0 & 1 & 0 & b'_2 \\ 0 & 0 & 1 & b'_3 \end{bmatrix}$$

In other words, we are reducing [A|b] to [I|b′]. MATLAB has a built-in function to do this, called **rref**. For example:

Note
The prime in b′ indicates that the numbers may have changed, but the systems are equivalent.

```
>> a = [1 3 0; 2 1 3; 4 2 3];
>> b = [1 6 3]';
>> ab = [a b];
>> rref(ab)
ans =
     1     0     0    -2
     0     1     0     1
     0     0     1     3
```

The solution is found from the last column, so $x_1 = -2$, $x_2 = 1$, and $x_3 = 3$. To get this in a column vector in MATLAB:

```
>> x = ans(:,end)
x =
    -2
     1
     3
```

14.5.2.3 Finding a Matrix Inverse by Reducing an Augmented Matrix

For a system of equations larger than a 2 × 2 system, one method of finding the inverse of a matrix A mathematically involves augmenting the matrix with an identity matrix of the same size, and then reducing it. The algorithm is:

- Augment the matrix with I: [A | I].
- Reduce it to the form [I | X]; X will be A^{-1}.

For example, in MATLAB, we can start with a matrix, augment it with an identity matrix, and then use the **rref** function to reduce it.

```
>> a = [1 3 0; 2 1 3; 4 2 3];
>> rref([a eye(size(a))])
ans =
    1.0000         0         0   -0.2000   -0.6000    0.6000
         0    1.0000         0    0.4000    0.2000   -0.2000
         0         0    1.0000         0    0.6667   -0.3333
```

In MATLAB, the **inv** function can be used to verify the result.

```
>> inv(a)
ans =
   -0.2000   -0.6000    0.6000
    0.4000    0.2000   -0.2000
         0    0.6667   -0.3333
```

14.6 SYMBOLIC MATHEMATICS

Symbolic mathematics means doing mathematics on symbols (not numbers!). For example, $a+a$ is $2a$. The symbolic math functions are in Symbolic Math Toolbox in MATLAB. Toolboxes contain related functions and are add-ons to MATLAB. (Therefore, this may or may not be part of your own system.) Symbolic Math Toolbox includes an alternative method for solving equations, and is therefore covered in this chapter.

14.6.1 Symbolic Variables and Expressions

MATLAB has a type called **sym** for symbolic variables and expressions; these work with strings. For example, to create a symbolic variable *a* and perform the addition just described, a symbolic variable would first be created by passing the string '*a*' to the **sym** function:

```
>> a = sym('a');
>> a+a
ans =
2*a
```

Symbolic variables can also store expressions. For example, the variables *b* and *c* store symbolic expressions:

```
>> b = sym('x^2');
>> c = sym('x^4');
```

All basic mathematical operations can be performed on symbolic variables and expressions (e.g., add, subtract, multiply, divide, and raise to a power, etc.). The following are examples:

```
>> c/b
ans =
x^2
>> b^3
ans =
x^6
>> c*b
ans =
x^6
>> b + sym('4*x^2')
ans =
5*x^2
```

It can be seen from the last example that MATLAB will collect like terms in these expressions, adding the x^2 and $4x^2$ to result in $5x^2$.

The following creates a symbolic expression by passing a string, but the terms are not collected automatically.

```
>> sym('z^3 + 2*z^3')
ans =
z^3 + 2*z^3
```

If, however, z is a symbolic variable to begin with, quotes are not needed around the expression, and the terms are automatically collected:

```
>> z = sym('z');
>> z^3 + 2*z^3
ans =
3*z^3
```

If using multiple variables as symbolic variable names is desired, the **syms** function is a shortcut instead of using **sym** repeatedly. For example,

```
>> syms x y z
```

is equivalent to

```
>> x = sym('x');
>> y = sym('y');
>> z = sym('z');
```

The built-in functions **sym2poly** and **poly2sym** convert from symbolic expressions to polynomial vectors and vice versa. For example:

```
>> myp = [1 2 -4 3];
>> poly2sym(myp)
ans =
x^3+2*x^2-4*x+3
>> mypoly = [2 0 -1 0 5];
>> poly2sym(mypoly)
ans =
2*x^4-x^2+5

>> sym2poly(ans)
ans =
     2     0    -1     0     5
```

14.6.2 Simplification Functions

There are several functions that work with symbolic expressions and simplify the terms. Not all expressions can be simplified, but the **simplify** function does whatever it can to simplify expressions, including gathering like terms. For example:

```
>> x = sym('x');
>> myexpr = cos(x)^2 + sin(x)^2
myexpr =
cos(x)^2+sin(x)^2

>> simplify(myexpr)
ans =
1
```

The functions **collect**, **expand**, and **factor** work with polynomial expressions. The **collect** function collects coefficients, such as the following:

```
>> x = sym('x');
>> collect(x^2 + 4*x^3 + 3*x^2)
ans =
4*x^2+4*x^3
```

The **expand** function will multiply out terms and **factor** will do the reverse:

```
>> expand((x+2)*(x–1));
ans =
x^2+x-2
>> factor(ans)
ans =
(x+2)*(x-1)
```

If the argument is not factorable, the original input argument will be returned unmodified.

The **subs** function will substitute a value for a symbolic variable in an expression. For example,

```
>> myexp = x^3 + 3*x^2 - 2
myexp =
x^3+3*x^2-2
>> subs(myexp,3)
ans =
    52
```

If there are multiple variables in the expression, one will be chosen by default for the substitution (in this case, x), or the variable for which the substitution is to be made can be specified:

```
>> syms a b x
>> varexp = a*x^2 + b*x;
>> subs(varexp,3)
ans =
9*a+3*b
 >> subs(varexp,'a',3)
ans =
3*x^2+b*x
```

With symbolic math, MATLAB works by default with rational numbers, meaning that results are kept in fractional forms. For example, performing the addition $1/3 + 1/2$ would normally result in a **double** value:

```
>> 1/3 + 1/2
ans =
     0.8333
```

However, by making the expression symbolic, the result is symbolic too. Any numeric function (e.g., **double)** could change that:

```
>> sym(1/3 + 1/2)
ans =
5/6
>> double(ans)
ans =
    0.8333
```

The **numden** function will return separately the numerator and denominator of a symbolic expression:

```
>> sym(1/3 + 1/2)
ans =
5/6
>> [n, d] = numden(ans)
n =
5
d =
6

>> [n, d] = numden((x^3 + x^2)/x)
n =
x^2*(x+1)
d =
x
```

14.6.3 Displaying Expressions

The **pretty** function will display symbolic expressions using exponents. For example:

```
>> b = sym('x^2')
b =
x^2
>> pretty(b)
                                                   2
                                                  x
```

There are several plot functions in MATLAB with names beginning with "ez" that perform the necessary conversions from symbolic expressions to numbers and plot them. For example, the function **ezplot** will draw a two-dimensional plot in the x-range from -2π to 2π, with the expression as the title (in pretty form). The expression

```
>> ezplot('x^3 + 3*x^2 - 2')
```

produces the figure that is shown in Fig. 14.5.

The domain for the **ezplot** function can also be specified; for example, to change the x-axis to the range 0 to π, the minimum and maximum values of the range are specified as a vector.

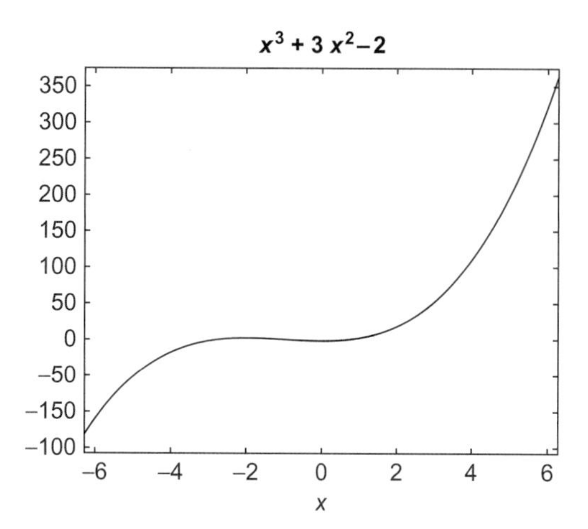

FIGURE 14.5
Plot produced using **ezplot**.

```
>> ezplot('cos(x)', [0  pi])
```

14.6.4 Solving Equations

We've already seen several methods for solving simultaneous linear equations using a matrix representation. MATLAB can also solve sets of equations using symbolic math.

The function **solve** solves an equation and returns the solution(s) as symbolic expressions. The solution can be converted to numbers using any numeric function, such as **double**:

```
>> x = sym('x');
>> solve('2*x^2 + x = 6')
ans =
   -2
  3/2
>> double(ans)
ans =
   -2.0000
    1.5000
```

If an expression is passed to the **solve** function rather than an equation, the **solve** function will set the expression equal to 0 and solve the resulting equation. For example, the following will solve $3x^2+x=0$:

```
>> solve('3*x^2 + x')
ans =
     0
  -1/3
```

If there is more than one variable, MATLAB chooses which one to solve for. In the following example, the equation $ax^2+bx=0$ is solved. There are three variables. As can be seen from the result, which is given in terms of *a* and *b*, the equation was solved for *x*. MATLAB has rules built-in that specify how to choose which variable to solve for. For example, *x* will always be the first choice if it is in the equation or expression.

```
>> solve('a*x^2 + b*x')
ans =
     0
 -b/a
```

However, it is possible to specify which variable to solve for:

```
>> solve('a*x^2 + b*x', 'b')
ans =
-a*x
```

MATLAB can also solve sets of equations. In this example, the solutions for *x*, *y*, and *z* are returned as a structure consisting of fields for *x*, *y*, and *z*. The individual solutions are symbolic expressions stored in fields of the structure.

```
>> solve('4*x-2*y+z=7', 'x+y+5*z=10', '-2*x+3*y-z=2')
ans =
    x: [1x1 sym]
    y: [1x1 sym]
    z: [1x1 sym]
```

To refer to the individual solutions, which are in the structure fields, the dot operator is used.

```
>> x = ans.x
x =
124/41
>> y = ans.y
y =
121/41
>> z = ans.z
z =
33/41
```

The **double** function can then be used to covert the symbolic expressions to numbers, and store the results from the three unknowns in a vector.

```
>> double([x y z])
ans =
    3.0244    2.9512    0.8049
```

PRACTICE 14.8

For each of the following expressions, show what the MATLAB result would be. Assume that all expressions are typed SEQUENTIALLY.

```
x = sym('x');
a = sym(x^3 - 2*x^2 + 1);
b = sym(x^3 + x^2);
res = a+b

p = sym2poly(res)

polyval(p,2)

sym(1/2 + 1/4)

solve('x^2 - 16')
```

14.7 CALCULUS: INTEGRATION AND DIFFERENTIATION

MATLAB has functions that perform common calculus operations on a mathematical function f(x), such as ***integration*** and ***differentiation***.

14.7.1 Integration and the Trapezoidal Rule

The integral of a function f(x) within the limits given by $x=a$ and $x=b$ is written as

$$\int_a^b f(x)dx$$

and is defined as the area under the curve f(x) from a to b, as long as the function is above the x-axis. Numerical integration techniques involve approximating this.

One simple method of approximating the area under a curve is to draw a straight line from f(a) to f(b) and calculate the area of the resulting trapezoid as

$$(b-a)\frac{f(a)+f(b)}{2}$$

In MATLAB, this could be implemented as a function.

THE PROGRAMMING CONCEPT

Here is a function to which the function handle and limits *a* and *b* are passed:

trapint.m

```
function int = trapint(fnh, a, b)
% trapint approximates area under a curve f(x)
%   from a to b using a trapezoid
% Format: trapint(handle of f, a, b)
int = (b-a) * (fnh(a) + fnh(b))/2;
end
```

To call it, for example, for the function $f(x) = 3x^2 - 1$, an anonymous function is defined and its handle is passed to the *trapint* function.

```
>> f = @ (x) 3 * x . ^2 - 1;
approxint = trapint(f, 2, 4)
approxint =

    58
```

THE EFFICIENT METHOD

MATLAB has a built-in function **trapz** that will implement the trapezoidal rule. Vectors with the values of *x* and $y = f(x)$ are passed to it. For example, using the anonymous function defined previously:

```
>> x = [2 4];
>> y = f(x);
>> trapz(x,y)
ans =
    58
```

An improvement on this is to divide the range from *a* to *b* into *n* intervals, apply the trapezoidal rule to each interval, and sum them. For example, for the preceding if there are two intervals, you would draw a straight line from $f(a)$ to $f((a+b)/2)$ and then from $f((a+b)/2)$ to $f(b)$.

THE PROGRAMMING CONCEPT

The following is a modification of the previous function to which the function handle, limits, and the number of intervals are passed:

Continued

THE PROGRAMMING CONCEPT—CONT'D

trapintn.m

```
function intsum = trapintn(fnh, lowrange,highrange, n)
% trapintn approximates area under a curve f(x) from
%  a to b using trapezoids with n intervals
% Format: trapintn(handle of f, a, b, n)
intsum = 0;
increm = (highrange - lowrange)/n;
for a = lowrange: increm : highrange - increm
    b = a + increm;
    intsum = intsum + (b-a) * (fnh(a) + fnh(b))/2;
end
end
```

For example, this approximates the integral of the previous function *f* with two intervals:

```
>> trapintn(f,2,4,2)
ans =
    55
```

THE EFFICIENT METHOD

To use the built-in function **trapz** to accomplish the same thing, the *x* vector is created with the values 2, 3, and 4:

```
>> x = 2:4;
>> y = f(x)
>> trapz(x,y)
ans =
    55
```

In these examples, straight lines, which are first-order polynomials, were used. Other methods involve higher-order polynomials. The built-in function **quad** uses Simpson's method. Three arguments are normally passed to it: the handle of the function, and the limits *a* and *b*. For example, for the previous function:

```
>> quad(f,2,4)
ans =
    54
```

MATLAB has a function **polyint**, which will find the integral of a polynomial. For example, for the polynomial $3x^2 + 4x - 4$, which would be represented by the vector `[3 4-4]`, the integral is found by:

```
>> origp = [3  4  -4];
>> intp = polyint(origp)
intp =
     1     2     -4     0
```

which shows that the integral is the polynomial $x^3 + 2x^2 - 4x$.

14.7.2 Differentiation

The derivative of a function $y=f(x)$ is written as $\frac{dy}{dx}f(x)$ or $f'(x)$, and is defined as the rate of change of the dependent variable y with respect to x. The derivative is the slope of the line tangent to the function at a given point.

MATLAB has a function **polyder**, which will find the derivative of a polynomial. For example, for the polynomial $x^3 + 2x^2 - 4x + 3$, which would be represented by the vector `[1 2 -4 3]`, the derivative is found by:

```
>> origp = [1 2 -4 3];
>> diffp = polyder(origp)
diffp =
     3     4     -4
```

which shows that the derivative is the polynomial $3x^2 + 4x - 4$. The function **polyval** can then be used to find the derivative for certain values of x, such as for $x=1$, 2, and 3:

```
>> polyval(diffp, 1:3)
ans =
     3     16     35
```

The derivative can be written as the limit

$$f'(x) = \lim_{h \to 0} \frac{f(x+h) - f(x)}{h}$$

and can be approximated by a difference equation.

Recall that MATLAB has a built-in function, **diff**, which returns the differences between consecutive elements in a vector. For a function $y = f(x)$ where x is a vector, the values of $f'(x)$ can be approximated as **diff(y)** divided by **diff(x)**. For example, the equation $x^3 + 2x^2 - 4x + 3$ can be written as an anonymous function. It can be seen that the approximate derivative is close to the values found using **polyder** and **polyval**.

```
>> f = @ (x) x .^ 3 + 2 * x .^ 2 - 4 * x + 3;
>> x = 0.5 : 3.5
x =
    0.5000    1.5000    2.5000    3.5000
>> y = f(x)
```

```
y =
    1.6250    4.8750    21.1250    56.3750
>> diff(y)
ans =
   3.2500   16.2500   35.2500
>> diff(x)
ans =
   1    1    1
>> diff(y) ./ diff(x)
ans =
   3.2500   16.2500   35.2500
```

14.7.3 Calculus in Symbolic Math Toolbox

There are several functions in Symbolic Math Toolbox™ to perform calculus operations symbolically (e.g., **diff** to differentiate and **int** to integrate). To learn about the **int** function, for example, from the Command Window:

```
>> help sym/int
```

For instance, to find the indefinite integral of the function $f(x)=3x^2-1$:

```
>> syms x
>> int(3*x^2 - 1)
ans =
x^3-x
```

To instead find the definite integral of this function from $x=2$ to $x=4$:

```
>> int(3*x^2 - 1, 2, 4)
ans =
54
```

Limits can be found using the **limit** function. For example, for the difference equation:

```
>> syms x h
>> f
f =
     @ (x) x .^3 + 2 .*x.^2 - 4 .* x + 3

>> limit((f(x+h)-f(x))/h,h,0)
ans =
3*x^2-4+4*x
```

To differentiate, instead of the anonymous function we write it symbolically:

```
>> syms x f
>> f = x^3 + 2*x^2 - 4*x + 3
f =
x^3+2*x^2-4*x+3

>> diff(f)
ans =
3*x^2-4+4*x
```

PRACTICE 14.9

For the function $3x^2 - 4x + 2$:

- Find the indefinite integral of the function.
- Find the definite integral of the function from $x = 2$ to $x = 5$.
- Approximate the area under the curve from $x = 2$ to $x = 5$.
- Find its derivative.
- Approximate the derivative for $x = 2$.

Explore Other Interesting Features

Investigate the **corrcoef** function, which returns correlation coefficients.

Investigate filtering data, for example using the **filter** function.

Investigate the moving statistical functions introduced in R2016a **movmean**, **movsum**, etc.

Investigate the flag introduced in R2015a to omit NaN in functions such as **max**, **min median**, etc.

Investigate the index vectors returned by the set functions.

Investigate the use of 'R2012a' to see future changes to the set functions, versus the use of 'legacy' to preserve the previous values.

Investigate passing matrices to the set functions, using the 'rows' specifier.

Investigate the **interp1** function, which does a table look-up to interpolate or extrapolate.

Investigate the **fminsearch** function, which finds local minima for a function.

Investigate the **fzero** function, which attempts to find a zero of a function near a specified *x* value.

Investigate linear algebra functions, such as **rank** for the rank of a matrix, or **null**, which returns the null space of a matrix.

Investigate the **blkdiag** function, which will create a block diagonal matrix.

Investigate the functions that return eigenvalues and eigenvectors, such as **eig** and **eigs**.

Investigate the **norm** function to find a vector or matrix norm.

Investigate the Ordinary Differential Equation (ODE) solve functions, such as **ode23** and **ode45**, which use the Runge-Kutta integration methods.

In the Command Window, type "odeexamples" to see some ODE example codes.

Investigate some of the other numerical integration functions, such as **integral**, **integral2** for double integrals, and **integral3** for triple integrals.

Investigate the **poly** function, which finds the characteristic equation for a matrix, and the **polyeig** function, which solves a polynomial eigenvalue problem of a specified degree. ■

SUMMARY

COMMON PITFALLS

- Forgetting that **max** and **min** return the index of only the first occurrence of the maximum or minimum value.
- Not realizing that a data set has outliers that can drastically alter the results obtained from the statistical functions.
- Extrapolating too far away from the data set
- Forgetting that the **fprintf** function by default only prints the real part of a complex number
- Forgetting that to augment one matrix with another, the number of rows must be the same in each

PROGRAMMING STYLE GUIDELINES

- Remove the largest and smallest numbers from a large data set before performing statistical analyses, in order to handle the problem of outliers.
- The better the curve fit, the more accurate interpolated and extrapolated values will be.
- When working with symbolic expressions, it is generally easier to make all variables symbolic to begin with.

MATLAB Functions and Commands			
mean	polyfit	issymmetric	factor
var	complex	istril	subs
std	real	istriu	numden
mode	imag	isbanded	pretty
median	isreal	inv	ezplot
union	conj	det	solve
intersect	angle	rref	trapz
unique	polarplot	sym	quad
setdiff	diag	syms	polyint
setxor	trace	sym2poly	polyder
ismember	eye	poly2sym	int
issorted	triu	simplify	limit
roots	tril	collect	
polyval	isdiag	expand	

Exercises

1. In a marble manufacturing plant, a quality control engineer randomly selects eight marbles from each of the two production lines and measures the diameter of each marble in millimeters. For the each data set here, determine the mean, median, mode, and standard deviation using built-in functions.

```
Prod. line A:15.94 15.98 15.94 16.16 15.86 15.86 15.90 15.88
Prod. line B:15.96 15.94 16.02 16.10 15.92 16.00 15.96 16.02
```

   Suppose the desired diameter of the marbles is 16 mm. Based on the results you have, which production line is better in terms of meeting the specification? (Hint: think in terms of the mean and the standard deviation.)
2. Write a function *mymin* that will receive any number of arguments, and will return the minimum. Note: the function is not receiving a vector; rather, all of the values are separate arguments.
3. Write a script that will do the following. Create two vectors with 20 random integers in each; in one the integers should range from 1 to 5, and in the other, from 1 to 500 (inclusive). For each vector, would you expect the mean and median to be approximately the same? Would you expect the standard deviation of the two vectors to be approximately the same? Answer these questions, and then use the built-in functions to find the minimum, maximum, mean, median, standard deviation, and mode of each. Do a histogram for each in a subplot. Run the script a few times to see the variations.
4. Write a function that will return the mean of the values in a vector, not including the minimum and maximum values. Assume that the values in the vector are unique. It is okay to use the built-in **mean** function. To test this, create a vector of 10 random integers, each in the range from 0 to 50, and pass this vector to the function.
5. A moving average of a data set $x = \{x_1, x_2, x_3, x_4, \ldots, x_n\}$ is defined as a set of averages of subsets of the original data set. For example, a moving average of every two terms would be $1/2 * \{x_1+x_2, x_2+x_3, x_3+x_4, \ldots, x_{n-1}+x_n\}$. Write a function that will receive a vector as an input argument, and will calculate and return the moving average of every two elements.

   Eliminating or reducing noise is an important aspect of any signal processing. For example, in image processing noise can blur an image. One method of handling this is called median filtering.
6. A median filter on a vector has a size, for example, a size of 3 means calculating the median of every three values in the vector. The first and last elements are left alone. Starting from the second element to the next-to-last element, every element of a vector vec(*i*) is replaced by the median of [vec(*i*-1) vec(*i*) vec(*i*+1)]. For example, if the signal vector is

```
signal = [5 11 4 2 6 8 5 9]
```

the median filter with a size of 3 is

```
medianFilter3 = [5 5 4 4 6 6 8 9]
```

Write a function to receive the original signal vector and return the median filtered vector.

7. What is the difference between the mean and the median of a data set if there are only two values in it?
8. A student missed one of four exams in a course and the professor decided to use the "average" of the other three grades for the missed exam grade. Which would be better for the student: the mean or the median if the three recorded grades were 99, 88, and 95? What if the grades were 99, 70, and 77?
9. Write a function *allparts* that will read in lists of part numbers for parts produced by two factories. These are contained in data files called *xyparts.dat* and *qzparts.dat*. The function will return a vector of all parts produced, in sorted order (with no repeats). For example, if the file *xyparts.dat* contains

```
123  145  111  333  456  102
```

and the file *qzparts.dat* contains

```
876  333  102  456  903  111
```

calling the function would return the following:

```
>> partslist = allparts
partslist =
   102    111    123    145    333    456    876    903
```

10. The set functions can be used with cell arrays of strings. Create two cell arrays to store (as strings) course numbers taken by two students. For example,

```
s1 = {'EC 101', 'CH 100', 'MA 115'};
s2 = {'CH 100', 'MA 112', 'BI 101'};
```

Use a set function to determine which courses the students have in common.

11. A vector *v* is supposed to store unique random numbers. Use set functions to determine whether or not this is true.
12. A program has a vector of structures that stores information on experimental data that has been collected. For each experiment, up to 10 data values were obtained. Each structure stores the number of data values for that experiment, and then the data values. The program is to calculate and print the average value for each experiment. Write a script to create some data in this format and print the averages.
13. Express the following polynomials as row vectors of coefficients:

$$2x^3 - 3x^2 + x + 5$$
$$3x^4 + x^2 + 2x - 4$$

14. Find the roots of the equation f(*x*) = 0 for the following function. Also, create *x* and *y* vectors and plot this function in the range from –3 to 3 to visualize the solution.

$$f(x) = 3x^2 - 2x - 5$$

15. Evaluate the polynomial expression $3x^3 + 4x^2 + 2x - 2$ at $x=4$, $x=6$, and $x=8$.
16. What is a danger of extrapolation?
17. Write a script that will generate a vector of 10 random integers, each in the inclusive range from 0 to 100. If the integers are evenly distributed in this range, then when arranged in order from lowest to highest, they should fall on a straight line. To test this, fit a straight line through the points and plot both the points and the line with a legend.
18. Write a function that will receive data points in the form of *x* and *y* vectors. If the lengths of the vectors are not the same, then they cannot represent data points, so an error message should be printed. Otherwise, the function will fit a polynomial of a random degree through the points, and will plot the points and the resulting curve with a title specifying the degree of the polynomial. The degree of the polynomial must be less than the number of data points *n*, so the function must generate a random integer in the range from 1 to *n*-1 for the polynomial degree.
19. Write a function *mirror* that will receive one input vector consisting of *y* coordinates of data points. The function will fit a second-order polynomial through the points. The function will plot on one graph, the original data points (using green *s), the curve (using blue and with enough points so that it is very smooth), and also "mirror image" points (in red *s). The "mirror image" points are, for every *x* coordinate, equidistant from the curve as the original data point. So, if the original data point is 2 above the curve, the mirror image point will be 2 below it.

Data on the flow of water in rivers and streams is of great interest to civil engineers, who design bridges, and to environmental engineers, who are concerned with the environmental impact of catastrophic events such as flooding

20. The Mystical River's water flow rate on a particular day is shown in the table below. The time is measured in hours and the water flow rate is measured in cubic feet per second. Write a script that will fit polynomials of degree 3 and 4 to the data and create a subplot for the two polynomials. Plot also the original data as black circles in both plots. The titles for the subplots should include the degree of the fitted polynomial. Also, include appropriate *x* and *y* labels for the plots.

Time	0	3	6	9	12	15	18	21	24
Flow Rate	800	980	1090	1520	1920	1670	1440	1380	1300

21. Write a function that will receive *x* and *y* vectors representing data points. The function will create, in one Figure Window, a plot showing these data points as circles and also in the top part a second-order polynomial that best fits these points and on the bottom a third-order polynomial. The top plot will have a line width of 3 and will be a gray color. The bottom plot will be blue, and have a line width of 2. For example, the Figure Window might look like Fig. 14.6.

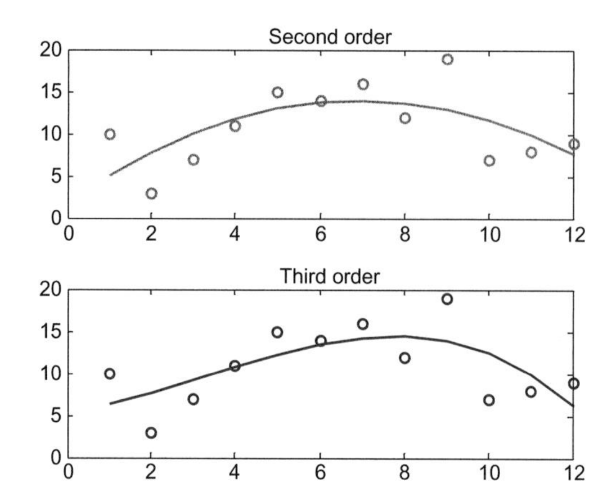

FIGURE 14.6
Subplot of second- and third-order polynomials with different line properties.

The axes are the defaults. Note that changing the line width also changes the size of the circles for the data points. You do not need to use a loop.

22. Store the following complex numbers in variables, and print them in the form a + bi.
    3 - 2i
    $\sqrt{-3}$
23. Create the following complex variables
    c1 = 2 - 4i;
    c2 = 5 + 3i;
    Perform the following operations on them:
    - add them
    - multiply them
    - get the complex conjugate and magnitude of each
    - put them in polar form
24. Represent the expression $z^3 - 2z^2 + 3 - 5i$ as a row vector of coefficients, and store this in a variable *compoly*. Use the **roots** function to solve $z^3 - 2z^2 + 3 - 5i = 0$. Also, find the value of *compoly* when $z = 2$ using **polyval**.
25. What is the value of the trace of an $n \times n$ identity matrix?
26. For an $m \times n$ matrix, what are the dimensions of its transpose?
27. What is the transpose of a diagonal matrix A?
28. When is a square matrix both, an upper triangular and lower triangular matrix?
29. Is the transpose of an upper triangular matrix also upper triangular? If not, how would you characterize it?
30. Given the following colormap matrix:

```
mycmap = [1 1 1;  1 0 0;  0 1 0;  0 0 1;  1 1 0];
```

Write code that will generate the following 4×4 "image" matrix, using the colormap method:

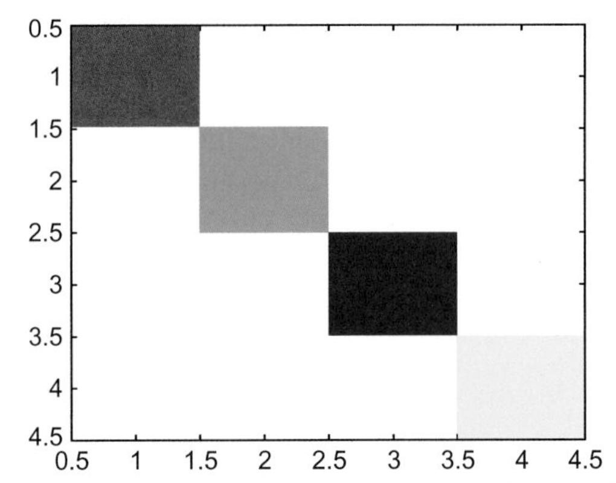

Note: the axes are the defaults, and you only need 3 lines of code to accomplish this.

31. Write a function *myupp* that will receive an integer argument *n*, and will return an $n \times n$ upper triangular matrix of random integers.
32. Analyzing electric circuits can be accomplished by solving sets of equations. For a particular circuit, the voltages V_1, V_2, and V_3 are found through the system:

$$V_1 = 5$$
$$-6\,V_1 + 10\,V_2 - 3\,V_3 = 0$$
$$-V_2 + 51\,V_3 = 0$$

Put these equations in matrix form and solve in MATLAB.

33. Rewrite the following system of equations in matrix form:

$$4x_1 - x_2 + 3x_4 = 10$$
$$-2x_1 + 3x_2 + x_3 - 5x_4 = -3$$
$$x_1 + x_2 - x_3 + 2x_4 = 2$$
$$3x_1 + 2x_2 - 4x_3 = 4$$

Set it up in MATLAB and use any method to solve.

34. Solve the simultaneous equations $x - y = 2$ and $x^2 + y = 0$ using **solve**. Plot the corresponding functions, $y = x - 2$ and $y = -x^2$, on the same graph with an *x*-range from -5 to 5.
35. For the following set of equations,

$$2x_1 + 2x_2 + x_3 = 2$$
$$x_2 + 2x_3 = 1$$
$$x_1 + x_2 + 3x_3 = 3$$

write it in symbolic form and solve using the **solve** function. From the symbolic solution, create a vector of the numerical (**double**) equivalents.

36. The reproduction of cells in a bacterial colony is important for many environmental engineering applications such as wastewater treatments. The formula

$$\log(N) = \log(N_0) + t/T \log(2)$$

can be used to simulate this, where N_0 is the original population, N is the population at time t, and T is the time it takes for the population to double. Use the solve function to determine the following: if $N_0=10^2$, $N=10^8$, and $t=8$ h, what will be the doubling time T? Use **double** to get your result in hours.

37. Using the symbolic function **int**, find the indefinite integral of the function $4x^2+3$, and the definite integral of this function from $x=-1$ to $x=3$. Also, approximate this using the **trapz** function.
38. Use the **quad** function to approximate the area under the curve $4x^2+3$ from -1 to 3. First, create an anonymous function and pass its handle to the **quad** function.
39. Use the **polyder** function to find the derivative of $2x^3-x^2+4x-5$.
40. Examine the motion or *trajectory* of a *projectile* moving in air. Assume that it has an initial height of 0, and neglect the air resistance for simplicity. The projectile has an initial velocity v_0, an angle of departure θ_0, and is subject to the gravity constant $g=9.81$ m/s^2. The position of the projectile is given by x and y coordinates, where the origin is the initial position of the projectile at time $t=0$. The total horizontal distance that the projectile travels is called its *range* (the point at which it hits the ground) and the highest peak (or vertical distance) is called its *apex*. Equations for the trajectory can be given in terms of the time t or in terms of x and y. The position of the projectile at any time t is given by:

    $$x=v_0\cos(\theta_0)t$$

    $$y=v_0\sin(\theta_0)t-{}^1/_2gt$$

    For a given initial velocity v_0 and angle of departure θ_0, describe the motion of the projectile by writing a script to answer the following:

    - What is the range?
    - Plot the position of the projectile at suitable x values
    - Plot the height versus time.
    - How long does it take to reach its apex?
41. Write a GUI function that creates four random points. Radio buttons are used to choose the order of a polynomial to fit through the points. The points are plotted along with the chosen curve.

APPENDIX I

MATLAB® Functions (not including those listed in the "Explore Other Interesting Features" sections)

abs absolute value
all true if all elements in the input argument are true
angle angle of a complex number
any true if any element in the input argument is true
area filled two-dimensional area plot
asin arcsine in radians
asind arcsine in degrees
asinh inverse hyperbolic sine in radians
axis sets limits on axes for a plot
bar two-dimensional bar chart
bar3 three-dimensional bar chart
bar3h three-dimensional horizontal bar chart
barh two-dimensional horizontal bar chart
blanks creates a string of all blank spaces
cast casts a variable to a specified type
categorical converts cell array to categorical array
categories lists categories from a categorical array
ceil rounds toward infinity
cell creates a cell array
celldisp displays contents of a cell array
cellplot displays contents of a cell array in boxes
cellstr converts from a character matrix to a cell array of strings
char creates a character matrix
checkcode displays Code Analyzer results for code files
class returns the type or class of the input argument
clear clears variable(s) from the workspace
clf clears the figure window
clock stores the current date and time in a vector
collect collects like terms in a symbolic math expression
colorbar displays a color scale in a plot

colormap returns the current colormap, or sets a matrix to be the current colormap
comet animated two-dimensional plot
comet3 three-dimensional animated plot
complex creates a complex number
conj complex conjugate
cross cross product
cummax cumulative, or running, maximum of a vector or columns of a matrix
cummin cumulative, or running, minimum of a vector or columns of a matrix
cumprod cumulative, or running, product of a vector or columns of a matrix
cumsum cumulative, or running, sum of a vector or columns of a matrix
cylinder returns three-dimensional data vectors to create a cylinder
date stores the current date as a string
dbcont continue executing code in debug mode
dbquit quit debug mode
dbstep step through code in debug mode
dbstop set a breakpoint in debug mode
deblank gets rid of trailing blanks in a string
deg2rad converts from degrees to radians
demo shows MATLAB Examples in the Help Browser
det finds the determinant of a matrix
diag returns the diagonal of a matrix, or creates a diagonal matrix
diff finds differences between consecutive elements; used to approximate derivatives
disp simple display (output)
doc brings up a documentation page
dot dot product
double converts to the type double
echo toggle; displays all statements as they are executed
end ends control statements and functions; refers to last element
error displays an error message
eval evaluates a string as a function or command
exit quits MATLAB
exp exponential function
expand expands a symbolic math expression
eye creates an identity matrix
ezplot simple plot function that plots a function without the need for data vectors
factor factors a symbolic math expression
factorial factorial of an integer *n*, is 1*2*3*...**n*
false equivalent to logical(0); creates an array of false values
fclose closes an open file
feof true if the specified file is at the end-of-file

feval evaluates a function handle on a string as a function call
fgetl low-level input function reads one line from a file as a string
fgets same as fgetl but does not remove newline characters
fieldnames returns the names of fields in a structure as a cell array of strings
figure create or refer to Figure Windows
find returns indices of an array for which a logical expression is true
fix rounds toward zero
flip flips an array, either left to right or up to down
fliplr flips columns of a matrix from left to right
flipud flips rows of a matrix up to down
floor rounds toward negative infinity
fopen low-level file function; opens a file for a specified operation
format many options for formatting displays
fplot plots a function passed as a function handle
fprintf formatted display (output); writes either to a file or to the screen (the default)
fscanf low-level file input function; reads from a file into a matrix
func2str converts from a function handle to a string
fzero attempts to find a zero of a function, given the function handle
gca handle to the current axes
gcf handle to the current figure
get gets properties of a plot object
getframe gets a movie frame, which is a snapshot of the current plot
ginput gets graphical coordinates from a mouse click
grid plot toggle; turns grid lines on or off
groot handle of graphics root object (screen)
gtext allows the user to place a string on a plot in location of a mouse click
help displays help information for built-in or user-defined functions, or scripts
histogram plot function: plots a histogram
hold plot toggle; freezes plot in Figure Window so the next will be superimposed
i constant for the square root of negative one
im2double converts an image matrix to type double
imag imaginary part of a complex number
image displays an image matrix
imread reads in an image matrix
imshow displays an image
imwrite writes a matrix in an image format
inf constant for infinity
input prompts the user and reads user's input
int symbolic math integration
int16 converts a number to a 16-bit signed integer
int2str converts from an integer to a string storing the integer

int32 converts a number to a 32-bit signed integer
int64 converts a number to a 64-bit signed integer
int8 converts a number to an 8-bit signed integer
intersect set intersection
intmax largest value possible in a specified integer type
intmin smallest value possible in a specified integer type
inv inverse of a matrix
isa true if the input argument is the specified class
isbanded true if the input matrix is banded
iscellstr true if the input argument is a cell array storing only strings
ischar true if the input argument is a string, or character vector
isdiag true if the input matrix is a diagonal matrix
isempty true if the input argument is an empty vector or empty string
isequal true if two array arguments are equal element-by-element
isfield true if a string is the name of a field within a structure
iskeyword true if the string input argument is the name of a keyword
isletter true if the input argument is a letter of the alphabet
ismember set function receives two sets; true for every member of first set also in second
isreal true if input argument is a real number (not complex)
issorted true if the input vector is sorted in ascending order
isspace true if the input argument is a white space character
isstrprop true if the string argument is a specified property
isstruct true if the input argument is a structure
issymmetric true if the input matrix is symmetric
istril true if the input matrix is lower triangular
istriu true if the input matrix is upper triangular
j constant for the square root of negative one
jet returns all or part of the 64 colors in the jet colormap
legend displays a legend on a plot
length length, or number of elements, in a vector; largest dimension for a matrix
limit computes limit of a symbolic math expression
line graphics primitive object that creates a line
linspace creates a vector of linearly-spaced values
load inputs a file into a matrix, or reads variables from a.mat file (the default)
log natural logarithm
log10 base 10 logarithm
log2 base 2 logarithm
logical converts numbers to the type logical
loglog plot function that uses logarithmic scales for x and y axes
logspace creates a vector of logarithmically spaced values
lookfor looks for a string in the H1 comment line in files

lower converts letters to lower-case in a string
max the maximum value in a vector, or for every column in a matrix
mean the mean (average) of values in a vector, or every column in a matrix
median the median (middle) value in a sorted vector, or for every column in a matrix
menu displays a menu of push buttons and returns number of choice (not recommended)
mesh three-dimensional mesh surface plot
meshgrid creates *x* and *y* vectors to be used in images or as function arguments
methods displays methods of a class
min the minimum value in a vector, or for every column in a matrix
mod modulus after division
mode the maximum value in a vector, or for every column in a matrix
movegui moves a Figure Window within the screen
movie plays a movie, or sequence of screen shots
namelengthmax the maximum length of identifier names
NaN mathematics constant for "Not a Number"
nargin number of input arguments passed to a function
nargout number of output arguments expected to be returned by a function
nthroot *n*th root of a number
num2str converts a real number to a string containing the number
numden symbolic math function, separates the numerator and denominator of a fraction
numel total number of elements in a vector or matrix
ones creates a matrix of all ones
parula returns all or part of the 64 colors in the parula colormap
patch graphics primitive object that creates a filled-in two-dimensional polygon
pi constant for π
pie creates a two-dimensional pie chart
pie3 creates a three-dimensional pie chart
pink returns all or part of the 64 colors in the pink colormap
plot simple plot function, plots 2D points; markers, color, etc. can be specified
plot3 simple three-dimensional plot function, plots three-dimensional points
plus the functional form of the addition operator
polarplot plot function for complex numbers, plots the magnitude and angle
poly2sym converts a vector of coefficients of a polynomial to a symbolic expression
polyder derivative of a polynomial
polyfit fits a polynomial curve of a specified degree to data points
polyint integral of a polynomial
polyval evaluates a polynomial at specified value(s)
pretty displays a symbolic expression using exponents

print prints or saves a figure or image
prod the product of all values in a vector, or of every column in a matrix
profile toggle; the Profiler generates reports on execution time of code
properties displays properties of a class
quad integration using Simpson's method
quit quits MATLAB
rad2deg converts from radians to degrees
rand generates uniformly distributed random real number(s) in the open interval (0,1)
randi generates random integer(s) in the specified range
randn generates normally distributed random real numbers
real real part of a complex number
rectangle graphics primitive to create a rectangle; curvature can vary
rem remainder after division
repelem replicates elements in a matrix; creates $m \ x \ n$ copies of each
repmat replicates a matrix; creates $m \ x \ n$ copies of the matrix
reshape changes dimensions of a matrix to any matrix with the same number of elements
rgb2gray converts an RGB image matrix to grayscale
rmfield removes a field from a structure
rng random number generator, sets the seed for random functions and gets the state
roots roots of a polynomial equation
rot90 rotates a matrix 90 degrees counter-clockwise
round rounds a real number toward the nearest integer
rref puts an augmented matrix in reduced row echelon form
save writes a matrix to a file or saves variables to a.mat file
semilogx plot function, uses a scale for logarithmic x and a linear scale for y
semilogy plot function, uses a linear scale for x and a logarithmic scale for y
set sets properties of a plot object
setdiff set function, returns elements that are in one vector, but not in another
setxor set exclusive or, returns the elements that are not in the intersection of two sets
sign signum, returns −1, 0, or 1
simplify simplifies a symbolic math expression
sin sine in radians
sind sine in degrees
single converts a number to the type single
sinh hyperbolic sine in radians
size returns the dimensions of a matrix
solve symbolic math function to solve an equation or simultaneous equations
sort sorts the elements of a vector (default is ascending order)
sortrows sorts the rows of a matrix; for strings result in an alphabetical sort

sound sends a sound signal (vector of amplitudes) to an output device
sphere returns three-dimensional data vectors to create a sphere
spiral creates a square matrix of integers spiraling from 1 in the middle
sprintf creates a formatted string
sqrt square root
std standard deviation
stem two-dimensional stem plot
stem3 three-dimensional stem plot
str2double converts from a string containing a number to a double number
str2func converts a string to a function handle
str2num converts from a string containing number(s) to a number array
strcat horizontal string concatenation
strcmp string compare, used instead of equality operator for strings
strcmpi string compare, ignoring case
strfind find a substring within a longer string
strjoin concatenates strings in a cell array into a long string
strncmp string compare the first *n* characters of strings
strncmpi string compare the first *n* characters, ignoring case
strrep replace all occurrences of one substring with another within a longer string
strsplit splits a string into elements in a cell array
strtok breaks one longer string into two shorter strings, with all characters retained
strtrim deletes both leading and trailing blanks from a string
struct creates a structure by passing pairs of field names and values
subplot creates a matrix of plots in the Figure Window
subs substitutes a value into a symbolic math expression
sum the sum of the values in a vector or of every column in a matrix
summary shows variables and statistics for a table
surf three-dimensional surface plot
sym creates a symbolic variable or expression
sym2poly converts a symbolic expression to a vector of coefficients for a polynomial
syms creates multiple symbolic variables
table creates a table data structure
text graphics primitive object to put a string on a plot
textscan file input function, reads from a file into a cell array of column vectors
tic/toc used to time code
timeit times a function execution
title writes a string as a title on a plot
trace the trace (sum of values on the diagonal) of a matrix
trapz trapezoidal rule to approximate the area under a curve
tril converts a matrix to a lower triangular matrix

triu converts a matrix to an upper triangular matrix
true equivalent to logical(1), creates a matrix of all true values
type display the contents of a file in the Command Window
uibuttongroup groups together button objects
uicontrol basic function to create graphical user interface objects of different styles
uint16 converts a number to a 16-bit unsigned integer
uint32 converts a number to a 32-bit unsigned integer
uint64 converts a number to a 64-bit unsigned integer
uint8 converts a number to an 8-bit unsigned integer
uipanel groups together graphical user interface objects
union set function, the union of two sets
unique returns all of the unique values within a set (vector)
upper converts all letters to upper-case
var variance
varargin built-in cell array to store input arguments
varargout built-in cell array to store output arguments
who displays variables in the base workspace
whos displays more information on the variables in the base workspace
xlabel puts a string as a label on the *x* axis of a plot
xlsread reads from a spreadsheet with filename.xls
xlswrite writes to a spreadsheet with filename.xls
xor exclusive or, true if only one argument is true
ylabel puts a string as a label on the *y* axis of a plot
zeros creates a matrix of all zero values
zlabel puts a string as a label on the *z* axis of a three-dimensional plot

APPENDIX II

MATLAB® and Simulink Toolboxes

In addition to the many functions included in MATLAB, there are additional Toolboxes that can be added. These toolboxes have groups of related functions that can be used for more advanced computations and data processing. The MathWorks, Inc. also has a family of simulation software called Simulink; it too, can be augmented with additional Toolboxes. For more detailed information, see the website www.mathworks.com.

Some of the more common Toolboxes include functions for advanced image processing, control system design, signal processing, curve fitting, parallel computing, and optimization.

Here is a list of some of these Toolboxes:

Symbolic Math Toolbox (some used in this text)
Statistics Toolbox (mentioned in this text)
Curve Fitting Toolbox
Optimization Toolbox
Partial Differential Equation Toolbox
Image Processing Toolbox
Image Acquisition Toolbox
Data Acquisition Toolbox
Instrument Control Toolbox
Signal Processing Toolbox
Control System Toolbox
Parallel Computing Toolbox
Aerospace Toolbox
Neural Network Toolbox

Index

Note: Page numbers followed by "*f*" indicate figures, "*t*" indicate tables, and "*b*" indicate boxes.

D

E

F

G

H

I

T

U